Statistics and Data Visualization in Climate Science with R and Python

A comprehensive overview of essential statistical concepts, useful statistical methods, data visualization, machine learning, and modern computing tools for the climate sciences and many others such as geography and environmental engineering. It is an invaluable reference for students and researchers in climatology and its connected fields who wish to learn data science, statistics, R and Python programming. The examples and exercises in the book empower readers to work on real climate data from station observations, remote sensing, and simulated results. For example, students can use R or Python code to read and plot the global warming data and the global precipitation data in netCDF, csv, txt, or JSON; and compute and interpret empirical orthogonal functions. The book's computer code and real-world data allow readers to fully utilize the modern computing technology and updated datasets. Online supplementary resources include R code and Python code, data files, figure files, tutorials, slides, and sample syllabi.

Samuel S. P. Shen is Distinguished Professor of Mathematics and Statistics at San Diego State University, and Visiting Research Mathematician at Scripps Institution of Oceanography, University of California – San Diego. Formerly, he was McCalla Professor of Mathematical and Statistical Sciences at the University of Alberta, Canada, and President of the Canadian Applied and Industrial Mathematics Society. He has held visiting positions at the NASA Goddard Space Flight Center, the NOAA Climate Prediction Center, and the University of Tokyo. Shen holds a B.Sc. degree in Engineering Mechanics and a Ph.D. degree in Applied Mathematics.

Gerald R. North is University Distinguished Professor Emeritus and former Head of the Department of Atmospheric Science at Texas A&M University. His research focuses on modern and paleoclimate analysis, satellite remote sensing, climate and hydrology modeling, and statistical methods in atmospheric science. He is an elected fellow of the American Geophysical Union and the American Meteorological Society. He has received several awards including the Harold J. Haynes Endowed Chair in Geosciences of Texas A&M University, the Jules G. Charney medal from the American Meteorological Society, and the Scientific Achievement medal from NASA. North holds both B.Sc. and Ph.D. degrees in Physics.

'*Statistics and Data Visualization in Climate Science with R and Python* by Sam Shen and Jerry North is a fabulous addition to the set of tools for scientists, educators and students who are interested in working with data relevant to climate variability and change … I can testify that this book is an enormous help to someone like me. I no longer can simply ask my grad students and postdocs to download and analyze datasets, but I still want to ask questions and find data-based answers. This book perfectly fills the 40-year gap since I last had to do all these things myself, and I can't wait to begin to use it … I am certain that teachers will find the book and supporting materials extremely beneficial as well. Professors Shen and North have created a resource of enormous benefit to climate scientists.'

Dr Phillip A. Arkin, University of Maryland

'This book is a gem. It is the proverbial fishing rod to those interested in statistical analysis of climate data and visualization that facilitates insightful interpretation. By providing a plethora of actual examples and R and Python scripts, it lays out the "learning by doing" foundation upon which students and professionals alike can build their own applications to explore climate data. This book will become an invaluable desktop reference in Climate Statistics.'

Professor Ana P. Barros, University of Illinois Urbana-Champain

'A valuable toolkit of practical statistical methods and skills for using computers to analyze and visualize large data sets, this unique book empowers readers to gain physical understanding from climate data. The authors have carried out fundamental research in this field, and they are master teachers who have taught the material often. Their expertise is evident throughout the book.'

Professor Richard C. J. Somerville, University of California, San Diego

'This book is written by experts in the field, working on the frontiers of climate science. It enables instructors to "flip the classroom", and highly motivated students to visualize and analyze their own data sets. The book clearly and succinctly summarizes the applicable statistical principles and formalisms and goes on to provide detailed tutorials on how to apply them, starting with very simple tasks and moving on to illustrate more advanced, state-of-the-art techniques. Having this book readily available should reduce the time required for advanced undergraduate and graduate students to achieve sufficient proficiency in research methodology to become productive scientists in their own right.'

Professor John M. Wallace, University of Washington

Statistics and Data Visualization in Climate Science with R and Python

SAMUEL S. P. SHEN

San Diego State University

GERALD R. NORTH

Texas A&M University

CAMBRIDGE
UNIVERSITY PRESS

Shaftesbury Road, Cambridge CB2 8EA, United Kingdom

One Liberty Plaza, 20th Floor, New York, NY 10006, USA

477 Williamstown Road, Port Melbourne, VIC 3207, Australia

314–321, 3rd Floor, Plot 3, Splendor Forum, Jasola District Centre, New Delhi – 110025, India

103 Penang Road, #05–06/07, Visioncrest Commercial, Singapore 238467

Cambridge University Press is part of Cambridge University Press & Assessment,
a department of the University of Cambridge.

We share the University's mission to contribute to society through the pursuit of
education, learning and research at the highest international levels of excellence.

www.cambridge.org
Information on this title: www.cambridge.org/9781108842570

DOI: 10.1017/9781108903578

© Cambridge University Press & Assessment 2023

First published 2023

Printed in the United Kingdom by CPI Group Ltd, Croydon CR0 4YY

A catalogue record for this publication is available from the British Library.

Library of Congress Cataloging-in-Publication Data
Names: Shen, Samuel S., author. | North, Gerald R., author.
Title: Statistics and data visualization in climate science with R and Python / Samuel S. P. Shen,
Gerald R. North.
Description: Cambridge ; New York : Cambridge University Press, 2023. | Includes bibliographical
references and index.
Identifiers: LCCN 2023029571 | ISBN 9781108842570 (hardback) | ISBN 9781108829465 (paperback) |
ISBN 9781108903578 (ebook)
Subjects: LCSH: Climatology – Statistical methods. | Climatology – Data processing. | Information
visualization. | R (Computer program language) | Python (Computer program language)
Classification: LCC QC874.5 .S48 2023 | DDC 551.63/3–dc23/eng/20230816
LC record available at https://lccn.loc.gov/2023029571

ISBN 978-1-108-84257-0 Hardback

Give the pupils something to do, not something to learn; and doing is of such a nature as to demand thinking; learning naturally results.

— John Dewey

Contents

Preface

What Can You Get Out of This Book?

The learning goal of this book is to master the commonly used statistical methods in climate science and use them for real climate data analyses and visualization using computers. We hope that this book will quickly help you improve your skills of data science and statistics. You will feel comfortable to explore large datasets in various kinds of data formats, compute their statistics, generate statistical plots, visualize the data, interpret your results, and present your results. We appreciate the psychological value of instant gratification, which is adopted in the design of most video games. We want you to gain some useful skill after you have spent two or three hours reading the book and following the interaction procedures described by running and modifying our R code or Python code, and by exploring the climate data visualization technologies, such as 4-dimensional visual delivery (4DVD) of big climate data (`www.4dvd.org`). You might have noticed that we use the word "skill" rather than "knowledge" gained from our book. We believe that you need skills and your knowledge comes from your skills and your practice of the skills. We emphasize "do!" Our book is designed in such a way that you will use your computer to interact with the book, for example running R code in R Studio or Python code in Jupyter Notebook and Colab to reproduce the figures in the book, or modifying the code for the analyses and plots of your own datasets. Your interaction with the book is essential to effectively improve your data science and statistics skills, which can be applied not only to climate data but also to other datasets of your interest. We expect that your skill improvement will make you more efficient when handling the data in your study or job, and hence will bring you more opportunities. Specifically, you will be able to plot big data in the file formats of `.nc`, `.csv`, `.json`, `.txt`, compute statistical parameters (e.g., mean, standard deviation, quantiles, test statistics, empirical orthogonal functions, principal components, and multivariate regressions), and generate various kinds of figures (e.g., surface air temperature for January 1942, atmospheric carbon dioxide data of Mauna Loa, pie chart, box plot, histogram, periodogram, and chi-square fit to monthly precipitation data). If you are already sophisticated with data analysis and statistics, you may appreciate our explanation of the assumptions and limitations of statistical methods, such as the four assumptions for a linear regression, the test of serial correlation, and intuition of the theory behind covariances and spectral analyses.

To help you quickly learn your desired skills, we try our best to make each chapter self-contained as much as we can. This allows you to read the chapter of your interest without reading the previous chapters or with minimum references. We are able to do so because of your interaction with the book through your computer and the Internet with some help from learning resources freely available at our book website www.climatestatistics.org.

Who should read this book? We wrote this book for the following groups of people: (i) undergraduate and graduate students majoring in atmospheric sciences, oceanic sciences, geography, and climate science in general to help them quickly gain the data handling skills for research, for supporting course work, or for job interviews; (ii) climate science professionals who may wish to check whether they have correctly applied statistical methods or answer questions from their junior colleagues, such as how to make inference on regression; (iii) college students in other majors who wish to use climate data as examples to learn data science skills and R or Python programming so that they can compete well in the job market; (iv) college professors who wish to modernize their courses or curricula amid the development of the digital economy and who wish to find a textbook that allows them to effectively engage students to learn data science skills; and (v) senior or retired scientists and engineers who wish to use an interactive book to sharpen their mind by mimicking computer programs and generating beautiful figures.

Why Did We Write This Book?

A simple answer is that the need is wide and urgent. Almost all climate scientists wonder at one time or another if a statistics method they or their students have used in a project is sound. What are the assumptions and limitations of the method? They may wish to tell a better story about the method and its results, which are appreciated and agreed upon by professional statisticians. Students who take statistics courses taught by faculty from the Department of Statistics often learn very little about real climate datasets, and find that what they have learned is completely disjoint from the real climate science problems. For example, a statistics course never offers chances to analyze a dataset of more than 2.0 Gigabytes in a .nc file, or to plot 2D or 3D graphics or maps. This is so because traditional books are theory based or based on small datasets because of limited access to laptop computers. Some books are more advanced for research, but they still use the figures adopted from publications, not easily reproducible by readers. Consequently, most students of statistics courses forget what they have learned after the exams, and can hardly tell a story about the usage of statistical methods. We wrote this book to fill this gap that links assumptions and computing tools to real world data and practical problems. We intend to train you with useful skills through your interactions with the book so that you will not feel disjoint and unmotivated, because you have a chance to interact with the book and work on your own datasets.

Both authors have taught climate statistics courses many times in the USA, Canada, Japan, and China. In the earlier years, we taught using the old method, although we once in a while used some real climate data. We learned the shortcomings of the traditional method of teaching: one-directional instruction and lecture from professors to students. One of us

(S. S. P. S.) changed the one-directional method to the interaction method of instruction in 2015, when it was feasible to require every student to bring a laptop computer to the classroom. This interactive learning has helped students effectively in the job market. The method has already been used by another book authored by S. S. P. S. with Richard C. J. Somerville, *Climate Mathematics: Theory and Applications*, also published by Cambridge University Press in 2019.

We wrote this statistics and data visualization book partly because of our experience in research and teaching. The book is built on class notes we wrote at Texas A&M University (eight times taught as a semester-long course by G. R. N.), University of Alberta (twice taught as a semester-long course by S. S. P. S.), University of Tokyo (once taught as a short course on error estimation by S. S. P. S.), the NOAA National Centers of Environmental Information (once taught as a short course mainly about R and its use in analyzing climate data with different formats and visualizing the analysis results by S. S. P. S.), and the Chinese Academy of Sciences (once taught as a short course mainly about data visualization by R by S. S. P. S.). The courses were at graduate level and the audiences were graduate students not only majoring in atmospheric and oceanic sciences, but also in engineering, mathematics, and statistics. Some audiences were climate research professionals. The main purpose of the courses was to prepare students to correctly and efficiently use statistical theory and methods to analyze climate data, visualize the data, and interpret data analysis results.

Another motivation for writing the current book is the overwhelming problems in the traditional statistics courses for today's climate science. Climate science students may find traditional statistics courses inadequate, as they often focus on the mathematical power of the statistical theories, but not much on the practical data analysis examples, and even less or not at all on the visualization of modern climate data. Such a course is not engaged with the students' research work and their career. For example, students need to plot data in both 2D and 3D, and compute the empirical orthogonal functions for large data matrices. Our book fills the gap and provides both computer codes and data that allow students to make analysis and visualization, and to modify our codes for their research and jobs. Our book may be considered as a toolbox for climate data. Because we purposely limited the book size to avoid the audience being overwhelmed by a thick book, we thus leave some computer codes on our book's website www.climatestatistics.org, including tutorials for R and Python, and some examples. The freely available toolbox on our website will be convenient to use and will be updated with new data, computer codes, additional exercise problems, and other resources to empower your interaction with the book. The toolbox also allows a reader to try our learning-by-doing method first, before purchasing our book.

What Is the Philosophy of Our Pedagogy?

We follow the education theory of learning-by-doing, which in our case means using your computer and our R or Python code to interact with our book and modifying our computer codes and websites to analyze your own data and generate corresponding figures. Learning-by-doing is the core methodology of the progressive education of John Dewey (1859–1952), an American philosopher and educator. Dewey felt that the experience of students and teachers together yields extra value for both. Instructors are not just to lecture and project authority, instead they are to collaborate with students and guide students to gain experience of solving problems of their interest. Although Dewey's education theory was established initially for schoolchildren, we feel that the same is applicable to undergraduate and graduate students. Our way of learning-by-doing is to enable students to use R or Python code and other resources in the book and its website to reproduce the figures and numerical results in the book. Further, students can modify the computer code and solve their own problems, such as visualizing the climate data in a similar format and of their own interest, or analyzing their own data using a similar or a modified method. Thus, audience interaction is the main innovative feature of our book, allowing the audience to gain experience of practicing, thinking, applying, and consequently understanding. The ancient Chinese educator Confucius (551–479 BC) said, "I hear, and I forget; I see, and I remember; and I do, and I understand." Although John Dewey and Confucius were more than 2,000 years apart, they had a similar philosophy of learning-by-doing.

As illustrated in Figure 0.1, our pedagogy has three stages: do, reflect, and apply. Coauthor S. S. P. S. practiced this pedagogy in recent years. He presents a question or a problem at the beginning of a class. Then he asks students to orally ask the same question, or describe the same problem, or answer his question in their own words. For example, why is the tiny amount of carbon dioxide in the atmosphere important for climate change? Next, he and students search and re-search for data and literature, work on the observed carbon dioxide data at Mauna Loa using computer data visualization, and discuss the structure of greenhouse gasses whose molecules have three or more atoms. To understand the data better, they use more data analysis methods, such as the time series decomposition. Next, he encourages his students to share this experience with their grandparents, other family members, or friends. Finally, students apply the skills gained to solve their own problems with their own data, by doing homework, working on projects, finding employment, or making innovations. In this cycle, students have gathered experience and skills to improve their life and to engage with the community. In the short term, students trained in this progressive cycle of learning-by-doing have a better chance of becoming good problem solvers, smooth story narrators, and active leaders in a research project in a lab or an internship company, and consequently to become competitive in the job market. In the long term, students trained in this cycle are likely to become to life-time learners and educators. John Dewey said: "Education is not preparation for life but life itself." We would like to modify this to "Education is not only preparation for a better life, but also is life itself."

Dewey's progressive education theory is in a sharp contrast to the traditional learning process based on the logic method, which aims at cultivating high-achieving scholars. The commonly used pedagogy of lecture-read-homework-exam often uses the logic-based approach. The climax of logic-based education is that the instructors have the pleasure of presenting their method and theory, while students are so creative that they will produce a new or a better theory. Many outstanding scholars went through their education this way, and many excellent textbooks were written for this approach. However, our book does not

The Cycle of Learning–by–Doing

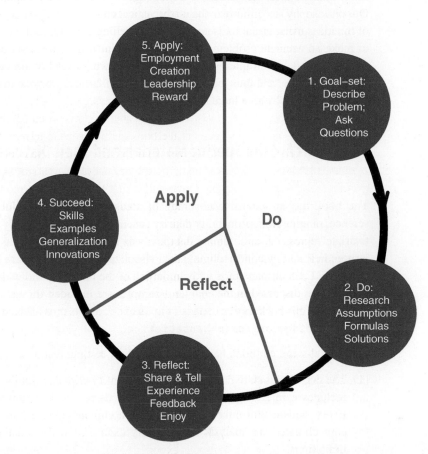

Figure 0.1 John Dewey's pedagogy of learning-by-doing: A cycle from a problem to skill training via solving a problem, to reflection, to success and applications, and to reward and new problems. It is a cycle of understanding and progress. The size of each piece of the "Do, Reflect, Apply" pie approximates the proportion of the learning effort. "Do" is emphasized here, indicated by the largest piece of the pie.

follow this approach, since our book is written for the general population of students, not just for the elite class or even scholars-to-be. If you wish to be such an ambitious scholar, then you may use our book very differently: you can read the book quickly and critically for gaining knowledge instead of skills, and skip our reader–book interaction functions.

Our pedagogy is result-oriented. If using car-driver and car-mechanic as metaphors, our book targets the 99% or more who are car-drivers, and provides some clues for the less than 1% of the audience who wish to become car-mechanics. Good drivers understand the limits of a car, are sensitive to abnormal noise or motion of a car, and can efficiently and safely get from A to B. Therefore, this book emphasizes (i) assumptions of a method, (ii) core and concise formulas, (iii) product development by a computer, and (iv) result interpretation. This is in contrast with traditional books, which are often in an expanded format of (ii) with many mathematical derivations, and challenge the mathematical capability of most readers. Our philosophy is to minimize the mathematical challenge, but to deepen the understanding of the ideas using visual tools and using storytelling experience. Our audience will be able to make an accurate problem statement, ask pointed questions, set up statistical models, solve the models, and interpret solutions. Therefore, instead of aiming our training at the few "mechanics" and thus incurring a high risk of failure, we wish to train a large number of good "drivers" with a large probability of success.

What Are the Specific Materials and Their Distinct Features?

The book has an extensive coverage of statistical methods useful in modern climate science, ranging from probability density functions, machine learning basics, modern multivariate regression, and climate data analysis examples, to a variety of computer codes written in R and Python for climate data visualization. The details are listed in the Table of Contents. Each chapter starts with an outline of the materials and ends with a summary of the methods discussed in the chapter. Examples are included for each major concept. The website for this book www.climatestatistics.org is cross-linked with the CUP site www.cambridge.org/climatestatistics.

Material-wise, our book has the following four distinct features:

(1) The book treats statistics as a language for every climate scientist to master. The book includes carefully selected statistical methods for modern climate science students. Every statistics formula in the book has a climate science interpretation, and every core climate data analysis is rigorously examined with statistical assumptions and limitations.

(2) The book describes a complete procedure for statistical modeling for climate data. The book discusses random fields, their meaning in statistical models of climate data, model solutions and interpretations, and machine learning.

(3) The book includes free computer codes in both R and Python for readers to conveniently visualize the commonly used climate data. The book has computer codes to analyze NCEP/NCAR Reanalysis data for EOF patterns, and has examples to analyze the NOAAGlobalTemp data for global warnings. The book website contains many online resources, such as an instructors' guide, PowerPoint slides, images, and tutorials for R and Python coding. Readers can use the given computer codes to reproduce and modify all the figures in the book, and generate high-quality figures from their own datasets for reports and publications.

(4) This textbook has a large number of exercise problems and has solution manuals. Every chapter has at least 20 exercise problems for homework, tests, and term projects. A solutions manual with both R and Python codes is available for instructors from the Cambridge University Press.

Acknowledgments

We thank our employers San Diego State University, University of Alberta, Texas A&M University, and NASA Goddard Space Flight Center for their support of our teaching and research on the materials included in this book. We appreciate that our employers gave us freedom to lecture using the many non-traditional materials included in this book. Of course, we are indebted to the students in our classes and labs who motivated us to teach better. The interactions with students gave us much joy when developing materials for this book and some students contributed directly. For example, Kaelia Okamura led the Python code development based on the original R code written by Samuel Shen; Kaelia Okamura, Thomas Bui, Danielle Lafarga, Joaquin Stawsky, Ian Ravenscroft, Elizabeth Reddy, Matthew Meier, Tejan Patel, Yiyi Li, Julien Pierret, Ryan Lafler, Anna Wade, Gabriel Smith, Braandon Gradylagrimas, and many other students contributed to the development of the computer code, book website, solutions manuals, and datasets. Several students in this group were awarded fellowships by the NOAA Education Partnership Program with Minority Serving Institutions, USA. We thank NOAA for this support.

We thank the friendly editorial team of Cambridge University Press (CUP) for their professional and high-quality work. In particular, we thank our book editor Matt Lloyd for his maximum patience with our writing progress, his professional guidance on the figure styles, and his advice on the book audience. We thank Ursula Acton for her excellent text copy-editing. We thank Reshma Xavier and Sapphire Duveau for their efficient management of the production of our book. We thank Sarah Lambert and Rowan Groat for their clear instructions for us to prepare the supplementary materials for the book. Our interactions with CUP were always pleasant and productive.

How to Use This Book

Prerequisites for Reading This Book

The course prerequisite for this book is one semester of calculus and one semester of linear algebra. These two prerequisite courses are included in the courses of a standard undergraduate curriculum of atmospheric science, oceanography, or climate science.

In lieu of these two courses for satisfying the prerequisite, you can also use chapters 3 and 4, and appendices of the book *Climate Mathematics: Theory and Applications* by Samuel S. P. Shen and Richard C. J. Somerville, Cambridge University Press, 2019, 391pp.

Besides, more than 50% of this book, including Chapters 1–4, and most of Chapters 7 and 9 requires only high school mathematics. These materials are sufficient for a one-semester course of computer programming and data science for students in majors that do not require calculus or linear algebra. Often geography, environmental science, and sustainability are such majors.

How Each Audience Group Can Make the Best Use of This Book

The book can be used as the textbook for a variety of courses, such as Statistical Methods [for Climate Science], Statistics and Data Visualization [for Climate Science], Climate Data Analysis, Climate Statistics, and Climate Data Visualization. The course can be taught at graduate student, advanced undergraduate student, and climate science professional levels. The book can also be used as a reference manual or a toolkit for a lab. The potential audience of this book and their purposes may be classified into the following groups.

> User Group I: Textbook for an upper division undergraduate course with a title like Climate Statistics, or Data Analysis, or Climate Data Science, or Computer Programming and Data Analysis, or Statistical Methods in Atmospheric Sciences. The course can use most of the materials in the book but exclude more advanced topics, such as Sections 4.2, 4.3, 6.4, 6.5, 7.5–7.7, 8.3, and 8.4.

How to Use This Book

User Group II: Textbook for a graduate course with some undergraduate seniors enrolled. The course title may be named as in Group I, or with the word "Advanced," e.g., Advanced Climate Data Science. The instructor might choose more advanced topics and omit some elementary sections that the students are already familiar with or can be quickly reviewed by themselves. The elementary sections include Sections 2.1, 4.1, and 5.1–5.4.

User Group III: Data science skills may help students in some traditional climate-related humanity majors to find employment. Data science courses, such as Python Programming and Data Science or Python Programming for Environmental Data, have been listed as electives. These students may not have mastered the methods of calculus and linear algebra. These courses may use materials from Chapters 1 to 4, and most of Chapters 7 and 9.

User Group IV: Textbook for a short course or workshop for climate scientists with a focus on data analysis and visualization. The instructor may select materials based on the purpose of the short course. For example, if the course is The Basics of Machine Learning for Climate Science, then use Chapter 9; if it is The Basics of Climate Data Visualization, then use Chapter 1; and if it is Python Coding for Climate Data, then select examples in Chapters 1, 4, 8, and 9.

Users Group V: Reference manual for climate professionals who wish to use this book to train their junior colleagues in their labs for data analysis, or use this book as a toolkit for their research. For example, when you wish to use the Kendall tau test for statistical significance, you can find a numerical example and its Python or R code in this book; when your students wish to analyze and plot a certain dataset, you may refer them to this book to find proper examples and computer codes. These are to make your work easy and efficient, which is a practical purpose we bore in mind when writing this book.

Users Group VI: Some senior or retired scientists and engineers may wish to keep their minds sharp by playing with R or Python codes and climate data, such as the CO_2 concentration data shown in Figure 7.3. Instead of learning many statistical concepts and methods, this book will allow them to play with a few computer codes and generate some interesting figures (e.g., Figs. 7.3 and 8.4) to discuss with their friends.

An Interactive Textbook to Support the *Learning-By-Doing*

This is an interactive textbook designed for readers in the digital age. The book contains many R and Python codes, which can also be found at the book's website `www.climatestatistics.org`. The relevant climate data used in the computer codes can also be downloaded directly from the website. An efficient way of learning from this book is by doing: to reproduce the figures and numerical results using the computer codes from either the book or its website. You can modify the code to alter the figures for your purposes and your datasets. These figures and computer codes can help you practice what you have learned and better understand the ideas and theory behind each method. We emphasize the value of practice in acquiring quantitative analysis skills.

For instructors of this textbook, we have tried to make this book user-friendly for both students and instructors. As an instructor, you can simply follow the book when teaching in a classroom, and you can assign the exercise problems at the end of each chapter. A solutions manual in R and Python with answers for all the problems is available through the publisher to qualified instructors.

Basics of Climate Data Arrays, Statistics, and Visualization

People talk about climate data frequently, read or imagine climate data, and yet rarely play with or use climate data, because they often think it takes a computer expert to do that. However, that is changing. With today's technology, anyone can use a computer to play with climate data, such as a sequence of temperature values of a weather station at different observed times or a matrix of data for a station for temperature, air pressure, precipitation; wind speed and wind direction at different times; and an array of temperature data on a 5-degree latitude–longitude grid for the entire world for different months. The first is a vector. The second is a variable-time matrix, and the third a space-time 3-dimensional array. When considering temperature variation in time at different air pressure levels and different water depths, we need to add one more dimension: altitude. The temperature data for the ocean and atmosphere for the Earth is a 4-dimensional array, with 3D space and 1D time. This chapter attempts to provide basic statistical and computing methods to describe and visualize some simple climate datasets. As the book progresses, more complex statistics and data visualization will be introduced.

We use both R and Python computer codes in this book for computing and visualization. Our method description is stated in R. A Python code following each R code is included in a box with a light yellow background. You can also learn the two computer languages and their applications to climate data from the book *Climate Mathematics: Theory and Applications* (Shen and Somerville 2019) and its website `www.climatemathematics.org`.

The climate data used in this book are included in the `data.zip` file downloadable from our book website `www.climatestatistics.org`. You can also obtain the updated data from the original data providers, such as `www.esrl.noaa.gov` and `www.ncei.noaa.gov`.

After studying this chapter, a reader should be able to analyze simple climate datasets, compute data statistics, and plot the data in various ways.

1.1 Global Temperature Anomalies from 1880 to 2018: Data Visualization and Statistical Indices

In a list of popular climate datasets, the global average annual mean surface air temperature anomalies might be on top. Here, *anomalies* means the temperature departures from the normal temperature or the *climatology*. Climatology is usually defined as the mean of temperature data in a given period of time, such as from 1971 to 2020. Thus, temperature anomaly data are the differences of the temperature data minus the climatology.

This section will use the global average annual mean surface air temperature anomaly dataset as an example to describe some basic statistical and computing methods.

1.1.1 The NOAAGlobalTemp Dataset

The 1880–2018 global average annual mean surface air temperature (SAT) anomaly data are shown as follows:

```
 [1]  -0.37 -0.32 -0.32 -0.40 -0.46 -0.47 -0.45 -0.50 -0.40 -0.35 -0.57
 [12] -0.49 -0.55 -0.56 -0.53 -0.47 -0.34 -0.36 -0.50 -0.37 -0.31 -0.39
 [23] -0.49 -0.58 -0.66 -0.53 -0.46 -0.62 -0.69 -0.68 -0.63 -0.68 -0.58
 [34] -0.57 -0.39 -0.32 -0.54 -0.56 -0.45 -0.45 -0.46 -0.39 -0.47 -0.46
 [45] -0.50 -0.39 -0.31 -0.40 -0.42 -0.54 -0.34 -0.32 -0.36 -0.49 -0.35
 [56] -0.38 -0.36 -0.26 -0.27 -0.26 -0.15 -0.05 -0.09 -0.09  0.04 -0.08
 [67] -0.25 -0.30 -0.30 -0.30 -0.41 -0.26 -0.22 -0.15 -0.36 -0.38 -0.44
 [78] -0.19 -0.13 -0.18 -0.22 -0.16 -0.15 -0.13 -0.39 -0.32 -0.27 -0.26
 [89] -0.27 -0.15 -0.21 -0.32 -0.22 -0.08 -0.32 -0.24 -0.32 -0.05 -0.13
[100] -0.02  0.02  0.06 -0.06  0.10 -0.10 -0.11 -0.01  0.13  0.13  0.05
[111]  0.19  0.16  0.01  0.03  0.09  0.21  0.08  0.27  0.39  0.20  0.18
[122]  0.30  0.35  0.36  0.33  0.41  0.37  0.36  0.30  0.39  0.45  0.33
[133]  0.38  0.42  0.50  0.66  0.70  0.60  0.54
```

The data are part of the dataset named as the NOAA Merged Land Ocean Global Surface Temperature Analysis (NOAAGlobalTemp) V4. The dataset was generated by the NOAA National Centers for Environmental Information (NCEI) (Smith et al. 2008; Vose et al. 2012). Here, NOAA stands for the United States National Oceanic and Atmospheric Administration.

The anomalies are with respect to the 1971–2000 climatology, i.e., 1971–2000 mean. An anomaly of a weather station datum is defined by the datum minus the station climatology. The first anomaly datum, -0.37 °C, indexed by [1] in the above data table, corresponds to 1880 and the last to 2018, a total of 139 years. The last row is indexed from [133] to [139].

One might be interested in various kinds of statistical characteristics of the data, such as mean, variance, standard deviation, skewness, kurtosis, median, 5th percentile, 95th percentile, and other quantiles. Is the data's probabilistic distribution approximately normal? What does the box plot look like? Are there any outliers? What is a good graphic representation of the data, i.e., popularly known as a climate figure?

When considering global climate changes, why do scientists often use anomalies, instead of the full values directly from the observed thermometer readings? This is because the observational estimates of the global average annual mean surface temperature are less accurate than the similar estimates for the changes from year to year. There is a concept of characteristic spatial correlation length scale for a climate variable, such as surface temperature. The length scale is often computed from anomalies.

The use of anomalies is also a way of reducing or eliminating individual station biases. A simple example of such biases is that due to station location, which is usually fixed in a long period of time. It is easy to understand, for instance, that a station located in the valley of a mountainous region might report surface temperatures that are higher than the true average surface temperature for the entire region and cannot be used to describe the behavior of climate change in the region. However, the anomalies at the valley station may synchronously reflect the characteristics of the anomalies for the region. Many online materials give justifications and examples on the use of climate data anomalies, e.g., NOAA NCEI (2021).

The global average annual mean temperature anomalies quoted here are also important for analyzing climate simulations. When we average over a large scale, many random errors cancel out. When we investigate the response of such large scale perturbations as the variations of Sun brightness or atmospheric carbon dioxide, these averaged data can help validate and improve climate models. See the examples in the book by Hennemuth et al. (2013) that includes many statistical analyses of both observed and model data.

1.1.2 Visualize the Data of Global Average Annual Mean Temperature

Many different ways have been employed to visualize the global average annual mean temperature anomalies. The following three are popular ones appearing in scientific and news publications: (a) a simple point-line graph, (b) a curve of staircase steps, and (c) a color bar chart, as shown in Figures 1.1–1.3. This subsection shows how to generate these figures by R and Python computer programming languages. The Python codes are in yellow boxes. To download and run the codes, visit www.climatestatistics.org.

1.1.2.1 Plot a Point-Line Graph of Time Series Data

Figure 1.1 is a simple line graph that connects all the data points, denoted by the small circles, together to form a curve showing the historical record of global temperature

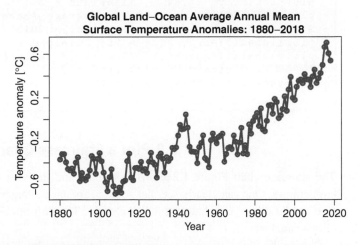

Figure 1.1 Point-line graph of the 1880–2018 global average annual mean temperature anomalies with respect to the 1971–2000 climatology, based on the NOAAGlobalTemp V4 data.

anomalies. It is plotted from the NOAAGlobalTemp V4 data quoted in Section 1.1.1. The
figure can be generated by the following computer code.

```
#R plot Fig. 1.1: A simple line graph of data
# go to your working directory
setwd("/Users/sshen/climstats")
# read the data file from the folder named "data"
NOAAtemp = read.table(
  "data/aravg.ann.land_ocean.90S.90N.v4.0.1.201907.txt",
  header=FALSE) #Read from the data folder
dim(NOAAtemp)# check the data matrix dimension
#[1] 140    6 #140 years from 1880 to 2019
#2019 will be excluded since data only up to July 2019
#col1 is year, col2 is anomalies, col3-6 are data errors
par(mar=c(3.5,3.5,2.5,1), mgp=c(2,0.8,0))
plot(NOAAtemp[1:139,1], NOAAtemp[1:139,2],
     type ="l", col="brown", lwd=3, cex.lab=1.2, cex.axis=1.2,
     main ="Global Land-Ocean Average Annual Mean
     Surface Temperature Anomalies: 1880-2018",
     xlab="Year",
     ylab=expression(
       paste("Temperature Anomaly [", degree,"C]")))
```

```
#Python plot Fig. 1.1:  A simple line graph of data
# Go to your working directory
os.chdir("/Users/sshen/climstats")
# read the data file from the folder named "data"
NOAAtemp = read_table(\
"data/aravg.ann.land_ocean.90S.90N.v4.0.1.201907.txt",
    header = None, delimiter = "\s+")
# check the data matrix dimension
print("The dimension of our data table is:", NOAAtemp.shape)
x = np.array(NOAAtemp.loc[0:138, 0])
y = np.array(NOAAtemp.loc[0:138, 1])
plt.plot(x, y, 'brown', linewidth = 3);
plt.title("Global Land-Ocean Average Annual Mean \n \
Surface Temperature Anomaly: 1880 - 2018", pad = 15)
plt.xlabel("Year", size = 25, labelpad = 20)
plt.ylabel("Temperature Anomaly [$\degree$C]",
           size = 25, labelpad = 20)
plt.show() # display on screen
```

1.1.2.2 Plot a Staircase Chart

The staircase chart Figure 1.2 shows both data and the year-to-year annual changes of
temperature. One can clearly see that some years have a larger change from the previous
year, while other years have a smaller change. This information can be used for further
climate analysis.

Denote the global average annual mean temperature by $T(t)$ where t stands for time in
year, the 1971–2000 climatology by C, and the anomalies by $A(t)$. We have

$$T(t) = C + A(t). \tag{1.1}$$

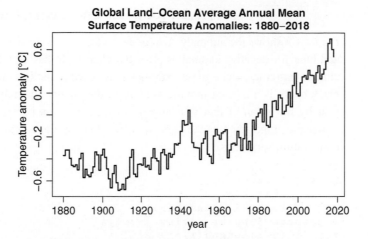

Figure 1.2 Staircase chart of the 1880–2018 global average annual mean temperature anomalies.

The temperature change from its previous year is

$$\Delta T = T(t) - T(t-1) = A(t) - A(t-1) = \Delta A. \tag{1.2}$$

This implies that an anomaly change is equal to its corresponding temperature change. We do not need to evaluate the 1971–2000 climatology C of the global average annual mean, when studying changes, as discussed earlier.

Figure 1.2 can be generated by the following computer code.

```
#R plot Fig. 1.2: Staircase chart of data
plot(NOAAtemp[1:139,1], NOAAtemp[1:139,2],
     type="s", #staircase curve for data
     col="black", lwd=2,
     main="Global␣Land-Ocean␣Average␣Annual␣Mean
␣␣␣␣␣Surface␣Temperature␣Anomalies:␣1880-2018",
     cex.lab=1.2,cex.axis=1.2,
     xlab="year",
     ylab=expression(paste(
       "Temperature␣Anomaly␣[", degree,"C]"))
)
```

The corresponding Python code is in the following box.

```
#Python plot Fig. 1.2: A staircase chart of data
# keyword arguments
kwargs = {'drawstyle' : 'steps'}
plt.plot(x, y, 'black', linewidth = 3, **kwargs);
plt.title("Global␣Land-Ocean␣Average␣Annual␣Mean␣\n␣␣\
Surface␣Temperature␣Anomaly:␣1880␣-␣2018", pad = 15)
plt.xlabel("Year", size = 25, labelpad = 20)
plt.ylabel("Temperature␣Anomaly␣[$\degree$C]",
           size = 25, labelpad = 20)
plt.show()
```

1.1.2.3 Plot a Bar Chart with Colors

Figure 1.3 shows the anomaly size for each year by a color bar: red for positive anomalies and blue for negative anomalies. This bar chart style has an advantage when visualizing climate extremes, since these extremes may leave a distinct impression on viewers. The black curve is a 5-point moving average of the annual anomaly data, computed for each year by the mean of that year, together with two years before and two years after. The moving average smooths the high-frequency fluctuations to reveal the long-term trends of temperature variations.

```
#R plot Fig. 1.3: A color bar chart of data
x   <-  NOAAtemp[,1]
y   <-  NOAAtemp[,2]
z   <-  rep(-99, length(x))
# compute 5-point moving average
for (i in 3:length(x)-2) z[i] =
  mean(c(y[i-2],y[i-1],y[i],y[i+1],y[i+2]))
n1 <- which(y>=0); x1 <- x[n1]; y1 <- y[n1]
n2 <- which(y<0); x2 <- x[n2]; y2 <- y[n2]
x3 <- x[2:length(x)-2]
y3 <- z[2:length(x)-2]
plot(x1, y1, type="h", #bars for data
     xlim = c(1880,2016), lwd=3,
     tck = 0.02,  #tck>0 makes ticks inside the plot
     ylim = c(-0.7,0.7), xlab="Year", col="red",
     ylab = expression(paste(
       "Temperature Anomaly [", degree ,"C]")),
     main ="Global Land-Ocean Average Annual Mean
      Surface Temperature Anomalies: 1880-2018",
     cex.lab = 1.2, cex.axis = 1.2)
lines(x2, y2, type="h",
     lwd = 3, tck = -0.02, col = "blue")
lines(x3, y3, lwd = 2)
```

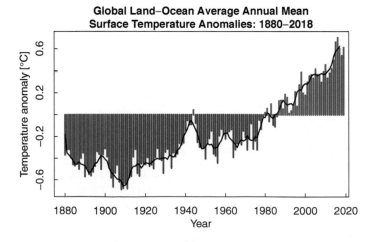

Figure 1.3   Color bar chart of the 1880–2018 global average annual mean temperature anomalies.

```python
#Python plot Fig. 1.3: A color bar chart of data
# define an array z with y number of ones
z = np.ones(y.size)
z[0] = -99
z[1] = -99
# computer 5-point moving average
for i in range(2, z.size - 2):
    z[i] = np.mean(y[i-2:i+3])
z[z.size - 2] = -99
z[z.size - 1] = -99
# define variables based on range
y1 = [y[i] for i in range(y.size) if y[i] >= 0]
x1 = [x[i] for i in range(y.size) if y[i] >= 0]
y2 = [y[i] for i in range(y.size) if y[i] < 0]
x2 = [x[i] for i in range(y.size) if y[i] < 0]
y3 = z[2:x.size-2]
x3 = x[2:x.size-2]

# plot the moving average
plt.plot(x3, y3, 'black', linewidth = 3)
plt.title("Global␣Land-Ocean␣Average␣Annual␣Mean␣\n␣\
Surface␣Temperature␣Anomaly:␣1880␣-␣2018", pad = 20)

# create bar chart
plt.bar(x1, y1, color = 'red');
plt.bar(x2, y2, color = 'blue');
plt.xlabel("Year", size = 25, labelpad = 20)
plt.ylabel("Temperature␣Anomaly␣[$\degree$C]",
           size = 25, labelpad = 20)
plt.show()
```

1.1.3 Statistical Indices

The commonly used basic statistical indices include mean, variance, standard deviation, skewness, kurtosis, and quantiles. We first use R to calculate these indices for the global average annual mean temperature anomalies. Then we describe their mathematical formulas and interpret the numerical results.

```r
#R code for computing statistical indices
setwd("/Users/sshen/climstats")
NOAAtemp = read.table(
  "data/aravg.ann.land_ocean.90S.90N.v4.0.1.201907.txt",
  header=FALSE)
temp2018=NOAAtemp[1:139,2] #use the temp data up to 2018
head(temp2018) #show the first six values
#[1] -0.370221 -0.319993 -0.320088 -0.396044 -0.458355 -0.470374
mean(temp2018) #mean
#[1] -0.1858632
sd(temp2018) #standard deviation
#[1] 0.324757
var(temp2018) #variance
```

```
#[1] 0.1054671
library(e1071)
#This R library is needed to compute the following parameters
#install.packages("e1071") #if it is not in your computer
skewness(temp2018)
#[1] 0.7742704
kurtosis(temp2018)
#[1] -0.2619131
median(temp2018)
#[1] -0.274434
quantile(temp2018,probs= c(0.05, 0.25, 0.75, 0.95))
#       5%          25%          75%          95%
#  -0.5764861 -0.4119770  0.0155245  0.4132383
```

We use $x = \{x_1, x_2, \ldots, x_n\}$ to denote the sampling data for a time series. The statistical indices computed by this R code are based on the following mathematical formulas for mean, variance, standard deviation, skewness, and kurtosis:

$$\text{Mean: } \mu(x) = \frac{1}{n} \sum_{k=1}^{n} x_k, \tag{1.3}$$

$$\text{Variance by unbiased estimate: } \sigma^2(x) = \frac{1}{n-1} \sum_{k=1}^{n} (x_k - \mu(x))^2, \tag{1.4}$$

$$\text{Standard deviation: } \sigma(x) = (\sigma^2(x))^{1/2}, \tag{1.5}$$

$$\text{Skewness: } \gamma_3(x) = \frac{1}{n} \sum_{k=1}^{n} \left(\frac{x_k - \mu(x)}{\sigma} \right)^3, \tag{1.6}$$

$$\text{Kurtosis: } \gamma_4(x) = \frac{1}{n} \sum_{k=1}^{n} \left(\frac{x_k - \mu(x)}{\sigma} \right)^4 - 3. \tag{1.7}$$

A *quantile* cuts a sorted sequence of data. For example, the 25th quantile, also called 25th percentile or 25% quantile, is the value at which 25% of the sorted data is smaller than this value and 75% is larger than this value. The 50th percentile is also known as the median.

```
#Python code for computing statistical indices
temp2018 = np.array(NOAAtemp.loc[0:138, 1]) # data string
# arithmetic average of the data
print("The␣Mean␣is␣%f.\n" % stats.mean(temp2018))
# standard deviation of the data
print("The␣Standard␣Deviation␣is␣%f.\n" %
      stats.stdev(temp2018))
# variance of the data
print("The␣Variance␣is␣%f.\n" % stats.variance(temp2018))
# skewness of the data
print("The␣Skewness␣is␣%f.\n" % skewness(temp2018))
# kurtosis of the data
print("The␣Kurtosis␣is␣%f.\n" % kurtosis(temp2018))
# median of the data
print("The␣Median␣is␣%f.\n" % stats.median(temp2018))
```

```
# percentiles
print("The 5th, 25th, 75th and 95th percentiles are:")
probs = [5, 25, 75, 95]
print([round(np.percentile(temp2018, p),5) for p in probs])
print()
```

The meaning of these indices may be explained as follows. The mean is the simple average of samples. The variance of climate data reflects the strength of variations of a climate system and has units of the square of the data entries, such as $[°C]^2$. You may have noticed the denominator $n - 1$ instead of n, which is for an estimate of unbiased sample variance. The standard deviation describes the spread of the sample entries and has the same units as the data. A large standard deviation means that the samples have a broad spread.

Skewness is a dimensionless quantity. It measures the asymmetry of sample data. Zero skewness means a symmetric distribution about the sample mean. For example, the skewness of a normal distribution is zero. Negative skewness denotes a skew to the left, meaning the existence of a long tail on the left side of the distribution. Positive skewness implies a long tail on the right side.

The words "kurtosis" and "kurtic" are Greek in origin and indicate peakedness. Kurtosis is also dimensionless and indicates the degree of peakedness of a probability distribution. The kurtosis of a normal distribution[1] is zero when 3 is subtracted as in Eq. (1.7). Positive kurtosis means a high peak at the mean, thus the distribution shape is slim and tall. This is referred to as leptokurtic. "Lepto" is Greek in origin and means thin or fine. Negative kurtosis indicates a low peak at the mean, thus the distribution shape is fat and short, referred to as platykurtic. "Platy" is also Greek in origin and means flat or broad.

For the 139 years of the NOAAGlobalTemp global average annual mean temperature anomalies, mean is $-0.1959°C$, which means that the 1880–2018 average is lower than the average during the climatology period 1971–2000. During the climatology period, the temperature anomaly average is approximately zero, and can be computed by the R command `mean(temp2018[92:121])`.

The variance of the data in the 139 years from 1880 to 2018 is $0.1055 [°C]^2$, and the corresponding standard derivation is $0.3248 [°C]$. The skewness is 0.7743, meaning skewed with a long tail on the right, thus with more extreme high temperatures than extreme low temperatures, as shown by Figure 1.4. The kurtosis is -0.2619, meaning the distribution is flatter than a normal distribution, as shown in the histogram Figure 1.4.

The median is $-0.2744°C$ and is a number characterizing a set of samples such that 50% of the sample values are less than the median, and another 50% are greater than the median. To find the median, sort the samples from the smallest to the largest. The median is then the sample value in the middle. If the number of the samples is even, then the median is equal to the mean of the two middle sample values.

---

[1] Chapter 2 will have a detailed description of the normal distribution and other probabilistic distributions.

Quantiles are defined in a way similar to median by sorting. For example, 25-percentile (also called the 25th percentile) $-0.4120°C$ is a value such that 25% of sample data are less than this value. By definition, 60-percentile is thus larger than 50-percentile. Here, percentile is a description of quantile relative to 100. Obviously, 100-percentile is the largest datum, and 0-percentile is the smallest one. Often, a box plot is used to show the typical quantiles (see Fig. 1.5).

The 50-percentile (or 50th percentile) $-0.2744°C$ is equal to the median. If the distribution is symmetric, then the median is equal to the mean. Otherwise, these two quantities are not equal. If the skew is to the right, then the mean is on the right of the median: the mean is greater than the median. If the skew is to the left, then the mean is on the left of the median: the mean is less than the median. Our 139 years of temperature data are right skewed and have mean equal to $-0.1959°C$, greater than their median equal to $-0.2969°C$.

1.2 Commonly Used Climate Statistical Plots

We will use the 139 years of NOAAGlobalTemp temperature data and R to illustrate some commonly used statistical figures: histogram, boxplot, Q-Q plot, and linear regression trend line.

1.2.1 Histogram of a Set of Data

A histogram of the NOAAGlobalTemp global average annual mean temperature anomalies data from 1880 to 2018 is shown in Figure 1.4, which can be generated by the following computer code.

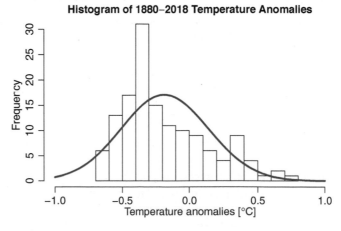

Figure 1.4 Histogram and its normal distribution fit of the global average annual mean temperature anomalies from 1880 to 2018. Each small interval in the horizontal coordinate is called a bin. The frequency in the vertical coordinate is the number of temperature anomalies in a given bin. For example, the frequency for the bin $[-0.5, -0.4]°C$ is 17.

```
#R plot Fig. 1.4: Histogram and its fit
par(mar=c(3.5,3.5,2.5,1), mgp=c(2,0.8,0))
h <- hist(NOAAtemp[1:139, 2],
        main="Histogram␣of␣1880-2018␣Temperature␣Anomalies",
        xlab=expression(paste(
              "Temperature␣anomalies␣[", degree, "C]")),
        xlim=c(-1,1), ylim=c(0,30),
        breaks=10, cex.lab=1.2, cex.axis=1.2)
xfit <- seq(-1, 1, length=100)
areat <- sum((h$counts)*diff(h$breaks[1:2]))#Normalization area
yfit <- areat*dnorm(xfit,
                    mean=mean(NOAAtemp[1:139,2]),
                    sd=sd(NOAAtemp[1:139,2]))
#Plot the normal fit on the histogram
lines(xfit, yfit, col="blue", lwd=3)
```

```
#Python plot Fig. 1.4: Histogram and its fit
mu, std = scistats.norm.fit(y)
# create an evenly spaced sequence of 100 points in [-1,1]
points = np.linspace(-1, 1, 100)
plt.hist(y, bins = 20, range=(-1,1), color ='white',
        edgecolor = 'k', density = True);
plt.plot(points, scistats.norm.pdf(points, mu, std),
        color = 'b', linewidth = 3)
plt.title(r"Histogram␣of␣1880␣-␣2018␣Temperature␣Anomalies",
        pad = 20)
plt.xlabel("Temperature␣Anomalies␣[$\degree$C]",
        size = 25, labelpad = 20)
plt.ylabel("Frequency", size = 25, labelpad = 20)
plt.show()
```

The shape of the histogram agrees with the characteristics predicted by the statistical indices in the previous subsection:

(i) The distribution is asymmetric and skewed to the right with skewness equal to 0.7743.

(ii) The distribution is platykurtic with a kurtosis equal to 0.2619, i.e., it is flatter than the standard normal distribution indicated by the blue curve.

1.2.2 Box Plot

Figure 1.5 is the box plot of the 1880-2018 NOAAGlobalTemp global average annual mean temperature anomaly data, and can be made from the following R command.

```
#R plot Fig. 1.5: Box plot
boxplot(NOAAtemp[1:139, 2], ylim = c(-0.8, 0.8),
        ylab=expression(paste(
          "Temperature␣anomalies␣[", degree, "C]")),
        width=NULL, cex.lab=1.2, cex.axis=1.2)
```

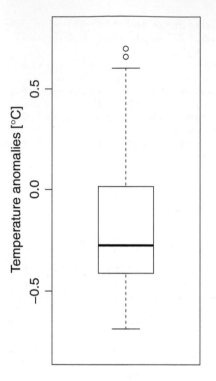

The rectangular box's mid line indicates the median, which is 0.2744°C. The rectangular box's lower boundary is the first quartile, i.e., 25th percentile 0.4120°C. The box's upper boundary is the third quartile, i.e., the 75th percentile 0.0155°C. The box's height is the third quartile minus the first quartile, and is called the interquartile range (IQR). The upper "whisker" is the third quartile plus 1.5 IQR. The lower whisker is supposed to be at the first quartile minus 1.5 IQR. However, this whisker would then be lower than the lower extreme. In this case, the lower whisker takes the value of the lower extreme, which is −0.6872°C. The points outside the two whiskers are considered outliers. Our dataset has two outliers, which are 0.6607

Figure 1.5 Box plot of the global average annual mean temperature anomalies from 1880 to 2018.

and 0.7011°C, and are denoted by two small circles in the box plot. The two outliers occurred in 2015 and 2016, respectively.

```
#Python plot Fig. 1.5: Box plot
medianprops = dict(linestyle='-', linewidth=2.5, color='k')
whiskerprops = dict(linestyle='--',linewidth=2.0, color='k')
plt.boxplot(y, medianprops = medianprops,
            whiskerprops = whiskerprops);
plt.title("Boxplot␣of␣1880␣-␣2018␣Temperature␣Anomalies",
          pad = 20)
plt.ylabel("Temperature␣Anomalies␣[$\degree$C]",
           size = 25, labelpad = 20)
y_ticks = np.linspace(-0.5,0.5,3)
plt.yticks(y_ticks)
plt.show()
```

1.2.3 Q-Q Plot

Figure 1.6 shows quantile-quantile (Q-Q) plots, also denoted by q-q plots, qq-plots, or QQ-plots.

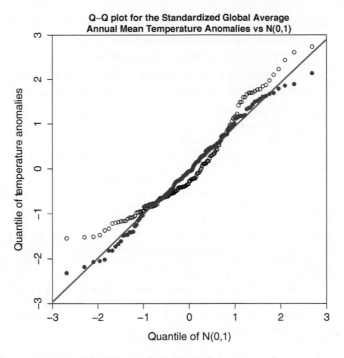

Figure 1.6 Black empty-circle points are the Q-Q plot of the standardized global average annual mean temperature anomalies versus standard normal distribution. The purple points are the Q-Q plot for the data simulated by `rnorm(139)`. The red is the distribution reference line of $N(0,1)$.

The function of a Q-Q plot is to compare the distribution of a given set of data with a specific reference distribution, such as a standard normal distribution with zero mean and standard deviation equal to one, denoted by $N(0,1)$. A Q-Q plot lines up the percentiles of data on the vertical axis and the same number of percentiles of the specific reference distribution on the horizontal axis. The pairs of the quantiles $(x_i, y_i), i = 1, 2, \ldots, n$ determine the points on the Q-Q plot. Here, x_i and y_i correspond to the same cumulative percentage or probability p_i for both x and y variables, where p_i monotonically increases from approximately 0 to 1 as i goes from 1 to n. A red Q-Q reference line is plotted as if the vertical axis values are also the quantiles of the given specific distribution. Thus, the Q-Q reference line should be diagonal.

The black empty circles in Figure 1.6 compare the quantiles of the standardized global average annual mean temperature anomalies marked on the vertical axis with those of the standard normal distribution marked on the horizontal axis. The standardized anomalies are equal to anomalies divided by the sample standard deviation. The purple dots shows a Q-Q plot of a set of 139 random numbers simulated by the standard normal distribution. As expected, the simulated points are located close to the red diagonal line, which is the distribution reference line of $N(0,1)$. On the other hand, the temperature Q-Q plot shows a considerable degree of scattering of the points away from the reference line. We may intuitively conclude that the global average annual temperature anomalies from 1880 to 2018 are not exactly distributed according to a normal (also known as Gaussian) distribution.

However, we may also conclude that the distribution of these temperatures is not very far away from the normal distribution either, because the points on the Q-Q plot are not very far away from the distribution reference line, and also because even the simulated $N(0,1)$ points are noticeably off the reference line for the extremes.

Figure 1.6 can be generated by the following computer code.

```
#R plot Fig. 1.6: Q-Q plot for the standardized
# global average annual mean temperature anomalies
temp2018 <- NOAAtemp[1:139,2]
tstand <- (temp2018 - mean(temp2018))/sd(temp2018)
set.seed(101)
qn <- rnorm(139) #simulate 139 points by N(0,1)
qns <- sort(qn) # sort the points
qq2 <- qqnorm(qns,col="blue", lwd = 2)

setEPS() #Automatically saves the eps file
postscript("fig0106.eps", height=7, width=7)
par(mar = c(4.5,5,2.5,1), xaxs = "i", yaxs = "i")
qt = qqnorm(tstand,
  main = "Q-Q plot for the Standardized Global Average
  Annual Mean Temperature Anomalies vs N(0,1)",
    ylab="Quantile of Temperature Anomalies",
    xlab="Quantile of N(0,1)",
  xlim=c(-3,3), ylim = c(-3,3),
          cex.lab = 1.3, cex.axis = 1.3)
qqline(tstand, col = "red", lwd=3)
points(qq2$x, qq2$y, pch = 19,
       col ="purple")
dev.off()
```

In the R code, we first standardize (also called normalize) the global average annual mean temperature data by subtracting the data mean and dividing by the data's standard deviation. Then, we use these 139 years of standardized global average annual mean temperature anomalies to generate a Q-Q plot, which is shown in Figure 1.6.

```
#Python plot Fig. 1.6: Q-Q plot for the standardized
# global average annual mean temperature anomalies
NOAAtemp = read_table(
     "data/aravg.ann.land_ocean.90S.90N.v4.0.1.201907.txt",
                     header = None, delimiter = "\s+")
x = np.array(NOAAtemp.loc[0:138, 0])
y = np.array(NOAAtemp.loc[0:138, 1])
line = np.linspace(-3, 3, y.size)
tstand = np.sort((y - np.mean(y))/np.std(y))
# simulate 139 points following N(0,1)
qn = np.random.normal(size = y.size)
qns = np.sort(qn) # sort the points
qq2 = sm.qqplot(qns)
fig = plt.figure(figsize=(12,12)) # set up figure
sm.qqplot(tstand, color = "k", linewidth = 1)
# plot diagonal line
plt.plot(line, line, 'r-', linewidth = 3)
```

```
# Q-Q plot of standard normal simulations
plt.plot(line, qns, 'mo')
plt.tick_params(length=6, width=2, labelsize=20)
plt.title("Q-Q␣plot␣for␣the␣Standardized␣Global␣\n␣\
Temperature␣Anomalies␣vs␣N(0,1)", pad = 20)
plt.xlabel("Quantile␣of␣N(0,1)", size = 25, labelpad = 20)
plt.ylabel("Quantile␣of␣Temperature␣Anomalies", size = 25)
plt.show()
```

1.2.4 Plot a Linear Trend Line

Climate data analysis often involves plotting a linear trend line for time series data, such as the linear trend for the global average annual mean surface temperature anomalies, shown in Figure 1.7. The R code for plotting a linear trend line of data sequence y and time sequence t is `abline(lm(y ~ t))`.

Figure 1.7 can be generated by the following computer code.

```
#R plot Fig. 1.7: Data line graph with a linear trend line
par(mar=c(3.5,3.5,2.5,1), mgp=c(2,0.8,0))
plot(NOAAtemp[1:139,1], NOAAtemp[1:139,2],
    type="l", col="brown", lwd=3,
    main="␣Global␣Land-Ocean␣Average␣Annual␣Mean
Surface␣Temperature␣Anomaly:␣1880-2018",
    cex.lab=1.2,cex.axis=1.2,
    xlab="Year",
    ylab=expression(paste(
      "Temperature␣Anomaly␣[", degree,"C]"))
)
abline(lm(NOAAtemp[1:139,2] ~ NOAAtemp[1:139,1]),
    lwd=3, col="blue")
```

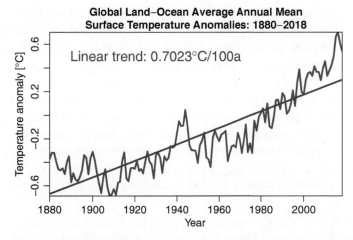

Global Land–Ocean Average Annual Mean Surface Temperature Anomalies: 1880–2018

Linear trend: 0.7023°C/100a

Figure 1.7 Linear trend line with the 1880–2018 global average annual mean surface temperature based on the NOAAGlobalTemp V4.0 dataset.

```
lm(NOAAtemp[1:139,2] ~ NOAAtemp[1:139,1])
#        (Intercept)   NOAAtemp[1:139, 1]
#-13.872921            0.007023
#Trend 0.7023 degC/100a
text(1930, 0.5,
 expression(paste("Linear trend: 0.7023",
                  degree,"C/100a")),
    cex = 1.5, col="blue")
```

```
#Python plot Fig. 1.7: Data line graph with a trend line
trend = np.array(np.polyfit(x, y, 1))
abline = trend[1] + x*trend[0]
plt.plot(x, y, 'k-',
         color = 'tab:brown', linewidth = 3);
plt.plot(x, abline, 'k-', color = 'b', linewidth = 3);
plt.title("Global Land-Ocean Average Annual Mean \n\
Surface Temperature Anomaly: 1880 - 2018", pad = 20);
plt.text(1880, 0.5, r"Linear trend: 0.7023$\degree$C/100a",
         color= 'b', size = 28)
plt.xlabel("Year", size = 25, labelpad = 20)
plt.ylabel("Temperature Anomaly [$\degree$C]",
           size = 25, labelpad = 20)
```

1.3 Read netCDF Data File and Plot Spatial Data Maps

1.3.1 Read netCDF Data

Climate data are at spatiotemporal points, such as at the grid points on the Earth's surface and at a sequence of time. NetCDF (Network Common Data Form) is a popular file format for modern climate data with spatial locations and temporal records. The gridded NOAA-GlobalTemp data have a netCDF version, and can be downloaded from
`www.esrl.noaa.gov/psd/data/gridded/data.noaaglobaltemp.html`

The data are written into a 3D array, with 2D latitude–longitude for space, and 1D for time. R and Python can read and plot the netCDF data. We use the NOAAGlobalTemp as an example to illustrate the netCDF data reading and plotting. Figure 1.8 displays a temperature anomaly map for the entire globe for December 2015.

Figure 1.8 can be generated by the following computer code.

```
#R read the netCDF data: NOAAGlobalTemp
setwd("/Users/sshen/climstats")
#install.packages("ncdf4")
library(ncdf4)
nc = ncdf4::nc_open("data/air.mon.anom.nc")
nc # describes details of the dataset
Lat <- ncvar_get(nc, "lat")
```

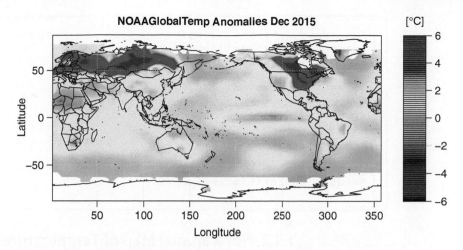

Figure 1.8 The surface temperature anomalies of December 2015 with respect to the 1971–2000 climatology (data source: The NOAAGlobalTemp V4.0 gridded monthly data).

```
Lat # latitude data
#[1] -87.5 -82.5 -77.5 -72.5 -67.5 -62.5
Lon <- ncvar_get(nc, "lon")
Lon # longitude data
#[1]    2.5   7.5  12.5  17.5  22.5  27.5
Time <- ncvar_get(nc, "time")
head(Time) # time data in Julian days
#[1] 29219 29250 29279 29310 29340 29371
library(chron) # convert Julian date to calendar date
nc$dim$time$units # .nc base time for conversion
#[1] "days since 1800-1-1 00:00:0.0"
month.day.year(29219,c(month = 1, day = 1, year = 1800))
#1880-01-01 # the beginning time of the dataset
tail(Time)
#[1] 79988 80019 80047 80078 80108 80139
month.day.year(80139,c(month = 1, day = 1, year = 1800))
#2019-06-01 # the end time of the dataset

# extract anomaly data in (lon, lat, time) coordinates
NOAAgridT <- ncvar_get(nc, "air")
dim(NOAAgridT) # dimensions of the data array
#[1]   72   36 1674 #5-by-5, 1674 months from Jan 1880-Jun 2019
```

```
#Python read the netCDF data: NOAAGlobalTemp
import netCDF4 as nc # import netCDF data reading package
# go to the working directory
os.chdir('/Users/sshen/climstats')
# read a .nc file from the folder named "data"
nc = nc.Dataset('data/air.mon.anom.nc')
# get the detailed description of the dataset
print(nc)
```

```
# extract latitude, longitude, time and temperature
lon = nc.variables['lon'][:]
lat = nc.variables['lat'][:]
time = nc.variables['time'][:]
air = nc.variables['air']
# covert Julian date to calendar date
from netCDF4 import num2date
from datetime import datetime, date, timedelta
from matplotlib.dates import date2num, num2date
units = nc.variables['time'].units
print(units)
dtime = num2date(time)
```

1.3.2 Plot a Spatial Map of Temperature

The NOAAGlobalTemp anomaly data on a 5-degree latitude–longitude grid for a given time can be represented by a color map. Figure 1.8 shows the temperature anomaly map for December 2015, an El Niño month. The eastern tropical Pacific has positive anomalies, which is a typical El Niño signal. That particular month also exhibited a very large anomaly across Europe, and the eastern United States and Canada. The white areas over the high latitude regions lack data.

The figure can be generated by the following computer code.

```
#R plot Fig. 1.8: Dec 2015 surface temp anomalies map
library(maps)# requires maps package
mapmat=NOAAgridT[,,1632]
# Column of NOAAgridT 1632 corresponds to Dec 2015
mapmat=pmax(pmin(mapmat,6),-6) # put values in [-6, 6]
int=seq(-6, 6, length.out=81)
rgb.palette=colorRampPalette(c('black','blue',
   'darkgreen', 'green', 'yellow','pink','red','maroon'),
                              interpolate='spline')
par(mar=c(3.5, 4, 2.5, 1), mgp=c(2.3, 0.8, 0))
filled.contour(Lon, Lat, mapmat,
               color.palette=rgb.palette, levels=int,
  plot.title=title(main="NOAAGlobalTemp Anomalies Dec 2015",
          xlab="Latitude", ylab="Longitude", cex.lab=1.2),
  plot.axes={axis(1, cex.axis=1.2, las=1);
     axis(2, cex.axis=1.2, las=2); map('world2', add=TRUE); grid()},
  key.title=title(main=expression(paste("[", degree, "C]"))),
  key.axes={axis(4, cex.axis=1.2)})
```

```
#Python plot Fig. 1.8: Dec 2015 surface temp anomalies map
dpi = 100
fig = plt.figure(figsize = (1100/dpi, 1100/dpi), dpi = dpi)
ax  = fig.add_axes([0.1, 0.1, 0.8, 0.9])
```

```
# create map
dmap = Basemap(projection = 'cyl', llcrnrlat = min(lat),
               urcrnrlat = max(lat), resolution = 'c',
               llcrnrlon = min(lon), urcrnrlon = max(lon))
# draw coastlines, state and country boundaries, edge of map
dmap.drawcoastlines()
dmap.drawstates()
dmap.drawcountries()
# convert latitude/longitude values to plot x/y values
x, y = dmap(*np.meshgrid(lon, lat))
# draw filled contours
cnplot = dmap.contourf(x, y, mapmat, clev, cmap=myColMap)
# tick marks
ax.set_xticks([0, 50, 100, 150, 200, 250, 300, 350])
ax.set_yticks([-50,0,50])
ax.tick_params(length=6, width=2, labelsize=20)
# add colorbar
# pad: distance between map and colorbar
cbar = dmap.colorbar(cnplot, pad = "4%",drawedges=True,
         shrink=0.55, ticks = [-6,-4,-2,0,2,4,6])
# add colorbar title
cbar.ax.set_title('[$\degree$C]', size= 17, pad = 10)
cbar.ax.tick_params(labelsize = 15)
# add plot title
plt.title('NOAAGlobalTemp␣Anomalies␣Dec␣2015',
         size = 25, fontweight = "bold", pad = 15)
# label x and y
plt.xlabel('Longitude', size = 25, labelpad = 20)
plt.ylabel('Latitude', size = 25, labelpad = 10)
# display on screen
plt.show()
```

1.3.3 Panoply Plot of a Spatial Map of Temperature

You can also use the Panoply software package to plot the map (see Fig. 1.9). This is a very powerful data visualization tool developed by NASA specifically for displaying netCDF files. The software package is free and can be downloaded from www.giss.nasa.gov/tools/panoply.

To make a Panoply plot, open Panoply and choose Open from the File menu. Open dropdown, which will allow you to go to the right directory to find the netCDF file you wish to plot. In our case, the file is air.mon.anom.nc. Choose the climate parameter air. Click on Create Plot. A map will show up. Then you have many choices to modify the map, ranging from the data Array, Scale, and Map, etc., and finally produce the figure. You can then tune the figure by choosing different graphics parameters underneath the figure, such as Array(s) to choose which month to plot, Map to choose the map projection types and the map layout options, and Labels to type proper captions and labels.

To learn more about Panoply, please use an online Panoply tutorial, such as the NASA Panoply help page www.giss.nasa.gov/tools/panoply/help/.

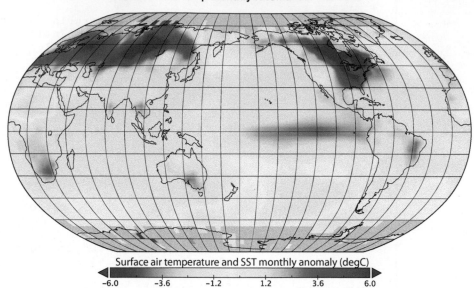

NOAA GlobalTemp Monthly Anomalies: Dec 2015

Surface air temperature and SST monthly anomaly (degC)

−6.0   −3.6   −1.2   1.2   3.6   6.0

Figure 1.9   A Panoply plot of Robinson projection map for the surface temperature anomalies of December 2015.

Compared with the R or Python map Figure 1.8, the visualization effect of the Panoply map seems more appealing. However, R and Python have the advantage of flexibility and can deal with all kinds of data. For example, the Plotly graphing library in R `https://plotly.com/r/` and in Python `https://plotly.com/python/` can even make interactive and 3D graphics. You may find some high-quality figures from the Intergovernmental Panel on Climate Change (IPCC) report (2021) `www.ipcc.ch` and reproduce them using the computing tools described here.

1.4 1D-space-1D-time Data and Hovmöller Diagram

A very useful climate data visualization technique is the Hovmöller diagram. It displays how a climate variable varies with respect to time along a given line section of latitude, longitude, or altitude. It usually has the abscissa for time and ordinate for the line section. A Hovmöller diagram can conveniently show time evolution of a spatial pattern, such as wave motion from south to north or from west to east.

Figure 1.10 is a Hovmöller diagram for the sea surface temperature (SST) anomalies at a longitude equal to 240°, i.e., 120° W, in a latitude interval [30° S, 30° N], with time range from January 1989 to December 2018. When the red strips become strong from the south to north, a strong El Niño occurs, such as those in the 1997–1998 and 2015–2016 winters.

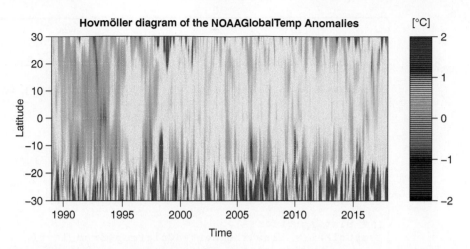

Figure 1.10   Hovmöller diagram for the gridded NOAAGlobalTemp monthly anomalies at longitude 120° W and a latitude interval [30° S, 30° N].

The Hovmöller diagram Figure 1.10 may be plotted by the following computer code.

```
#R plot Fig. 1.10: Hovmoller diagram
library(maps)
mapmat=NOAAgridT[30,12:24,1309:1668]
#Longitude= 240 deg, Lat =[-30 30] deg
#Time=Jan 1989-Dec 2018: 30 years
mapmat=pmax(pmin(mapmat,2),-2) # put values in [-2,2]
par(mar=c(4,5,3,0))
int=seq(-2,2,length.out=81)
rgb.palette=colorRampPalette(c('black','blue',
 'darkgreen','green', 'yellow','pink','red','maroon'),
                        interpolate='spline')
par(mar=c(3.5,3.5,2.5,1), mgp=c(2.4, 0.8, 0))
x = seq(1989, 2018, len=360)
y = seq(-30, 30, by=5)
filled.contour(x, y, t(mapmat),
      color.palette=rgb.palette, levels=int,
      plot.title=title(main=
   "Hovmoller␣diagram␣of␣the␣NOAAGlobalTemp␣Anomalies",
             xlab="Time",ylab="Latitude", cex.lab=1.2),
    plot.axes={axis(1, cex.axis=1.2);
               axis(2, cex.axis=1.2);
               map('world2', add=TRUE);grid()},
      key.title=title(main =
             expression(paste("[", degree, "C]"))),
      key.axes={axis(4, cex.axis=1.2)})
```

```
#Python plot Fig. 1.10: Hovmoller diagram
mapmat2 = NOAAgridT[1308:1668,11:23,29]
```

```python
# define values between -2 and 2
mapmat2 = np.array([[j if j < 2 else 2 for j in i]
                    for i in mapmat2])
mapmat2 = np.array([[j if j > -2 else -2 for j in i]
                    for i in mapmat2])
# find dimensions
mapmat2 = np.transpose(mapmat2)
print(mapmat2.shape)
lat3 = np.linspace(-30,30, 12)
print(lat3.shape)
time = np.linspace(1989, 2018, 360)
print(time.shape)

# plot functions
myColMap = LinearSegmentedColormap.from_list(
    name='my_list',
    colors=['black','blue','darkgreen','green','lime',
            'yellow', 'pink','red','maroon'], N=100)
clev2 = np.linspace(mapmat2.min(), mapmat2.max(), 501)
contf = plt.contourf(time, lat3, mapmat2,
                     clev2, cmap=myColMap);

plt.text(2019.2, 31.5,
         "[$\degree$C]", color='black', size = 23)
plt.title("Hovmoller diagram of the \n \
NOAAGlobalTemp Anomalies",
          fontweight = "bold", size = 25, pad = 20)
plt.xlabel("Time", size = 25, labelpad = 20)
plt.ylabel("Latitude", size = 25, labelpad = 12)
colbar = plt.colorbar(contf, drawedges=False,
                      ticks = [-2,-1,0,1,2])
```

1.5 4D netCDF File and Its Map Plotting

The 4D climate data means 3D spatial dimensions and 1D time. For example, the NCEP Global Ocean Data Assimilation System (GODAS) monthly water temperature data are at 40 depth levels ranging from 5 meters to 4478 meters and at 1/3 degree latitude by 1 degree longitude horizontal resolution and are from January 1980. The NOAA-CIRES 20th Century Reanalysis (20CR) monthly air temperature data are at 24 different pressure levels ranging from 1,000 mb to 10 mb and at 2-degree latitude and longitude horizontal resolution and are from January 1851.

GODAS data can be downloaded from NOAA ESRL,
www.esrl.noaa.gov/psd/data/gridded/data.godas.html.
The data for each year is a netCDF file and is about 140MB. The following R code can read the GODAS 2015 data into R.

```
#R read a 4D netCDF file: lon, lat, level, time
setwd("/Users/sshen/climstats")
library(ncdf4)
# read GODAS data 1-by-1 deg, 40 levels, Jan-Dec 2015
nc=ncdf4::nc_open("data/godas2015.nc")
nc
Lat <- ncvar_get(nc, "lat")
Lon <- ncvar_get(nc, "lon")
Level <- ncvar_get(nc, "level")
Time <- ncvar_get(nc, "time")
head(Time)
#[1] 78527 78558 78586 78617 78647 78678
library(chron)
month.day.year(78527,c(month = 1, day = 1, year = 1800))
# 2015-01-01
# potential temperature pottmp[lon, lat, level, time]
godasT <- ncvar_get(nc, "pottmp")
dim(godasT)
#[1] 360 418  40  12,
#i.e., 360 lon, 418 lat, 40 levels, 12 months=2015
t(godasT[246:250, 209:210, 2, 12])
#Dec level 2 (15-meter depth) water temperature [K] of
#a few grid boxes over the eastern tropical Pacific
#         [,1]      [,2]      [,3]      [,4]      [,5]
#[1,] 300.0655 299.9831 299.8793 299.7771 299.6641
#[2,] 300.1845 300.1006 299.9998 299.9007 299.8045
```

```
#Python read a netCDF data file
import netCDF4 as nc
# go to your working directory
os.chdir('/Users/sshen/climstats')
# read a .nc file from the folder named "data"
nc1 = nc.Dataset('data/godas2015.nc')
# get the detailed description of the dataset
print(nc1)
# dimensions of the 2015 pottem data array
godasT = nc1.variables['pottmp'][:]
print(godasT.shape)
# (12, 40, 418, 360)
```

Figure 1.11 shows the 2015 annual mean water temperature at 195 meters depth based on the GODAS data. At this depth level, the equatorial upwelling appears: The deep ocean cooler water in the equatorial region upwells and makes the equatorial water cool. The equatorial water is not the hottest anymore at this level, and is cooler than the water in some subtropical regions as shown in Figure 1.11.

Figure 1.11 can be generated by the following computer code.

```
#R plot Fig. 1.11: The ocean potential temperature
# the 20th layer from surface: 195 meters depth
# compute 2015 annual mean temperature at 20th layer
library(maps)
climmat=matrix(0,nrow=360,ncol=418)
```

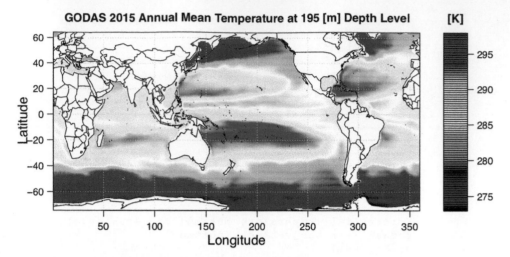

Figure 1.11   The 2015 annual mean water temperature at 195 meters depth based on GODAS data.

```
sdmat=matrix(0,nrow=360,ncol=418)
Jmon<-1:12
for (i in 1:360){
  for (j in 1:418){
    climmat[i,j] = mean(godasT[i,j,20,Jmon]);
    sdmat[i,j]=sd(godasT[i,j,20,Jmon])
                }
              }
int=seq(273,298,length.out=81)
rgb.palette=colorRampPalette(c('black','blue',
    'darkgreen','green', 'white','yellow',
    'pink','red','maroon'), interpolate='spline')
par(mar=c(3.5, 3.5, 2.5, 0), mgp=c(2, 0.8, 0))
filled.contour(Lon, Lat, climmat,
    color.palette=rgb.palette, levels=int,
    plot.title=title(main=
  "GODAS␣2015␣Annual␣Mean␣Temperature␣at␣195[m]␣Depth␣Level",
            xlab="Longitude",ylab="Latitude",
            cex.lab=1.3, cex.axis=1.3),
plot.axes={axis(1); axis(2); map('world2', add=TRUE);grid()},
            key.title=title(main="[K]"))
```

```
#Python plot Fig. 1.11: A spatial map from 4D data
climmat = np.zeros((360, 418))
for i in range(360):
    for j in range(418):
        climmat[i,j] = np.mean(godasT[:, 20, j, i])
climmat = np.transpose(climmat)
lat4 = np.linspace(-75, 65, 418)
long = np.linspace(0, 360, 360)
```

```
myColMap = LinearSegmentedColormap.from_list(
    name='my_list',
    colors=['black','blue','darkgreen','green',
            'white','yellow','pink','red','maroon'],
        N=100)
plt.figure(figsize=(18,12));
clev3 = np.arange(godasT.min(), godasT.max(), 0.1)
contf = plt.contourf(long, lat4, climmat,
                     clev3, cmap=myColMap);
plt.text(382, 66, "[K]", fontsize=23, color='black')
plt.tick_params(length=6, width=2, labelsize=20)
m = Basemap(projection='cyl', llcrnrlon=0,
    urcrnrlon=360, resolution='l', fix_aspect=False,
    suppress_ticks=False, llcrnrlat=-75, urcrnrlat=65)
m.drawcoastlines(linewidth=1)
plt.title("GODAS␣2015␣Annual␣Mean␣Temperature␣\n␣\
at␣195[m]␣Depth␣Level",
          size = 25, fontweight = "bold", pad = 20)
plt.xlabel("Longitude", size = 25, labelpad = 20)
plt.ylabel("Latitude", size = 25, labelpad = 15)
plt.tick_params(length=6, width=2, labelsize=30)
colbar = plt.colorbar(contf, drawedges=False,
                      ticks = [275,280,285,290,295])
```

1.6 Paraview, 4DVD, and Other Tools

Besides using R, Python, and Panoply to plot climate data, other software packages may also be used to visualize data for some specific purposes, such as Paraview for 3D visualization and 4DVD for fast climate diagnostics and data delivery.

1.6.1 Paraview

Paraview is an open-source data visualization software package, available at www.paraview.org/download. There are many online tutorials, such as www.paraview.org/tutorials.

You may use the software ParaView to plot the GODAS data in a 3D view as shown in Figure 1.12 for the December 2015 water temperature.

1.6.2 4DVD

4DVD (4-dimensional visual delivery of big climate data) is a fast data visualization and delivery software system at www.4dvd.org. It optimally harnesses the power of distributed computing, databases, and data storage to allow a large number of general public users to quickly visualize climate data. For example, teachers and students can use 4DVD to

Paraview Plot of 3D Ocean Water Temperature, December 2015

Figure 1.12   A 3D view of the December 2015 annual mean water temperature based on the GODAS data. The surface of the Northern Hemisphere and the cross-sectional map along the equator from surface to the 2,000 meters depth. The dark color indicates land; the red, yellow, and blue colors indicate temperature.

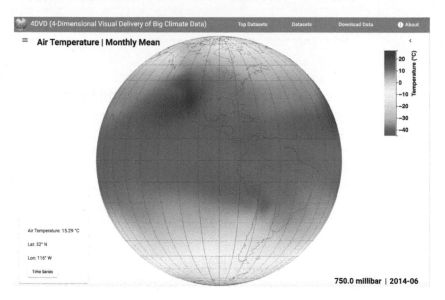

Figure 1.13   The 4DVD screenshot for the NOAA-CIRES 20th Century Reanalysis temperature at 750 millibar height level for June 2014.

visualize climate model data in classrooms and download the visualized data instantly. The 4DVD website has a tutorial for users.

Here, we provide an example of 4DVD visualization of the NOAA-CIRES 20CR climate model data for atmosphere. The 4DVD can display a temperature map at a given time and given pressure level, as shown in Figure 1.13 for January 1851 and 750 millibar pressure level, or one can obtain a time series of the monthly air temperature for a specific grid box from January 1851. In fact, 4DVD can show multiple time series for the same latitude-longitude location but at different pressure levels. The 4DVD not only allows a user to visualize the data but also to download the data for the figure shown in the 4DVD system. In this sense, 4DVD is like a machine that plays data, while the regular DVD player machine, popular for about 30 years since the 1980s, plays DVD discs for music and movies.

1.6.3 Other Online Climate Data Visualization Tools

Besides R, Python, Panoply, ParaView, and 4DVD, there are many other data visualization and delivery software systems. A few popular and free ones are listed as follows.

Nullschool `https://nullschool.net` is a beautiful data visualizer of wind, ocean flows, and many climate parameters. It is supported by the data from a global numerical weather prediction system, named the Global Forecast System (GFS) run by the United States' National Weather Service (NWS).

Ventusky `www.ventusky.com` has both website and smartphone app. It has an attractive and user-friendly interface that allows users to get digital weather information instantly around the globe.

Climate Reanalyzer `climatereanalyzer.org` is a comprehensive tool for climate data plotting and download. It has a user-friendly interface for reanalysis and historical station data. The data can be exported in either CSV (comma-separated values) format or JSON (JavaScript object notation) format.

Google Earth Engine `earthengine.google.com` provides visualization tools together with a huge multi-petabytes storage of climate data. Its modeling and satellite data sources are from multiple countries.

Giovanni `giovanni.gsfc.nasa.gov` is an online climate data plotting tool with an interface. It allows users to download the plotted figures in different formats, such as png. It is supported by various kinds of NASA climate datasets.

Climate Engine `climateengine.org` is a web application for plotting climate and remote sensing data. Similar to Giovanni, it also has a tabular interface for a user to customized data and maps.

NOAA Climate at a Glance `www.ncdc.noaa.gov/cag` is a data visualization tool mainly for visualizing the observed climate data over the United States. It has functions for both spatial maps and historical time series.

Web-based Reanalyses Intercomparison Tools (WRIT) `psl.noaa.gov/data/writ` is similar to Giovanni, in that WRIT also has an interface table for a user to enter plot parameters. The plot (in postscript format) and its data (in netCDF format) can be downloaded. WRIT is designed for the data from climate reanalysis models.

1.6.4 Use ChatGPT as a Study Assistant

ChatGPT (generative pre-trained transformer) is an artificial intelligence (AI) system that can help us develop computer codes, write essays, draft course syllabi, and more. For example, in your box of the ChatGPT interface, you can enter "Write a Python code to plot the cosine function as a blue thick dashed curve." ChatGPT will give you the following Python code:

```python
import numpy as np
import matplotlib.pyplot as plt

x = np.linspace(-np.pi, np.pi, 100)
y = np.cos(x)

plt.plot(x, y, linestyle='--', linewidth=3, color='blue')
plt.xlabel('x')
plt.ylabel('cos(x)')
plt.title('Plot of the cosine function')
plt.show()
```

You can copy the above code and paste it to your Jupyter Notebook cell, then run the code to produce the curve of the cosine function in the interval $[-\pi, \pi]$. With this sample and according to your needs, you can modify the code for different functions, intervals, colors, thicknesses, titles, axis labels, and more.

You can also ask ChatGPT to work on a dataset, such as "Write an R code to plot the NOAAGlobalTempts.csv data and its trend line." ChatGPT will give you an R code. You copy and paste the code to your RStudio. Set your R code in the correct directory that has the NOAAGlobalTemp annual time series data file, `NOAAGlobalTempts.csv`. Run your R code to plot the data and the linear trend line. Of course, you can modify the code according to your requirements. The R code is as follows.

```r
setwd('/Users/sshen/climmath/data')
library(tidyverse)

# Load the data
data <- read.csv("NOAAGlobalTempts.csv")

# Plot the data
ggplot(data, aes(x = Year, y = Tempts)) +
  geom_line() +
  geom_smooth(method = "lm", se = FALSE) +
  ggtitle("NOAA Global Temperature Data") +
  xlab("Year") +
  ylab("Temperature (Celsius)")
```

Another example is that you enter "Write an essay on global warming." ChatGPT will write an essay for you. Most likely, it will not reflect your ideas, but it will give you a few tips, such as the definition of global warming, evidences and data, consequences, main causes, how to slow down the warming, and a conclusion. Again, you can use this Chat-GPT essay as a sample and modify the essay using your own data and ideas, such as, for example, green technologies, solar farms, microclimate resources for smart cities, trends of the oil business, lifestyles, AI in self-driving electric cars, and education.

ChatGPT was released by the OpenAI Lab in November 2022. You may use this tool as your assistant while learning and working. It is only your assistant. It is not you! You can efficiently use it to give you hints and inspire your ideas. You still have to produce your own work.

1.7 Chapter Summary

This chapter has provided a brief introduction to useful statistical concepts and methods for climate science and has included the following material.

(i) Formulas to compute the most commonly used statistical indices:

- Mean as a simple average, median as a datum whose value is in the middle of the sorted data sequence, i.e., the median is larger than 50% of the data and smaller than the remaining 50%,
- Standard deviation as a measure of the width of the probability distribution,
- Variance as the square of the standard deviation,
- Skewness as a measure of the degree of asymmetry in distribution, and
- Kurtosis as a measure of the peakedness of the data distribution compared to that of the normal distribution.

(ii) The commonly used statistical plots:

- Histogram for displaying the probability distribution of data,
- Linear regression line for providing a linear model for data,
- Box plot for quantifying the probability distribution of data, and
- Q-Q plot for checking whether the data are normally distributed.

(iii) Read and plot the netCDF file for plotting a 2D map by R and Python. Other data visualization tools, such as Panoply, Paraview, 4DVD, Plotly, and Nullschool, were briefly introduced. The online tools like 4DVD and Nullschool may be used for classroom teaching and learning.

Computer codes and climate data examples are given to demonstrate these concepts and the use of the relevant tools and formulas. With this background, plus some R or Python programming skill, you will have sufficient knowledge to meet the needs of the basic statistical analysis for climate data.

[1] B. Hennemuth, S. Bender, K. Bulow, et al., 2013: *Statistical Methods for the Analysis of Simulated and Observed Climate Data, Applied in Projects and Institutions Dealing with Climate Change Impact and Adaptation.* CSC Report 13, Climate Service Center, Germany.

```
www.climate-service-center.de/imperia/md/content/csc/
projekte/csc-report13_englisch_final-mit_umschlag.pdf
```

> This free statistics recipe book outlines numerous methods for climate data analysis, collected and edited by climate science professionals, and has examples of real climate data.

[2] IPCC, 2021: *AR6 Climate Change 2021: The Physical Science Basis.* Contribution of Working Group I to the Sixth Assessment Report of the Intergovernmental Panel on Climate Change [V. Masson-Delmotte, P. Zhai, A. Pirani et al. (eds.)]. Cambridge University Press.

> This is the famous IPCC report available for free at the IPCC website, `www.ipcc.ch/report/ar6/wg1/`. It includes many high-quality figures plotted from climate data.

[3] NCEI, 2021: *Anomalies vs. Temperature.* Last accessed on 12 May 2021.

```
www.ncdc.noaa.gov/monitoring-references/dyk/
anomalies-vs-temperature
```

> This is an excellent site for education and gives an easy-to-understand justification of the use of anomalies. From this site, you can find many other educational resources for climate science, such as the concise description of climate extreme index and global precipitation percentile maps.

[4] S. S. P. Shen and R. C. J. Somerville, 2019: *Climate Mathematics: Theory and Applications.* Cambridge University Press.

> This book attempts to modernize mathematical education for undergraduate students majoring in atmospheric and oceanic sciences, or related fields. The book integrates six traditional mathematics courses (i.e., Calculus I, II, and III, Linear Algebra, Statistics, and Computer Programming) into a single course named Climate Mathematics. The course uses real climate data and can be taught in three semesters for freshmen and sophomores, or in one semester as an upper level or graduate class. Computer codes of R and Python and other learning resources are available at the book website `www.climatemathematics.org`.

[5] T. M. Smith, R. W. Reynolds, T. C. Peterson, and J. Lawrimore, 2008: Improvements to NOAA's historical merged land-ocean surface temperatures analysis (1880-2006). *Journal of Climate*, 21, 2283–2296, https://doi.org10.1175/ 2007 JCLI2100.1.

> These authors are among the main contributors who reconstructed sea surface temperature (SST). They began their endeavor in the early 1990s.

[6] R. S. Vose, D. Arndt, V. F. Banzon et al., 2012: NOAA's merged land-ocean surface temperature analysis. *Bulletin of the American Meteorological Society*, 93, 1677–1685.

> These authors are experts on the reconstruction of both SST and the land surface air temperature, and have published many papers in data quality control and uncertainty quantifications.

Exercises

1.1 The NOAA Merged Land Ocean Global Surface Temperature Analysis (NOAAGlobalTemp) V5 includes the global land-spatial average annual mean temperature anomalies as shown here:

```
1880   -0.843351   0.031336   0.009789   0.000850   0.020698
1881   -0.778600   0.031363   0.009789   0.000877   0.020698
1882   -0.802413   0.031384   0.009789   0.000897   0.020698
......
```

The first column is for time in years, and the second is the temperature anomalies in [Kelvin]. Columns 3–6 are data errors. This data file for the land temperature can be downloaded from the NOAAGlobalTemp website

`www.ncei.noaa.gov/data/noaa-global-surface-temperature/v5/access/`
`timeseries`

The data file named `aravg.ann.land.90S.90N.v5.0.0.202104.asc.txt` is also included in `data.zip` that can be downloaded from the book website `www.climatestatistics.org`

(a) Plot the anomalies against time from 1880 to 2020 using the point-line graph like Figure 1.1.

(b) Plot the anomalies against time from 1880 to 2020 using the staircase chart like Figure 1.2.

(c) Plot the anomalies against time from 1880 to 2020 using the color bar chart like Figure 1.3.

1.2 NOAAGlobalTemp V5 also has the global **ocean**-spatial average annual mean temperature anomalies contained in a data file named

`aravg.ann.ocean.90S.90N.v5.0.0.202104.asc.txt`

For this dataset, please plot the anomaly data in the same three styles (a), (b) and (c) as in the previous problem. You may download the data file from the NOAAGlobalTemp V5 website or from `data.zip` for this book.

1.3 NOAAGlobalTemp V5 also has the global **land**-spatial average monthly mean temperature anomalies from January 1880 to April 2021. The data file is named

`aravg.mon.land.90S.90N.v5.0.0.202104.asc.txt`

For this dataset, please plot the January anomaly data from January 1880 to January 2021 in the same three styles (a) - (c) as in the previous problem.

1.4 For the monthly global land data of NOAAGlobalTemp V5 in the previous problem,

(a) For the January data from 1880 to 2020, compute the following statistical indices: mean, standard deviation, variance, skewness, kurtosis, max, 75%-percentile, median, 25th percentile, and min.

(b) Do the same for February, March, ..., December.

(c) Aggregate all the results in (a) and (b) into a 10×12 matrix with the 10 statistical indices in rows and the 12 months in columns. Add proper column names, such as `Jan Feb Mar ...`, and also row names.

(d) Use 100–300 words to describe the differences of the statistical indices for different months.

1.5 For the monthly data CRUTEM4 (Climate Research Unit surface air temperature anomalies) in 4DVD, download the historical time series data for January of a location of your interest, and compute the following statistical indices: mean, standard deviation, variance, skewness, kurtosis, max, 75th percentile, median, 25th percentile, and min.

1.6 Do the same as the previous problem but for July. Comment on the differences between the statistical indices in January and July.

1.7 (a) Plot a histogram of the January global land temperature anomalies using the data file in the previous problem, i.e.,

`aravg.mon.land.90S.90N.v5.0.0.202104.asc.txt`

(b) Do the same for July.

(c) Use 100–200 words to describe the comparison of the two histograms.

1.8 (a) Plot a box plot for the January global land temperature anomalies

`aravg.mon.land.90S.90N.v5.0.0.202104.asc.txt`

(b) Do the same but for the July data.

(c) Put the two box plots next to each other in the same figure.

(d) Use 100–200 words to describe the comparison of the two box plots.

1.9 (a) Use the monthly data in the above problem to generate 12 Q-Q plots relative to the standard normal distribution for every month from January to December.

(b) From the 12 Q-Q plots, discuss which month's temperature anomalies are the closest to the normal distribution.

1.10 (a) Using the monthly data in the previous problem
`aravg.mon.land.90S.90N.v5.0.0.202104.asc.txt`
compute the linear trend of January temperature anomalies from 1880 to 2020. Output your result in the unit: [°C/century].

(b) Repeat (a) for each of the other 11 months.

(c) Generate a list for the 12 linear trends.

(d) Use 100–200 words to discuss the numerical values of the list.

1.11 Plot the December 1997 surface temperature anomalies using the NOAAGlobalTemp V5 data on a $5° \times 5°$ grid in the netCDF format:

`NOAAGlobalTemp_v5.0.0_gridded_s188001_e202104_c20210509T133251.nc`

The data can be downloaded from NOAAGlobalTemp V5 or extracted from `data.zip` for this book at the website `www.climatestatistics.org`.

1.12 Use Panoply to make the same December 1997 surface temperature anomalies plot as the previous problem, but with the *Robinson* map projection.

1.13 Use Panoply to make the same December 1997 surface temperature anomalies plot as the previous problem, but with the *Mercator* map projection.

1.14 Use Panoply to make the same December 1997 surface temperature anomalies plot as the previous problem, but with the *Orthographic* map projection.

1.15 Use Panoply to make the same December 1997 surface temperature anomalies plot as the previous problem, but with the *Stereographic (Two-Hemisphere)* map projection.

1.16 Plot a Hovmöller diagram like Figure 1.10 for the gridded NOAAGlobalTemp monthly anomalies at longitude 150° W and a latitude interval [50° S, 50° N] from January 1989 to December 2018 (i.e., 240 months).

1.17 Plot a Hovmöller diagram like Figure 1.10 for the gridded NOAAGlobalTemp monthly anomalies at longitude 140° W and a latitude interval [40° S, 40° N] from January 1971 to December 2000 (i.e., 360 months).

1.18 Plot a Hovmöller diagram like Figure 1.10 for the gridded NOAAGlobalTemp monthly anomalies on the equator with longitude from 160° E to 90° W from January 1971 to December 2000 (i.e., 360 months).

1.19 Plot a Hovmöller diagram like Figure 1.10 for the gridded NOAAGlobalTemp monthly anomalies at latitude 5° S with longitude from 160° E to 90° W from January 1971 to December 2000 (i.e., 360 months).

1.20 Use R or Python to plot four December 2015 potential temperature maps at the depth layers 5, 25, 105, and 195 meters based on the GODAS data. You can use the netCDF

data file godas2015.nc in data.zip for this book or download the netCDF GODAS data from a NOAA website, such as
www.esrl.noaa.gov/psd/data/gridded/data.godas.html

1.21 Use Panoply to plot the same maps as the previous problem but with the Robinson projection.

1.22 Use Plotly in R or Python to plot four December 2015 potential temperature maps at the depth layers 5, 25, 105, and 195 meters based on the GODAS data. Try to place the four maps together to make a 3D visualization.

1.23 Plot four June 2015 potential temperature maps at the depth layers 5, 25, 105, and 195 meters using the same netCDF GODAS data file as the previous problem.

1.24 Use Plotly in R or Python to plot four June 2015 potential temperature maps at the depth layers 5, 25, 105, and 195 meters based on the GODAS data. Try to place the four maps together to make a 3D visualization.

1.25 Plot three cross-sectional maps of the December 2015 potential temperature at 10° S, equator, and 10° N using the same netCDF GODAS data file as the previous problem. Each map is on a 40 × 360 grid, with 40 depth levels and 1-deg longitude resolution for the equator.

1.26 Do the same as the previous problem but for the June 2015 GODAS potential temperature data.

1.27 Use R or Python to plot a cross-sectional map of the December 2015 potential temperature along a meridional line at longitude at 170° E using the same netCDF GODAS data file as the previous problem.

1.28 Use R or Python to plot a cross-sectional map of the June 2015 potential temperature along a meridional line at longitude at 170° E using the same netCDF GODAS data file as the previous problem.

## 2

Elementary Probability and Statistics

This chapter describes the basic probability and statistics that are especially useful in climate science. Numerous textbooks cover the subjects in this chapter. Our focus is on the climate applications of probability and statistics. In particular, the emphasis is on the use of modern software that is now available for students and journeyman climate scientists. It is increasingly important to pull up a dataset that is accessible from the Internet and have a quick look at the data in various forms without worrying about the details of mathematical proofs and long derivations. Special attention is also given to the clarity of the assumptions and limitations of the statistical methods when applied to climate datasets. Several application examples have been included, such as the probability of dry spells and binomial distribution, the probability of the number of storms in a given time interval based on the Poisson distribution, the random precipitation trigger based on the exponential distribution, and the standard precipitation index and Gamma distribution.

2.1 Random Variables

2.1.1 Definition

For a conventional variable x in a function $y = f(x)$, we can assign a value to x, say, $x = 3$, or determine x value by a constraint, say, $2x = 1$, hence $x = 1/2$. A *random variable* (RV), usually indicated by an uppercase letter, such as X, is a much more complex concept, involving three things: (i) a random phenomenon, such as drought, (ii) a function or procedure that assigns a real value for each outcome (also called a trial or realization) of the random action, and (iii) a probability of the real value. The function assigns a real number for each outcome of a random event, e.g., 1 for a rainy day and 0 for a sunny day. We usually use a lowercase letter, such as x, to denote the real numbers for different outcomes. The probability for the outcome and the corresponding real value is also defined, such as the probability of a rainy day $P(X = 1) = 0.2$ where X is a random variable, 1 is in the place of x, and 0.2 is the probability of a rainy day. Another notation commonly used for this is $p_X(1) = 0.2$. The following example shows the meaning of an RV in real weather.

Example 2.1   Let X be the number of wet days in any two successive days in New York City, when the weather is classified as two types: *Wet* or W if the day has more than 0.2 mm/day precipitation, and *Dry* or D if otherwise. Consider two sequential days denoted by the outcomes D and W. There are only four possible outcomes $\{DD, DW, WD, WW\}$ and three possible values $\{0, 1, 2\}$ of x as shown in Table 2.1. Figure 2.1 shows precipitation data and the D and W days for New York City data (based on the Global Historical Climatology Network-Daily (GHCN-D) station USC00280907 at latitude 40°54′00″ N, longitude 74°24′00″ W) from June 1, 2019 to August 31, 2019. The probabilities in the fourth column of Table 2.1 are also computed from the daily station data.

Table 2.1 Random variable table: Wet and dry days

Random weather outcome	Wet days (RV X value)	Probability (independence)	Probability (observations)	Transitions (number)
DD	0	0.25	0.3626	33
DW	1	0.25	0.1978	18
WD	1	0.25	0.1978	18
WW	2	0.25	0.2418	22

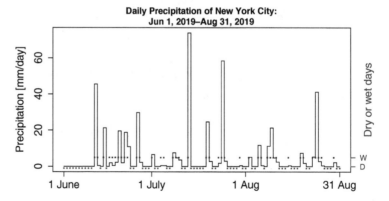

Figure 2.1   The daily precipitation data for New York City from June 1, 2019 to August 31, 2019, based on the GHCN-D station USC00280907 at latitude 40°54′00″ N, longitude 74°24′00″ W. Red dots indicate dry days and blue dots indicate wet days.

Figure 2.1 can be generated by the following computer code.

```
#R plot Fig. 2.1: Daily precipitation of New York City
# Go to your working directory
setwd("/Users/sshen/climstats")
nycDat=read.csv("data/NYCDailyDatasummer2019.csv", header=T)
dim(nycDat)
#[1] 7555    13
```

```
dayNum=92
nycP=nycDat[1:dayNum, c(3,6)]

dw=c()
for (i in 1:dayNum){if (nycP[i,2] >= 0.2){dw[i] = 1}
                    else {dw[i]=0} }
dw
n0=which(dw==0)
n1=which(dw==1)
m=dayNum - 1
par(mfrow=c(1,1))
par(mar=c(2.5,4.5,2.5,4))
plot(1:dayNum, nycP[,2], type="s",
     main="Daily␣Precipitation␣of␣New␣York␣City:
␣␣␣␣␣␣␣1␣Jun␣2019-␣31␣Aug␣2019",
     ylab="Precipitation␣[mm/day]",
     xaxt="n", xlab="",
     cex.lab=1.4, cex.axis=1.4)
axis(1, at=c(1, 30, 61, 92),
     labels=c("1␣June", "1␣July", "1␣Aug", "31␣Aug"),
     cex.axis=1.4)
par(new=TRUE)
plot(n0+1.0, dw[n0]-0.2, cex.lab=1.4,
     ylim=c(0,15), pch=15, cex=0.4, col="red",
     axes=FALSE, xlab="", ylab="")
points(n1+1, dw[n1], pch=15, col="blue", cex=0.4)
axis(4, col="blue", col.axis="blue",
     at=c(0,1), labels =c("D", "W"),
     las=2, cex.axis=1)
mtext("Dry␣or␣Wet␣Days",
      col="blue", side=4, line=1.5, cex=1.4)
```

```
#Python plot Fig. 2.1: Daily precipitation of New York City
os.chdir("/Users/sshen/climstats")
precipData = \
 np.array(read_csv("data/NYCDailyDatasummer2019.csv",
                              usecols=["DATE","PRCP"]))
date = precipData[:, 0]
precip = precipData[:,1]

#Detect wet or dry day
dayNum = 92
dw = np.array([1 if p > .2 else 0 for p in precip])
n0 = np.where(dw[0:dayNum] == 0)
n1 = np.where(dw[0:dayNum] == 1)
nd = n0[0]
nw = n1[0]

date = np.arange(0, dayNum)
precip = precipData[0:dayNum,1]
kwargs = {'drawstyle' : 'steps-post'}
plt.plot(date, precip, 'black', linewidth = 2, **kwargs);
plt.plot(nd, dw[nd]-1, 'ro', markersize = 4)
plt.plot(nw,dw[nw]*5, 'bo', markersize = 4)
```

```
plt.title("Daily precipitation of New York City: \
 \n 1 Jun 2019 - 31 Aug 2019", fontweight="bold", pad = 20)
plt.ylabel(r"Precipitation [mm/day]", size=25, labelpad=20)
plt.ylim(-5,80)
ax = plt.gca()
ax.text(96,5," W",size=20, color = 'b')
ax.text(96,-1.0," D",size=20, color = 'b')
ax.text(103, 10, "Dry or Wet Days", size = 25,
        color = 'b',rotation = 'vertical')
ax.spines['right'].set_color('b')
ax.set_xticks([0, 29, 60, 92])
ax.set_xticklabels(['1 June', '1 July', '1 Aug', '31 Aug'])
```

Here, X is a random variable that is a function mapping the weather event outcomes $\{DD, DW, WD, WW\}$ to real values $\{0, 1, 2\}$, and associates each real value with a probability. The RV takes value only when the outcome of the random event is known, such as $X = 2$ when the outcome of the weather event is WW. The probability of $X = 2$ is 0.25, denoted by $P(X = 2) = 0.25$, if the day-to-day weather is independent, i.e., unrelated. Similarly, $P(X = 0) = 0.25$. However, $P(X = 1) = 0.5$ because $X = 1$ corresponds to two outcomes $\{DW, WD\}$.

However, the real weather has a serial correlation and the day-to-day weather is not independent. The probability for each of the weather event outcomes $\{DD, DW, WD, WW\}$ is not the same. For this particular dataset, Table 2.1 indicates that the probabilities for the outcomes

$$\{DD, DW, WD, WW\}$$

are 0.3626, 0.1978, 0.1978, and 0.2418, respectively. They are computed from the number of transitions in the fifth column of Table 2.1. For example, $33/(33 + 18 + 18 + 22) = 0.3626$ is larger than the expected 0.25 under the assumption of independent weather events. This means that New York has a better chance of another dry day, when today is dry, than of a wet day. This is because the probability of $\{DD\}$ is 0.3626, while that of $\{DW\}$ is only 0.1978.

When the RV takes on only discrete real values, such as the number of wet days, then the RV is called a *discrete* RV. Otherwise, it is called a *continuous* RV, such as the amount of total precipitation at a station for two consecutive days. The RV Y can take any nonnegative real value and is a continuous RV. The probability $P(Y = 1.5)$ does not make sense anymore. Instead, the probability of Y in an interval, such as $P(1.0 < Y < 1.5)$, can be defined. Most RVs in climate science are continuous, such as temperature, pressure, and atmospheric moisture.

The data for a random variable consists of the x values of X based on the outcomes of events that have already occurred, such as the observed temperature or atmospheric pressure. The forecast or prediction for a random variable means conclusions on X based on the calculated x values and their associated probabilities for the events to occur, such as the weather forecast of precipitation for the next day: How much will it rain? What is the probability of rain? What is the probability of a given rain amount in a given interval? We will discuss more on these issues in the next chapter.

In a sense, understanding RV can help better understand weather data and weather forecasting. You should be cautioned that Table 2.1 only used the data for one time period, not the entire history of New York weather.

2.1.2 Probabilities of a Random Variable

Based on 92 days of New York City daily data, we consider the first 91 days and use the last day only as an indicator for transition. The number of dry days is 51, and that of wet days is 40. An estimate of the dry day probability is

$$p_D = \frac{51}{91} = 0.5604, \tag{2.1}$$

and the wet day probability is

$$p_W = \frac{40}{91} = 0.4396. \tag{2.2}$$

The probabilities of the events of the consecutive two days are shown in Table 2.1, such as,

$$P_{WW} = 0.2418. \tag{2.3}$$

Thus, we denote this as $P(X = 2) = 0.2418$ when we use the random variable formula to describe the probability. Similarly, $P(X = 0) = 0.3626$. However, $X = 1$ has two cases: DW and WD. Thus,

$$P(X = 1) = P_{DW} + P_{WD} = 0.1978 + 0.1978 = 0.3956. \tag{2.4}$$

2.1.3 Conditional Probability and Bayes' Theorem

Also related to this example is the probability of a wet day today based on the condition that yesterday was a dry day. This is called *conditional probability*, denoted by $P(W|D)$, in this case. It can be estimated by the number of DW transition days divided by the total number of dry days:

$$P(W|D) = \frac{18}{51} = 0.3529. \tag{2.5}$$

Similarly,

$$P(D|D) = \frac{33}{51} = 0.6471. \tag{2.6}$$

The event *DW* occurs when the first day is dry and has probability $P(D)$, and the second day changes to wet from dry: conditional probability $P(W|D)$. Thus,

$$P(DW) = P(D)P(W|D) = \frac{51}{91} \times \frac{18}{51} = \frac{18}{91} = 0.1978. \tag{2.7}$$

This agrees with the fourth column of Table 2.1.

In the long run, every D to W transition triggers a corresponding W to D transition. Thus,

$$P(WD) = P(W)P(D|W) = \frac{40}{91} \times \frac{18}{40} = \frac{18}{91} = 0.1978. \tag{2.8}$$

The equality $P(WD) = P(DW)$ implies

$$P(W)P(D|W) = P(D)P(W|D). \tag{2.9}$$

This is *Bayes' theorem*, sometimes called Bayes' formula, often written as in conditional probability:

$$P(D|W) = \frac{P(D)P(W|D)}{P(W)}. \tag{2.10}$$

Bayes' theorem holds for any two random events A and B:

$$P(A|B) = \frac{P(A)P(B|A)}{P(B)}. \tag{2.11}$$

This notation for the Bayes' theorem is most commonly used in books and the Internet.

2.1.4 Probability of a Dry Spell

Let p denote the conditional probability of the second day being dry when the first day is dry, i.e.,

$$p = P(D|D). \tag{2.12}$$

Let q denote the conditional probability of the second day being wet when the first day is wet, i.e.,

$$q = P(W|W). \tag{2.13}$$

In the case of the New York City example, $p = 0.3626$ and $q = 0.2418$.

The second day is either dry or wet, given that the first day is dry. The mathematical expression for this sentence is

$$P(D|D) + P(W|D) = 1, \tag{2.14}$$

or

$$p + P(W|D) = 1. \tag{2.15}$$

Similarly,

$$P(D|W) + P(W|W) = 1 \tag{2.16}$$

means that the second day is either dry or wet, given that the first day is wet.

The above leads to the following equations:

$$P(W|D) = 1 - P(D|D) = 1 - p, \quad P(D|W) = 1 - P(W|W) = 1 - q. \tag{2.17}$$

A dry spell of length n, denoted by RV D_n, is defined as n successive dry days followed by a wet day. Thus, the probability of a dry spell of length n is

$$P(D_n) = (1 - p)p^{n-1}. \tag{2.18}$$

A wet spell of length n is defined in a similar way, and has probability

$$P(W_n) = (1-q)q^{n-1}. \tag{2.19}$$

The cumulative probability of dry spells from one day to N days is

$$\text{CPD}_N = \sum_{n=1}^{N} P(D_n) = \sum_{n=1}^{N} (1-p)p^{n-1} = 1 - p^N. \tag{2.20}$$

The probability of a dry spell longer than N is thus

$$1 - \text{CPD}_N = 1 - (1-p^N) = p^N. \tag{2.21}$$

The mean length L_D of a dry spell is the expected value $E[n]$ when we treat n as an RV:

$$L_D = \sum_{n=1}^{\infty} n \times (1-p)p^{n-1} = \frac{1}{1-p}, \tag{2.22}$$

because

$$\sum_{n=1}^{\infty} nx^n = \frac{x}{(1-x)^2}. \tag{2.23}$$

This last summation formula can be found from the Internet.

In the case of the New York City example, $p = 0.3616$. The mean length of a dry spell in the summer is

$$L_D = \frac{1}{1-p} = \frac{1}{1-0.3616} = 1.57 \approx 2 \text{ [days]}. \tag{2.24}$$

This appears to be reasonable since the total number of wet days was 40, among the total 92 days in the dataset, and a wet day interrupts a dry spell.

One can similarly calculate the mean length of a wet spell in the summer for New York City. See the exercises at the end of this chapter.

2.1.5 Generate Random Numbers

The R command `rnorm(4)` generates four random numbers following the standard normal distribution $N(0, 1^2)$ with mean equal to 0 and standard deviation equal to 1:
`0.6341538 0.4210989 1.1772246 1.0102169`

```
#Generate random numbers by Python
np.random.normal(0, 1, 4)
#array([-0.96506567,  1.02827408,  0.22863013,
0.44513761])
```

To enable the generation of the same four random numbers, use `set.seed`:
```
set.seed(153)
rnorm(4)
```

Here, 153 is an arbitrary seed number you choose to assign. If you want to obtain the same four random numbers, it is necessary to apply `set.seed(153)` each time before running `rnorm(4)`.

```
#Python generates the same random numbers at each run
#Seed remembers the previous random numbers
np.random.seed(8) #Seed number 8
np.random.normal(0, 1, 4)
#array([ 0.09120472,  1.09128273, -1.94697031, -1.38634953])
```

The R command `rnorm(10, mean=8, sd=3)` generates ten random numbers according to the normal distribution $N(8, 3^2)$.

```
#Python generates normal random numbers with mean=8 and SD=3
np.random.normal(8, 3, 10) #Python code
```

In general, `rname` is the standard R command for generating random numbers with distribution `name`.

2.2 PDF and CDF

2.2.1 The Dry Spell Example

We use $F(N)$ to denote the probability of occurrence of a dry spell of length equal to one day, two days, three days, ..., or N days, i.e., $F(N) = P(n \leq N)$. Then, $F(N)$ is equal to the sum from $n = 1$ to N of the probabilities in (2.18):

$$F(N) = \sum_{n=1}^{N} P(D_n) = \sum_{n=1}^{N} (1-p)p^{n-1}$$

$$= (1-p) \sum_{n=1}^{N} p^{n-1} = (1-p) \times \frac{1-p^N}{1-p} = 1-p^N. \qquad (2.25)$$

When N is large, it should cover all the possibilities, and $F(N)$ should approach one, i.e.,

$$\lim_{N \to \infty} F(N) = \lim_{N \to \infty} (1-p^N) = 1, \qquad (2.26)$$

because $p < 1$ and p^N goes to zero as N goes to infinity.

The function $P(D_n)$ gives the probability for every value n and is called the *probability distribution function* (PDF), or probability density function (PDF), shown in Figure 2.2(a).

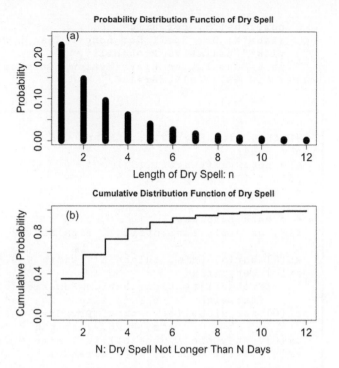

Probability Distribution Function of Dry Spell

Figure 2.2 (a) Probability distribution function (PDF) of the New York City dry spell. (b) Cumulative distribution function (CDF) of the New York City dry spell.

The function $F(N)$ gives the cumulative probability of $P(D_n)$ up to N and is called the cumulative distribution function (CDF), shown in Figure 2.2(b).

Strictly speaking, the PDF for a discrete RV should be called probability mass function (PMF), or discrete PDF. However, we usually do not make a clear distinction in climate science, since the problem itself defines if an RV is discrete or continuous.

Figure 2.2 can be plotted by the following computer code.

```
#R plot Fig. 2.2: PDF and CDF
p=0.6471
n=1:12
pdfn= (1-p)*p^n
par(mar=c(4.5,4.5,3,0.5))
plot(n, pdfn,
     ylim=c(0,0.25), type="h", lwd=15,
        cex.lab=1.5, cex.axis=1.5,
        xlab="Length of Dry Spell: n", ylab="Probability",
        main="Probability Distribution Function of Dry Spell")
text(1.5,0.25, "(a)",cex=1.5)

p=0.6471
N=1:12
cdfN= 1-p^N
par(mar=c(4.5,4.5,3,0.5))
plot(N, cdfN,
     ylim=c(0,1), type="s", lwd=3,
```

```
    cex.lab=1.5, cex.axis=1.5,
    xlab="N:␣Dry␣Spell␣Not␣Longer␣Than␣N␣Days",
    ylab="Cumulative␣Probability",
    main="Cumulative␣Distribution␣Function␣of␣Dry␣Spell")
text(1.5,0.95, "(b)",cex=1.5)
```

```
#Python plot Fig. 2.2: PDF and CDF
p = 0.6471
n = np.array(range(1,13))
pdfn = (1 - p)*p ** n
cdfn = 1 - p ** n

# create subplots
fig, ax = plt.subplots(2,1, figsize=(12,12))

ax[0].bar(n, pdfn, color='k', width = 0.5)
ax[0].set_title(
    'Probability␣Distribution␣Function␣of␣Dry␣Spell',
    fontweight = 'bold', size = 20, pad = 20)
ax[0].set_xlabel("Length␣of␣Dry␣Spell:␣n",
                size = 25, labelpad = 10)
ax[0].set_ylabel('Probability', size = 25, labelpad = 20)
ax[0].text(1.5,0.22,'(a)',size=20)

ax[1].plot(n, cdfn, 'k', **kwargs, linewidth = 3)
ax[1].set_title(
    'Cumulative␣Distribution␣Function␣of␣Dry␣Spell',
        fontweight = 'bold', size = 20, pad = 20)
ax[1].set_xlabel('N:␣Dry␣Spell␣not␣longer␣than␣N',
                size = 25, labelpad = 10)
ax[1].set_ylabel('Cumulative␣Probability',
                size = 25, labelpad = 20)
ax[1].text(1.5,0.95,'(b)',size=20)

fig.tight_layout(pad=3)
```

Usually, we use $f(x)$ to denote the PDF, and $F(x)$ for CDF. If X is a discrete RV, then

$$F(x) = \sum_{t=t_0}^{x} f(t), \tag{2.27}$$

where t_0 is the smallest x value allowed for the RV.

If X is a continuous RV, then

$$F(x) = \int_{t_0}^{x} f(t)\mathrm{d}t. \tag{2.28}$$

Here t_0 may be a real value or $-\infty$.

For a continuous RV, we also have

$$f(x) = \frac{\mathrm{d}F(x)}{\mathrm{d}x}. \tag{2.29}$$

2.2.2 Binomial Distribution

In general, any function $f(x) \geq 0$ can be a PDF if the integration or summation over its domain is 1. For example, if $p > 0, q > 0$, and $p + q = 1$, then

$$1 = p + q = (p + q)^n = \sum_{k=0}^{n} C_k^n p^k q^{n-k}, \tag{2.30}$$

where C_k^n is the combination of choosing k distinct elements out of n distinct elements

$$C_k^n = \frac{n!}{k!(n-k)!}. \tag{2.31}$$

Therefore,

$$f(k) = C_k^n p^k q^{n-k} \tag{2.32}$$

is a PDF, and

$$F(K) = \sum_{k=0}^{K} C_k^n p^k q^{n-k} \tag{2.33}$$

is a CDF. This is the binomial distribution, of which the random event has only two possible outcomes, such as dry day or wet day, success or failure, connected or disconnected, alive or dead, 0 or 1. If it is dry or wet, then $f(k)$ may be interpreted as the probability of k dry days among n days. These k dry days can be interwoven with wet days and is different from a dry spell of k days. The latter is a consecutive k dry days. Figure 2.3 shows the PDF and CDF of the binomial distribution for $n = 20, p = 0.3$.

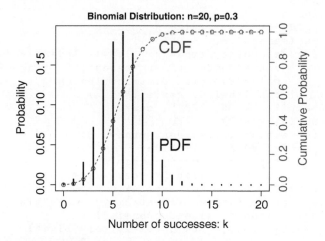

Figure 2.3   PDF and CDF of the binomial distribution for $n = 20, p = 0.3$.

When the dry and wet days are interwoven, there are C_k^n ways of selecting k dry days from n days (probability $C_k^n p^k$), and remaining $(n-k)$ days are wet (probability $q^{(n-k)}$). Thus, the probability of exactly k dry days is

$$C_k^n p^k q^{(n-k)} = f(k). \tag{2.34}$$

This is another way to explain the coefficients of the binomial expansion. The function $f(k)$ is the probability of k dry days among the total n days with the normalization property

$$\sum_{k=0}^{n} f(k) = 1. \tag{2.35}$$

This is intuitive since all the possibilities are exhausted when k runs from 0 to n.

We denote the binomial distribution by $B(p,n)$, and $f(k)$ is called the *probability mass function* (PMF).

Example 2.2   Tossing a coin is a binomial random variable. Its sample space is $\Omega = \{\text{heads, tails}\}$. *Sample space* in probability theory is defined as the set of all possible outcomes of an experiment. Here, the experiment is tossing a coin. A real-valued function is defined on the sample space and has only two values: $x = 1$ if it is a head, and $x = 0$ if it is a tail. The probabilities defined on the real values are $p_X(1) = 0.5$ and $p_X(0) = 0.5$. The PDF of the coin toss RV is

$$f(k) = \left(\frac{1}{2}\right)^n C_k^n, \tag{2.36}$$

where n is the total number tosses, i.e., the size of the sample space Ω.

Figure 2.3 can be generated by the following computer code.

```
#R plot Fig. 2.3: PDF and CDF of binomial distribution
n=20
x=0:n
pdfdx=dbinom(x, size=n, prob=0.3)
par(mar=c(4.5,4.5,2,4.5))
plot(x, pdfdx, type="h", lty=1, lwd=3,
     xlab="Number of successes: k", ylab="Probability",
     main="Binomial Distribution: n=20, p=0.3",
     cex.lab=1.4, cex.axis=1.4)
text(11, 0.05, "PDF", cex=2)
par(new=TRUE)
cdfx=pbinom(x, size=n, prob=0.3)
plot(x, cdfx, col="blue",
     type="o", lty=2, axes=FALSE,
     xlab="", ylab="", cex.axis=1.4)
axis(4, col="blue", col.axis = "blue", cex.axis=1.4)
mtext("Cumulative Probability", col="blue",
      side=4, line=3, cex=1.4)
text(11, 0.9, "CDF", cex=2, col="blue")
```

```
#Python plot Fig. 2.3: PDF and CDF of binomial distribution
import scipy.stats as stats
n = 20
p = 0.3
```

```
k = np.arange(0, n + 1)
pdf = stats.binom.pmf
cdf = stats.binom.cdf
fig, ax1 = plt.subplots(figsize=(12,8))

# plot a bar graph for pdf
## note: this is a probability plot, NOT a histogram
ax1.bar(k, pdf(k, n, p), color='k', width = 0.2)
ax1.set_xlabel("Number␣of␣successes:␣k",
               size = 23, labelpad = 15)
ax1.set_ylabel("Probability",
               size = 25, color = 'k', labelpad = 20)
ax1.set_title("Binomial␣Distribution:␣n=20,␣p␣=␣0.3",
               size = 25, pad = 20)
ax1.text(10, 0.17, "CDF", color = 'b', size = 26)
ax1.tick_params(length=6, width=2, labelsize=21)
ax1.set_xticks([0,5,10,15,20])
ax1.set_yticks([0.00, 0.05, 0.10, 0.15])

ax2 = ax1.twinx()

# plot cdf
ax2.plot(k, cdf(k, n, p), 'bo-');
ax2.set_ylabel("Cumulative␣Probabilty",
               size = 25, color='b', labelpad = 15)
ax2.tick_params(length=6, width=2,
               labelsize=21, color = 'b', labelcolor = 'b')
ax2.spines['right'].set_color('b')
ax2.set_ylim(0,)
ax2.text(10, 0.2, "PDF", color = 'k', size = 26)
plt.savefig("Figure0203.eps")

fig.tight_layout()
```

2.2.3 Normal Distribution

In the R code in Example 2.2, dbinom provides data for the PDF, and pbinom yields data
for the CDF. The command dname and pname apply to all the commonly used probability distributions, such as dnorm and pnorm for normal distributions. For example, if
x=seq(-5,5, len=101), then plot(x, dnorm(x)) plots the PDF of the standard normal distribution, as shown in Figure 2.4, which can be produced by the following computer
code.

```
#R plot Fig. 2.4: PDF and CDF of normal distribution
x=seq(-5,5, len=101)
y=dnorm(x)
plot(x, y,
    type="l", lty=1,
    xlab="x", ylab="Probability␣Density␣f(x)",
    main="Standard␣Normal␣Distribution",
```

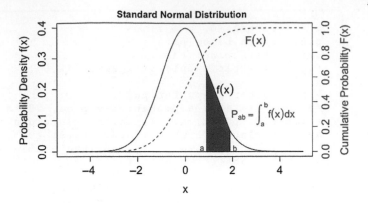

Figure 2.4   PDF and CDF of the standard normal distribution $N(0, 1)$.

```
         cex.lab=1.4, cex.axis=1.4)
lines(x,rep(0, 101))
xx=c(x[60], x[60], x[70], x[70])
yy=c(0, y[60], y[70], 0)
polygon(xx, yy, col="brown")
text(0.7, 0.01, "a", col="brown")
text(2.1, 0.01, "b", col="brown")
text(1.7,0.2, expression(f(x)), cex=1.4)
par(new=TRUE)
plot(x, pnorm(x), type="l", lty=2, col="blue",
     axes=FALSE, ylab="", xlab="", cex.axis=1.4)
axis(4, col="blue", col.axis = "blue", cex.axis=1.4)
text(3,0.9, expression(F(x)), cex=1.4, col="blue")
mtext("Cumulative␣Probability␣F(x)", col="blue",
      side=4, line=3, cex=1.4)
text(3.3,0.3, cex=1.3, col="brown",
     expression(P[ab]==integral(f(x)*dx, a,b)))
```

```
#Python plot Fig. 2.4: PDF and CDF of normal distribution
import scipy.stats as norm
dist = scistats.norm(0,1) # normal distribution
n = 101
b = 5
#Generate n evenly spaced numbers between [-b,b]
x = np.linspace(-b, b, n)
pdf = dist.pdf(x) # normal probability density function
cdf = dist.cdf(x) # normal cumulative density function

#Plot the figure
fig, ax1 = plt.subplots(figsize=(12,8))

# define the interval to be shaded
interval = np.array([True if i > .85 and i < 2 \
                     else False for i in x])
```

```
# plot probability density (f(x))
ax1.plot(x, pdf, 'k')

# fill red section
ax1.fill_between(x, np.zeros(x.size),
                 y1, where= interval, color ='red')
ax1.set_xlabel("$x$", size = 25, labelpad = 20)
ax1.set_ylabel("Probability␣Density␣f(x)",
               size = 22, labelpad = 20)
ax1.tick_params(length=6, width=2, labelsize=20)
ax1.set_yticks([0,0.1,0.2,0.3,0.4]);
ax1.text(1.3,.2, "$f(x)$", fontsize = 25)
ax1.text(1.7,.15, "$P_{ab}␣=␣\int_a^b␣f(x)\␣dx$",
         fontsize = 25, color ='red')
ax1.text(.6,.01, "$a$", fontsize = 20, color = 'red')
ax1.text(2,.01, "$b$", fontsize = 20, color ='red')
ax1.set_ylim(0,)

ax2 = ax1.twinx()

# plot cumulative probability (F(x))
ax2.plot(x, cdf, 'b--')
ax2.set_ylabel("Cumulative␣Probabilty␣$F(x)$",
               size = 22, color ='b', labelpad = 20)
ax2.tick_params(length=6, width=2, labelsize=20,
                color = 'b', labelcolor = 'b')
ax2.spines['right'].set_color('b');
ax2.text(1.7,.9, "$F(x)$", fontsize=25, color = 'blue')
ax2.set_ylim(0,)

plt.show()
```

Note that for a continuous RV, the probability at a single point $P(x)$ does not make any sense, but the probability in x's small neighborhood of length dx, denoted by $f(x)dx$, is meaningful. The probability of a continuous RV in an interval (a,b) is also meaningful and is defined as the area underneath the PDF:

$$P_{ab} = \int_a^b f(x)dx = F(b) - F(a). \tag{2.37}$$

For example, the standard normal distribution has the following PDF formula:

$$f(x) = \frac{1}{\sqrt{2\pi}} \exp\left(-\frac{x^2}{2}\right). \tag{2.38}$$

If $a = -1.96$ and $b = 1.96$, then

$$P_{ab} = \int_{-1.96}^{1.96} f(x)dx = 0.95. \tag{2.39}$$

These numbers will be used in the description of confidence level and confidence interval in Chapter 3.

2.2.4 PDF and Histogram

A histogram shows the distribution of data from a single realization, also called a sample. When it is normalized, a histogram may be considered as an approximation to the PDF of the random process, for which data are taken or generated. Figure 2.5 shows the histogram computed from 200 random numbers generated by R, and its PDF fit, where the PDF is subject to an area normalization, which is 200 for this figure.

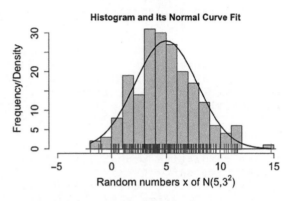

Histogram and Its Normal Curve Fit

Figure 2.5 Histogram and its PDF fit of the standard normal distribution $N(5, 2^2)$. The blue short vertical lines are the 200 random numbers x generated by the R code `rnorm(200, mean=5, sd=3)`.

Figure 2.5 can be generated by the following computer code.

```
#R plot Fig. 2.5: Histogram and PDF
par(mar=c(4.5, 4.5, 2.5, 0.5))
x<- rnorm(200, mean=5, sd=3)
h<-hist(x, breaks=15, col="gray",
        xlim=c(-5,15), ylim=c(0,30),
  xlab=expression("Random numbers x of N(5," * 3^2 *")"),
  ylab="Frequency/Density",
  main="Histogram and Its Normal Curve Fit",
  cex.lab=1.4, cex.axis=1.4)
xfit <- seq(min(x),max(x),length=40)
yfit <- dnorm(xfit, mean=mean(x), sd=sd(x))
yfit <- yfit*diff(h$mids[1:2])*length(x)
#diff(h$mids[1:2])*length(x) =200
#is the total histogram area
points(x, rep(0, 200), cex=0.8, pch=3, col="blue")
lines(xfit, yfit,  lwd=2)
```

```
#Python plot Fig. 2.5: Histogram and PDF
# generate 200 random numbers from N(5, 3^2)
x = np.random.normal(5, 3, 200)
```

```
fig, ax = plt.subplots(figsize=(12,8))

# plot histogram
counts, bins, bars = ax.hist(x, bins=17,
  range = (-5,15), color = 'gray', edgecolor = 'k');

# fit normal curve
xfit = np.linspace(x.min(), x.max(), 40)
yfit = norm.pdf(xfit, np.mean(xfit), np.std(xfit))
yfit = yfit * len(x) * ((bins[2] - bins[1]))


# plot normal curve
ax.plot(xfit, yfit,
        color = 'k', linewidth = 2);

ax.spines['right'].set_visible(False)
ax.spines['top'].set_visible(False)

ax.tick_params(length=6, width=2, labelsize=20)
#ax.set_yticks([0, .05, 0.10, 0.15]);
ax.set_xticks(np.round(np.linspace(-5, 15, 5), 2))
ax.set_title("Histogram␣and␣Its␣Normal␣Curve␣Fit",
              size = 22, pad = 20);
ax.set_xlabel("Random␣Numbers␣$x$␣of␣$N(5,3^2)$",
              size = 22, labelpad = 10);
ax.set_ylabel("Frequency/Density",
              size = 22, labelpad = 20);

# generate 200 zeros
y = np.zeros(200)

# plot random points
s = np.repeat(100,200)
plt.scatter(x, y-0.5, marker ='|',
        lw = 2, color = 'blue')

plt.show()
```

Please note that since the data is randomly generated, your histogram might differ from the one shown here.

2.3 Expected Values, Variances and Higher Moments of an RV

Chapter 1 showed numerical examples of estimating mean values, variances, skewness, and kurtosis from real climate datasets. This section presents the mathematical formulas related to them based on RV and its PDF.

2.3.1 Definitions

The *expected value*, also commonly referred to as mean value, average value, expectation, or mathematical expectation, of an RV is defined as

$$E[X] = \int_D x f(x) \, dx \equiv \mu_x, \tag{2.40}$$

where $f(x)$ is the PDF of X, and D is the domain of x, which is defined from the set of all the event outcomes, i.e., the sample space. The domain D is $(-\infty, \infty)$ for the standard normal distribution.

The expected value of a function of a random variable $g(X)$ is similarly defined:

$$E[g(X)] = \int_D g(x) f(x) \, dx. \tag{2.41}$$

The *variance* measures the dispersion property of an RV and is defined as follows:

$$\begin{aligned} \text{Var}[X] &= E[(X - \mu_x)^2] \\ &= \int_D (x - \mu_x)^2 f(x) \, dx. \end{aligned} \tag{2.42}$$

This expression is interpreted as the second centered moment of the distribution, while the expected value defined in Eq. (2.40) is the first moment. In general, the nth centered moment is defined by

$$\mu_n = E[(X - \mu_x)^n]. \tag{2.43}$$

Uncentered moments are defined by

$$\mu_n' = E[X^n]. \tag{2.44}$$

Another commonly used notation for variance is *var*, and many computer languages use `var(x)` as the command to calculate the variance of the data string x.

The *standard deviation*, abbreviated SD, is the square root of the variance:

$$\text{SD}[X] = \sqrt{\text{Var}[X]} \tag{2.45}$$

It has the same units as x. We also use *sd* to denote standard deviation and `sd(x)` is often a computer command to calculate the standard deviation of the data string x.

Integration cannot be applied to a discrete random variable and has to be replaced by summation. The expected value of a discrete random variable is defined by a summation as follows:

$$E[x_n] = \sum_n x_n p(x_n), \tag{2.46}$$

where $p(x_n)$ is the probability of x_n, i.e., p is the PDF of the discrete random variable.

For example, for a binomial distribution $B(n, p)$, the expected value of k is np and can be derived from the following formula:

$$E[K] = \sum_{k=0}^{n} k C_k^n p^k (1-p)^{n-k} = np. \tag{2.47}$$

The variance of a discrete random variable is also defined:

$$\mathrm{Var}[K] = \sum_{k=0}^{n} (k-np)^2 C_k^n p^k (1-p)^{n-k} = np(1-p). \tag{2.48}$$

2.3.2 Properties of Expected Values

Consider a linear function of a random variable Y expressed as

$$Y = a + bX, \tag{2.49}$$

where a and b are ordinary numbers and X is a random variable. The integral definition of the expected value implies that

$$\mathrm{E}[Y] = a + b\mathrm{E}[X]. \tag{2.50}$$

Similarly, the integral definition of variance yields

$$\mathrm{Var}[a+bX] = b^2 \mathrm{Var}[X]. \tag{2.51}$$

The sum $X+Y$ of two random variables X and Y has the following property:

$$\mathrm{E}[X+Y] = \mathrm{E}[X] + \mathrm{E}[Y]. \tag{2.52}$$

Two random variables X and Y are *independent* if and only if

$$P(X=x, Y=y) = P(X=x)P(Y=y). \tag{2.53}$$

If X and Y are independent random variables, then

$$\mathrm{E}[XY] = \mathrm{E}[X]\mathrm{E}[Y], \tag{2.54}$$

$$\mathrm{E}[X'Y'] = 0, \tag{2.55}$$

$$\mathrm{Var}[X+Y] = \mathrm{Var}[X] + \mathrm{Var}[Y], \tag{2.56}$$

where $X' = X - \mathrm{E}[X]$ and $Y' = Y - \mathrm{E}[Y]$ are anomalies.

2.4 Joint Distributions of X and Y

2.4.1 Joint Distributions and Marginal Distributions

When two random variables X and Y (also called *variates*) present, their PDF is a *joint distribution* $f(x,y)$, with

$$\iint_D f(x,y)\,\mathrm{d}x\mathrm{d}y = 1, \tag{2.57}$$

where D is a 2-dimensional domain, corresponding to the sample space of X and Y. The probability of an event in a subset of D is

$$P_{\mathrm{sub}} = \iint_{D_{\mathrm{sub}}} f(x,y)\,\mathrm{d}x\mathrm{d}y. \tag{2.58}$$

For a given x, the cross-sectional integral

$$f_1(x) = \int_{D_x} f(x,y)\,\mathrm{d}y \qquad (2.59)$$

is a PDF for X, and is called the *marginal distribution function*, where D_x is a cross section of the 2-dimensional domain D for a fixed x.

Figure 2.6 shows a simulated joint distribution: normal distribution $N(5,3^2)$ in x and uniform distribution $U(0,5)$ in y. The locations of the black dots are determined by the coordinates (x,y). The bar chart on top is the histogram of x data, and the bar chart on the right is the histogram of y data.

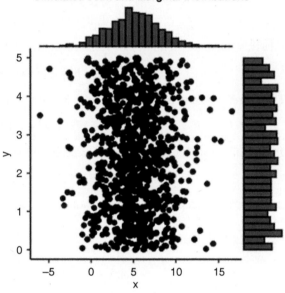

Simulated Joint and Marginal Distributions

Figure 2.6   Simulated joint and marginal distributions: Normal distribution in x and uniform distribution in y.

Figure 2.6 can be generated by the following computer code.

```
#R plot Fig. 2.6: Joint and marginal distributions
#install.packages("ggExtra")
library(ggplot2)
df <- data.frame(x = rnorm(1000, 5, 3), y = runif(1000, 0,5))
p <- ggplot(df, aes(x, y)) + geom_point() + theme_classic()
ggExtra::ggMarginal(p, type = "histogram")
```

```
#Python plot Fig. 2.6: Joint and marginal distributions
fig, ax = plt.subplots(2,2,
              gridspec_kw={'width_ratios': [4, 1],
                           'height_ratios': [1,4]},
                  figsize=(12,8))
```

```
# generate 1,000 samples from normal distribution
x = np.random.normal(5, 3, 1000)

# generate 1,000 samples from uniform distribution
y = np.random.uniform(0, 5, 1000)

# plot normal histogram
ax[0,0].hist(x, edgecolor = 'k',color = 'grey',
            bins =25, density=True)

# plot uniform histogram
ax[1,1].hist(y, edgecolor = 'k',color = 'grey',
            bins =25, density=True,
            orientation=u'horizontal', histtype='bar')

# plot points
ax[1,0].plot(x,y, 'ko', linewidth = 2);
ax[1,0].set_xlabel("$x$", size = 25, labelpad = 20);
ax[1,0].set_ylabel("$y$", size = 25, labelpad = 15);
ax[1,0].tick_params(length=6, width=2, labelsize=20);
ax[1,0].set_xticks(np.linspace(-5,15,5))

ax[0,0].axes.get_xaxis().set_visible(False)
ax[0,0].axes.get_yaxis().set_visible(False)
ax[0,0].spines['right'].set_visible(False)
ax[0,0].spines['top'].set_visible(False)
ax[0,0].spines['left'].set_visible(False)

ax[1,1].axes.get_xaxis().set_visible(False)
ax[1,1].axes.get_yaxis().set_visible(False)
ax[1,1].spines['right'].set_visible(False)
ax[1,1].spines['top'].set_visible(False)
ax[1,1].spines['bottom'].set_visible(False)

ax[0,1].axes.get_xaxis().set_visible(False)
ax[0,1].axes.get_yaxis().set_visible(False)
ax[0,1].spines['right'].set_visible(False)
ax[0,1].spines['top'].set_visible(False)
ax[0,1].spines['left'].set_visible(False)

fig.tight_layout(pad=-1)
```

We can also define a *conditional probability density* for x for a given y, denoted $f(x|y)$. If the marginal distribution $f_2(y)$ is also known, we have

$$f(x,y) = f(x|y)f_2(y). \tag{2.60}$$

We can state the independence from the joint distribution perspective. The RVs X and Y are *independent* (i.e., the occurrence of y does not alter the probability of the occurrence of x), if and only if

$$f(x,y) = f_1(x)f_2(y) \quad \text{(the condition of independence of } x \text{ and } y). \tag{2.61}$$

The example shown in Figure 2.6 has independent x and y. Chapters 3, 4, and 7 have more examples and discussions on independence and serial correlations.

2.4.2 Covariance and Correlation

The variance of the sum of two random variables X and Y is

$$\text{Var}[X+Y] = \text{Var}[X] + \text{Var}[Y] + 2\text{E}\left[(X - \text{E}[X])(Y - \text{E}[Y])\right]. \tag{2.62}$$

The last part of Eq. (2.62) is called the *covariance* of X and Y, denoted by

$$\text{Cov}[X,Y] = \text{E}\left[(X - \text{E}[X])(Y - \text{E}[Y])\right]. \tag{2.63}$$

A normalized version of the covariance is the *correlation coefficient*, or simply *correlation*, often denoted by ρ, r, $\text{cor}[X,Y]$, or $\text{Cor}[X,Y]$:

$$\text{Cor}[X,Y] = \frac{\text{E}\left[(X - \text{E}[X])(Y - \text{E}[Y])\right]}{\sqrt{\text{Var}[X] \cdot \text{Var}[Y]}}. \tag{2.64}$$

2.5 Additional Commonly Used Probabilistic Distributions in Climate Science

2.5.1 Poisson Distribution

Poisson distribution is a discrete probability distribution that shows the probability of the number of events occurring in a given time interval. For example, what is the probability of having ten rain storms over New York in June when the average number of June storms based on the historical data is five? The mean occurrence rate is denoted by λ. The PDF of the Poisson distribution is

$$f(k) = \frac{\text{e}^{-\lambda} \lambda^k}{k!}, \quad k = 0, 1, 2, \ldots \tag{2.65}$$

It can be proved that

$$\text{E}[X] = \lambda, \tag{2.66}$$

$$\text{Var}[X] = \lambda. \tag{2.67}$$

Figure 2.7 shows the PDF, also known as PMF, for a discrete RV, and CDF of the Poisson distribution. If the average number of June storms is five, then the probability of having ten storms in June is $f(10) = 0.0181$ or approximately 2%, which can be computed by the R command `dpois(10, lambda=5)`.

Figure 2.7 can be generated by the following computer code.

```
#R plot Fig. 2.7: Poisson distribution
k=0:20
y= dpois(k, lambda=5)
par(mar=c(4.5,4.5,2,4.5))
```

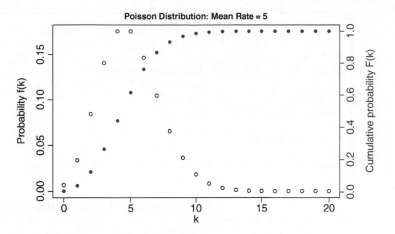

Figure 2.7   PDF and CDF of a Poisson distribution of mean rate equal to 5.

```
plot(x, y,
    type="p", lty=1, xlab="k", ylab="",
    main="Poisson Distribution: Mean Rate = 5",
    cex.lab=1.4, cex.axis=1.4)
mtext(side=2, line=3, cex=1.4, 'Probability f(k)')
par(new=TRUE)
plot(x, ppois(x, lambda=5), type="p", pch=16,
    lty=2, col="blue",
    axes=FALSE, ylab="", xlab="", cex.axis=1.4)
axis(4, col="blue", col.axis = "blue", cex.axis=1.4)
mtext("Cumulative Probability F(k)", col="blue",
    side=4, line=3, cex=1.4)
```

```
#Python plot Fig. 2.7: Poisson distribution
lam = 5 #mean rate for Poisson distribution
n = 20 #number of occurrence
x = np.arange(n+1)# integers from [0,n]
pdf = np.array(
 [(np.exp(-lam)*lam**k)/np.math.factorial(k) for k in x])
cdf = np.array([np.sum(pdf[:k]) for k in x])

fig, ax1 = plt.subplots(figsize=(12,8))
ax1.plot(x, pdf,'ko', ms = 10) #plot pdf
ax1.set_title('Poisson Distribution: Mean Rate=%1.1f'% lam,
            size = 22, pad = 20)
ax1.set_xlabel("k", size = 25, labelpad = 20)
ax1.set_ylabel("Probability f(k)",
            size = 22, color = 'k', labelpad = 20)
ax1.tick_params(length=6, width=2, labelsize=21)
ax1.set_xticks([0,5,10,15,20])
ax1.set_yticks([0.00,0.05, 0.10, 0.15])

ax2 = ax1.twinx()
```

```
ax2.plot(x, cdfdx, 'bo', ms = 10) #plot pdf
ax2.set_ylabel("Cumulative␣Probabilty␣$F(k)$",
               size = 22, color='b', labelpad = 20)
ax2.tick_params(length=6, width=2,
                labelsize=21, labelcolor = 'b')
ax2.spines['right'].set_color('b')
ax2.set_ylim(-0.02,1.1)

fig.tight_layout()
```

2.5.2 Exponential Distribution

The *exponential distribution* is a probability distribution of the time between events. It is a continuous distribution and its PDF is

$$f(t) = \lambda e^{-\lambda t}, \text{ for } t \geq 0, \tag{2.68}$$

where λ is the mean rate of occurrence, i.e., how many times of occurrence per unit time. Its CDF is

$$F(t) = 1 - e^{-\lambda t}. \tag{2.69}$$

The exponential distribution is a continuous distribution from 0 to ∞. The mean time of the exponential distribution is $1/\lambda$. This is reasonable since λ is the mean rate of occurrence, hence on average the time needed for an event to occur is $1/\lambda$. Based on this intuition, we have $F(0) = 0$ and $F(\infty) = 1$, because nothing can happen given zero time, and it is certain that an event occurs when waiting for an infinitely long time. Based on the CDF Eq. (2.69), the probability for an event to occur between t_1 and t_2 is $F(t_2) - F(t_1)$.

The variance of t is $1/\lambda^2$ and can be derived from the following formula:

$$\int_0^\infty \left(t - \frac{1}{\lambda}\right)^2 \lambda e^{-\lambda t} \, dt = \frac{1}{\lambda^2}. \tag{2.70}$$

Considering the precipitation process over a given grid box in a climate model, let Δt be the time step. Since the exponential distribution is for the time between two successive events, the probability for a nonzero precipitation event to occur in the time step Δt is

$$F(\Delta t) = 1 - e^{-\lambda \Delta t}. \tag{2.71}$$

The probability for no precipitation to occur in the time step Δt is thus

$$1 - F(\Delta t) = e^{-\lambda \Delta t}. \tag{2.72}$$

One can also relate this result to the Poisson distribution. According to the Poisson PDF, the probability of zero precipitation events occurring in the time interval $(t, t + \Delta t)$ is

$$p(0; \lambda \Delta t) = \frac{e^{-\lambda \Delta t}(\lambda \Delta t)^0}{0!} = e^{-\lambda \Delta t}. \tag{2.73}$$

Then, the probability for a nonzero precipitation event to occur in the time interval is

$$p_e(\Delta t) = 1 - p(0; \lambda \Delta t) = 1 - e^{-\lambda \Delta t}. \tag{2.74}$$

This agrees with the CDF of the exponential distribution $F(\Delta t)$.

One may use the exponential distribution to model the probability of the precipitation state transition from precipitation to nonprecipitation, or vice versa, from one time step to the next in a climate model. Between two neighboring time steps, the state of precipitation may switch from one to another, with 0 denoting no precipitation and 1 denoting precipitation. The transition probability is determined by the CDF of an exponential distribution

$$p_{m,n} = 1 - e^{-r_{m,n}\Delta t}, \tag{2.75}$$

where m, n take values of either 0 or 1, with $m, n = 1, 0$ denoting a transition from precipitation to nonprecipitation and $m, n = 0, 1$ for otherwise, and the rate parameter $r_{m,n}$ is a function of the precipitable water vapor concentration in the atmosphere. This provides a random precipitation trigger for a climate model, in contrast to the deterministic precipitation trigger that must make precipitation when the precipitable water vapor concentration reaches a critical level. The random trigger is more realistic. In real weather, precipitation may not happen in a very humid atmosphere, because the occurrence of precipitation is determined by a very complex process.

2.5.3 Mathematical Expression of Normal Distributions, Mean, and the Central Limit Theorem

The normal distribution $N(\mu, \sigma^2)$, by name, is a distribution we normally have in mind when we consider the distribution of random events in our daily life. Although many random events are not normally distributed, their mean is approximately normally distributed, as long as the mean is from a sufficiently large number of random variables. This is the conclusion of the *central limit theorem*.

The normal distribution is also called the Gaussian distribution after the German mathematician Carl Friedrich Gauss (1777–1855) and is a continuous distribution. The mathematical expression of the bell shape PDF for a normal RV $X \sim N(\mu, \sigma^2)$ is

$$f(x) = \frac{1}{\sqrt{2\pi}\sigma} e^{-\frac{1}{2}(x-\mu)^2/\sigma^2}, \quad -\infty < x < \infty. \tag{2.76}$$

Calculus can show that

$$\mathrm{E}[X] = \mu, \tag{2.77}$$

$$\mathrm{Var}[X] = \sigma^2. \tag{2.78}$$

If $X_1 \sim N(\mu_1, \sigma_1^2)$ and $X_2 \sim N(\mu_2, \sigma_2^2)$ are independent, then their mean is also a normal RV with

$$X = \frac{X_1 + X_2}{2} \sim N((\mu_1 + \mu_2)/2, (\sigma_1^2 + \sigma_2^2)/2^2). \tag{2.79}$$

Note that the mean has a smaller variance, i.e., is more accurate.

If the RVs $X_1, X_2, \ldots, X_n$ are independent and follow the same normal distribution $N(\mu, \sigma^2)$, then their mean is normally distributed $N(\mu, \sigma_A^2)$ with

$$\sigma_A^2 = \frac{\sigma^2}{n}. \tag{2.80}$$

This conclusion is particularly useful in science experiments. Repeating an experiment n times reduces the experimental error σ dramatically:

$$\sigma_A = \frac{\sigma}{\sqrt{n}}. \tag{2.81}$$

This is called the standard error of an experiment repeated n times, or n repeated measurements.

The $X_1, X_2, \ldots, X_n$ are independent RVs from a population with the same mean μ and finite variance σ^2, but may not be normally distributed. However, their average

$$X = \frac{\sum_{i=1}^n X_i}{n} \tag{2.82}$$

is still approximately normally distributed as long as n is sufficiently large with

$$X \sim N(\mu, \sigma^2/n). \tag{2.83}$$

This formula introduces the classical version of the *central limit theorem* (CLT), and supports our intuition that we will have a reduced standard error $\sigma/\sqrt{n}$, if we repeat our measurements and take their average.

The CLT states that so long as the variances of the terms are finite, the sum will be normally distributed as n goes to infinity, whether or not the RV of each term is normal. Nearly all variates in climate science satisfy this criterion.

One distribution that is rare but violates the finite variance assumption of CLT is the Cauchy distribution $C(x_0, \gamma)$ whose PDF is

$$f(x) = \frac{1}{\pi\gamma}\frac{1}{1+\left(\frac{x-x_0}{\gamma}\right)^2}, \quad -\infty < x < \infty, \tag{2.84}$$

where x_0 is called the location parameter and γ is called the scale parameter. The Cauchy distribution has mean zero, but has infinite variance. Figure 2.8 shows a fat tail of Cauchy distribution, compared to the normal distribution of the same peak.

Figure 2.8 can be generated by the following computer code.

```
#R plot Fig. 2.8: Normal and Cauchy distributions
x=seq(0,20, len=201)
ycauchy=dcauchy(x, location = 10, scale=1)
ygauss=dnorm(x, mean=10, sd=sqrt(pi/2))
plot(x, ycauchy, type="l", lwd=2,
     ylab="Density",
     main="Comparison between Cauchy and Gaussian Distributions",
     cex.lab=1.4, cex.axis=1.4)
legend(-1,0.35, legend="Cauchy distribution",
       cex=1.2, bty="n", lty=1, lwd=2)
lines(x, ygauss, lty=2)
legend(-1,0.32, legend="Normal distribution",
       cex=1.2, bty="n", lty=2)
```

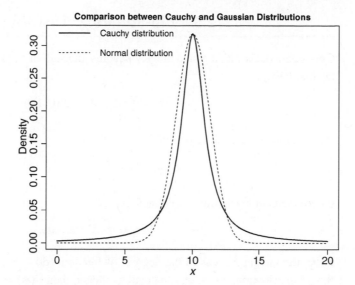

Figure 2.8   Compare the PDF of the normal distribution $N(10, \pi/2)$ and the Cauchy distribution $C(10,1)$.

```
#Python plot Fig. 2.8: Normal and Cauchy distributions
x = np.linspace(0, 20, 201)
pInv = 1/np.pi # inverse pi
# Normal density function
ygauss = np.array(
    [pInv*np.exp(-((i - 10)**2)*pInv) for i in x])
# Cauchy density function
ycauchy = np.array(
    [1/(np.pi*(1+(i - 10)**2)) for i in x])
#Plot
fig, ax = plt.subplots(figsize=(12,8))

# plot Cauchy distribution
ax.plot(x, ycauchy, 'k-',
        label="Cauchy␣Distribution", linewidth=3)
# plot Gaussian distribution
ax.plot(x, ygauss, 'k--',
    label="Normal␣(Gaussian)␣Distribution", linewidth=3)
ax.set_xlabel("$x$", size = 22, labelpad = 20)
ax.set_ylabel("Density", size = 22, labelpad = 20)
ax.tick_params(length=6, width=2, labelsize=21);
ax.set_xticks([0,5,10,15,20])
ax.set_yticks([0.00,0.10, 0.20, 0.30])
ax.set_title("Comparison␣between␣Cauchy␣and␣\
␣␣␣␣␣␣␣␣\n␣Gaussian␣Distributions",
            size = 23, pad = 20);
ax.legend(loc = 'upper␣left', prop={'size':16})
fig.tight_layout()
```

2.5.4 Chi-Square χ^2 Distribution

Chi-square deals with the sum of n RV squares; the sum of squares of independent standard normal RVs:

$$X = \sum_{n=1}^{k} Z_n^2, \ \ Z_n \sim N(0,1). \tag{2.85}$$

RV X satisfies the χ^2 distribution with a PDF as

$$f(x) = \frac{1}{2^{k/2}\Gamma(k/2)} x^{k/2-1} e^{-x/2}, \tag{2.86}$$

where the Gamma function is defined as

$$\Gamma(t) = \int_0^\infty y^{t-1} e^{-y} \, dy. \tag{2.87}$$

Here, the integer k is called the degrees of freedom (dof). We thus also use χ_k^2 to denote the χ^2 distribution. If $k = 2m$ is a positive integer, then $\Gamma(m) = (m-1)!$ and

$$f(x) = \frac{1}{2^m (m-1)!} x^{m-1} e^{-x/2}. \tag{2.88}$$

The expected value and variance of the chi-square χ^2 variable X are given as follows:

$$\mathrm{E}[X] = k, \tag{2.89}$$

$$\mathrm{Var}[X] = 2k. \tag{2.90}$$

The following simple R command can plot the probability density function (PDF) of a χ^2 variable:

```
x= seq(1,10, by=0.2)
plot(x,dchisq(x, df = 4))
```

As an example, the standardized anomalies of the NOAAGlobalTemp monthly temperature data from January 1880 to December 2015 over the 5-degree grid box centered at $(52.5° \text{ N}, 112.5° \text{ W})$ that covers Edmonton, Canada, are approximately $N(0,1)$. This data stream has 1,632 months and can be broken into 272 6-month periods. Then, compute the 6-month mean squares of the standardized anomalies

$$w = \sum_{n=1}^{6} T_n^2. \tag{2.91}$$

The histogram of these w data fits the distribution χ_6^2 as shown in Figure 2.9.

The R code for generating Figure 2.9 is as follows.

```
#R plot Fig. 2.9: Chi-square data and fit
setwd("/Users/sshen/climstats")
da1 =read.csv("EdmontonT.csv", header=TRUE)
x=da1[,3]
m1 = mean(x)
s1 = sd(x)
xa = (x- m1)/s1
y = matrix(xa[1:1632], ncol=6, byrow=TRUE)
```

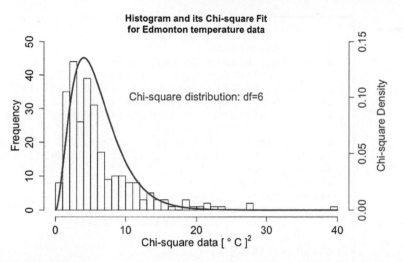

Histogram and its chi-square fit. The chi-square data are derived from Edmonton monthly surface air temperature anomaly data: Jan 1880–Dec 2015.

```
w = rowSums(y^2)
hist(w, breaks= seq(0,40,by=1),
    xlim=c(0,40), ylim=c(0,50),
    main="Histogram␣and␣its␣Chi-square␣Fit
for␣Edmonton␣temperature␣data",
    xlab=expression("Chi-square␣data␣["~degree~C~"]"^2),
    cex.lab=1.4, cex.axis=1.4)
par(new=TRUE)
density = function(x) dchisq(x, df=6)
x=seq(0,40, by=0.1)
plot(x, density(x), type="l", lwd=3,
    col="blue", ylim=c(0,0.15),
    axes=FALSE, bty="n", xlab="", ylab="")
axis(4, cex.axis=1.4, col='blue', col.axis='blue')
mtext("Chi-square␣Density", cex=1.4,
      side = 4, line = 3, col="blue")
text(20, 0.1, cex=1.4, col="blue",
    "Chi-square␣distribution:␣df=6")
```

```
#Python plot Fig. 2.9: Chi-square data and fit
os.chdir("/Users/sshen/climstats")
da1 = pd.read_csv('data/EdmontonT.csv')
x = da1.iloc[:,2]# all rows from 2nd column of data
m1 = np.mean(x)
s1 = np.std(x)
xa = (x - m1)/s1
y = np.array(xa[0:1632]).reshape((272, 6))#vector=>matrix
w = np.sum(y**2, axis = 1)# calculate the row sums
x = np.linspace(0,40,200)
cpdf = scistats.chi2.pdf(x, 6) # chi-square PDF
fig, ax = plt.subplots(figsize=(12,8))
```

```
# plot histogram
ax.hist(w, bins = 41, edgecolor = 'k',
        color = 'white', linewidth=3)
ax.set_title("Histogram␣and␣its␣Chi-square␣Fit␣\n\
for␣Emonton␣temperature␣data")
ax.set_xlabel("Chi-square␣data␣[$\degree$C]$^2$",
              size = 22)
ax.set_ylabel("Frequency", size = 22, labelpad = 20)
ax.tick_params(length=6, width=2, labelsize=21)
ax.set_xticks([0,10,20,30,40])
ax.set_yticks([0,10,20,30,40,50])
ax.spines['top'].set_visible(False)
ax.text(10, 35, "Chi-square␣distribution:␣df=6",
        size = 20, fontdict = font)

ax1 = ax.twinx()
# plot chi-square fit
ax1.plot(x,cpdf, color = 'b', linewidth = 3)
ax1.set_ylabel("Chi-square␣Density", size = 22,
               color='b', labelpad = 20)
ax1.set_ylim(0,)
ax1.tick_params(length=6, width=2, color = 'b',
                labelcolor = 'b', labelsize=21)
ax1.spines['top'].set_visible(False)
ax1.spines['right'].set_color('b')
ax1.set_yticks([0.00,0.05, 0.10, 0.15])

plt.show()
```

A popular application of the chi-square distribution in data analysis is to use the chi-square test to determine goodness of fit to a given distribution, continuous or discrete. It is sum of the squares of observed counts minus the expected counts from the given distribution:

$$\chi^2 = \sum_{i=1}^{n} \frac{(O_i - E_i)^2}{E_i}, \tag{2.92}$$

where O_i is the number of observations in the ith category, E_i is the expected number of observations in the ith category according to the given distribution, and n is the total number of observations. The perfect fit implies $\chi^2 = 0$, while a large χ^2 value means a not good fit.

2.5.5 Lognormal Distribution

The lognormal distribution is simply the case that the logarithm of the variable is distributed normally. Its PDF is

$$f(x) = \frac{1}{x\sigma\sqrt{2\pi}} e^{-\frac{(\ln x - \mu)^2}{2\sigma^2}}, \ \ 0 < x < \infty. \tag{2.93}$$

You can use calculus to show that

$$E[X] = e^{\mu + \sigma^2/2}, \tag{2.94}$$

$$\mathrm{Var}[X] = \left[e^{\sigma^2} - 1\right] e^{2\mu + \sigma^2}. \tag{2.95}$$

The distribution of weekly, 10-day or even monthly total precipitation of some stations may be approximated by the lognormal distribution. The June precipitation of Madison, Wisconsin, USA, from 1941 to 2019 fits lognormal distribution well. The R command for the fitting is

```
fitdist(x, distr = "lnorm", method = "mle")
```

where x is the Madison precipitation data stream, and `mle` stands for maximum likelihood estimation. The fitting results are that $\mu = 4.52$, and $\sigma = 0.64$. The theoretical expected value and variance from the fitted distribution are

$$E[X] = e^{4.52 + 0.64^2/2} = 112.81, \tag{2.96}$$

$$\mathrm{Var}[X] = \left[e^{0.64^2} - 1\right] e^{2 \times 4.52 + 0.64^2} = 6602.91. \tag{2.97}$$

This model expected value 112.81 [mm] is approximately equal to the sample mean of 79 years of June precipitation which is 109.79 [mm]. The model standard deviation is $\sqrt{6602.91} = 81.25$, much larger than the sample standard deviation 63.94 [mm] from the Madison precipitation data.

The aforementioned statistics results can be computed by the following computer code.

```
#R lognormal distribution of precipitation
setwd("/Users/sshen/climstats")
Madison = read.csv("data/MadisonP.csv", header=TRUE)
Madison[1:2,]#Take a look at the first two lines of the data
daP = matrix(Madison[,7], ncol=12, byrow=TRUE)
x = daP[,6] #The June precipitation data
n=length(x)
m1=mean(x)
A = log(m1)- sum(log(x))/n
c = (1/(4*A))*(1 + sqrt(1 + 4*A/3))
beta = c/m1
mu=mean(log(x))
sig = sd(log(x))
round(c(mu, sig), digits = 2)
#[1] 4.52 0.65
modelmean = exp(mu + 0.5*sig^2)
modelsd = sqrt((exp(sig^2)-1)*exp(2*mu + sig^2))
datamean = mean(x)
datasd = sd(x)
round(c(modelmean, datamean, modelsd, datasd), digits =2)
#[1] 112.81 109.79  81.26  63.94
```

If you wish to plot the fitted lognormal distribution, you may use the following simple R command.

```
#R plot lognormal distribution
x=seq(0, 600, by=0.1)
plot(x, dlnorm(x, mean = 4.52, sd = 0.64))
```

2.5.6 Gamma Distribution

The Gamma distribution is often used to describe the distribution of rain rates and aerosol sizes. It is more common in the social sciences, especially in considering incomes and other nonsymmetric distributions. Its PDF is

$$f(x) = \frac{1}{b\Gamma(c)} \left(\frac{x}{b}\right)^{c-1} e^{-x/b}, \ 0 < x < \infty. \tag{2.98}$$

Here, b is called the scale parameter, has the same dimension as x, and determines kurtosis, i.e., the distribution peakedness. The smaller b means large kurtosis, or a sharp peak. The parameter $\beta = 1/b$ is called the rate. A computer fitting often yields the rate parameter β.

The parameter c, the shape parameter, determines the shape of the distribution. A large c, say $c \geq 5$, makes the Gamma distribution approximate to a normal distribution, $N(bc, b^2c)$; $\Gamma(c)$ is the Gamma function defined earlier. Calculus can show that

$$E[X] = bc, \tag{2.99}$$

$$\mathrm{Var}[X] = b^2 c. \tag{2.100}$$

The monthly precipitation of stations or grid boxes can often be modeled by the Gamma distribution. For example, the June precipitation of Omaha, Nebraska, USA, from 1948 to 2019 fits the Gamma distribution well with rate $\beta = 0.0189$ and shape $c = 1.5176$. Data source: Climate Data Online (CDO), NOAA, www.ncdc.noaa.gov/cdo-web.

The R command is

```
fitdist(x, distr = "gamma", method = "mle")
```

where x is the June precipitation data stream of 72 years (i.e., 72 numbers) from 1948 to 2019.

The theoretical expected value from the fitted Gamma distribution is thus $c/\beta = 1.5176/0.0189 = 80.30$ [mm]. The sample mean is 80.33 [mm] and is approximately equal to the theoretical expectation 80.30 [mm]. The theoretical standard deviation is $\sqrt{c}/\beta = \sqrt{1.5176}/0.0189 = 65.18$ [mm]. The sample standard deviation 58.83 [mm] and is in the same order as the theoretical standard deviation, although the approximation is not very good.

The Gamma distribution has a popular application in meteorological drought monitoring. It is used to define the Standardized Precipitation Index (SPI), which is a widely used index on a range of timescales from a month to a year.

The parameters of Gamma distribution can be directly estimated from data using the following formulas:

$$c = \frac{1}{4A} \left(1 + \sqrt{1 + \frac{4A}{3}}\right), \tag{2.101}$$

$$b = \frac{\bar{x}}{c}, \tag{2.102}$$

where

$$A = \ln(\bar{x}) - \frac{\sum_{i=1}^{n} \ln(x_i)}{n}. \tag{2.103}$$

The cumulative distribution function (CDF) of the Gamma distribution is

$$G(x) = \frac{1}{\Gamma(c)} \int_0^x t^{c-1} e^{-t} dx. \tag{2.104}$$

The R command for computing $G(x)$ is `pgamma(x, shape, rate)`.

This CDF cannot take care of the zero precipitation case because $G(0) = 0$. We thus modify the CDF as

$$H(x) = q + (1-q)G(x), \tag{2.105}$$

where $q = m/n$ is approximately equal to the probability of zero precipitation, and m is the number of zero data records among the sample of size n.

The CDF of the standard normal distribution is

$$J(x) = \frac{1}{\sqrt{2\pi}} \int_{-\infty}^x e^{-t^2/2} dt. \tag{2.106}$$

For a given precipitation datum x_d, one can compute $H(x_d)$. Let

$$J(z_s) = H(x_d). \tag{2.107}$$

Solving this equation determines the value of z_s which is defined as SPI. The R command for the solution is `zs = qnorm(p)`, where $p = H(x_d)$ is the cumulative probability.

The Omaha June precipitation data are all positive. Thus, $q = 0$. The precipitation of June 2014 was $x_d = 5.6$ [mm]. Using R command

`pgamma(5.6, shape=1.5176, rate=0.0189)`

we can compute the following results: G(5.6) = 0.0231. Then we use R command `qnorm(0.0231)` to find $z_s = 1.9936$. Thus, the June 2014 SPI is approximately 2, which is considered to be an extreme drought.

2.5.7 Student's *t*-Distribution

If $x_1, x_2, \ldots, x_n$ are normally distributed data with a given population mean μ, an unknown standard deviation, and a small sample n, say, $n < 30$, then

$$t = \frac{\bar{x} - \mu}{s/\sqrt{n}} \tag{2.108}$$

follows a Student's *t*-distribution, or simply *t*-distribution, with $n - 1$ degrees of freedom (dof), where

$$\bar{x} = \frac{1}{n} \sum_{i=1}^n x_i \tag{2.109}$$

is the sample mean, and

$$s^2 = \frac{1}{n-1} \sum_{i=1}^n (x_i - \bar{x})^2 \tag{2.110}$$

is the sample variance.

The *t*-distribution theory was due to William Gosset (1876–1937), who published the *t*-distribution under the pseudonym "Student" while working at the Guinness Brewery in Dublin, Ireland. The shape of the *t*-distribution is symmetric and is not too much different from the standard normal distribution. When dof is infinity, the *t*-distribution is exactly the same as the standard normal distribution $N(0, 1)$. Even when dof $= 6$, the t-distribution is already very close to the standard normal distribution. Thus, t-distribution is meaningfully different from the standard normal distribution only when the sample size is small, say, $n = 5$.

The mathematical formula of the PDF for the t-distribution is quite complicated. The formula is rarely used and is not presented in this book.

2.6 Chapter Summary

This chapter has described the basics of a random variable, PDF or PMF, CDF, CLT, and some commonly used probability distributions in climate science. Some important concepts are as follows.

(i)   Definition of RV: When talking about an RV X, we mean three things: (a) A sample space that contains all the possible outcomes of the random events, (b) a function that maps the outcomes to real values, and (c) the probability associated with the real values in a discrete way or over an interval. Therefore, a random variable is different from a regular real-valued variable that can take a given value in its domain. A random variable takes a value only after the outcome of a random event is known, because RV is itself a function. When the value is taken, we then ask what the measure is of uncertainty of the value.

(ii)   Commonly used probability distributions: We have presented some distributions commonly used in climate science, including the normal distribution, binomial distribution, Poisson distribution, exponential distribution, chi-square distribution, lognormal distribution, and Gamma distribution. We have provided both R and Python codes to plot the PDFs and CDFs of these distributions. We have also fitted real climate data to appropriate distributions. Following our examples, you may wish to fit your own data to appropriate distributions.

(iii)   The central limit theorem (CLT) deals with the mean of independent RVs of finite variance. CLT claims that the mean is approximately normal if the number of RVs are many. The key assumption is the finite variance for each RV.

(iv)   Chi-square distribution χ_k^2 deals with the sum of squares of a few independent standard normal RVs. The chi-square distribution is very useful in climate science, such as analyzing goodness of fit for a model.

References and Further Reading

[1] L. M. Chihara and T. C. Hesterberg, 2018: *Mathematical Statistics with Resampling and R*. John Wiley & Sons.

> This is a textbook intended for an upper-undergraduate and graduate mathematical statistics course. The R code and the figures produced by the code are very helpful to readers.

[2] R. A. Johnson and G. K. Bhattacharyya, 1996: *Statistics: Principles and Methods*. 3rd ed., John Wiley & Sons.

> This is a popular textbook for a basic statistics course. It emphasizes clear concepts and includes descriptions of statistical methods and their assumptions. Many real-world data are analyzed in the book. The 8th edition was published in 2019.

[3] H. Von Storch and F. W. Zwiers, 2001: *Statistical Analysis in Climate Research*. Cambridge University Press.

> This is a comprehensive climate statistics book that helps climate scientists apply statistical methods correctly. The book has both rigorous mathematical theory and practical applications. Numerous published research results are discussed.

[4] D. S. Wilks, 2011: *Statistical Methods in the Atmospheric Sciences*. 3rd ed., Academic Press.

> This excellent textbook is easy to read and contains many simple examples of analyzing real climate data. It is not only a good reference manual for climate scientists but also a guide tool that helps scientists in other fields make sense of the data analysis results in climate literature.

Exercises

2.1 Use the data in Table 2.1 to calculate the mean length of wet spell in the summer for New York City, following the theory of conditional probability.

2.2 Use the data in Table 2.1 to calculate the probability of exactly n successive dry days in the summer for New York City for $n = 1, 2, 3, 4, 5, 6$, and 7. *Hint: The $(n+1)th$ day must be a wet day.*

2.3 Use the data in Table 2.1 to calculate the probability of exactly n successive wet days in the summer for New York City for $n = 1, 2, 3, 4, 5, 6$, and 7. *Hint: The $(n+1)th$ day must be a dry day.*

2.4 Write down the details of proving that the expected number of occurrence of a binomial event in $B(p, n)$ is

$$E[K] = \sum_{k=0}^{n} k C_k^n p^k (1-p)^{n-k} = np, \tag{2.111}$$

where C_k^n is

$$C_k^n = \frac{n!}{k!(n-k)!}. \tag{2.112}$$

2.5 The variance of the number of occurrence of a binomial event in $B(p, n)$ is $np(1-p)$. Write down the mathematical details to derive the following:

$$\mathrm{Var}[K] = \sum_{k=0}^{n} (k - np)^2 C_k^n p^k (1-p)^{n-k} = np(1-p). \tag{2.113}$$

2.6 Show that

$$\sum_{k=0}^{\infty} k \times \frac{e^{-\lambda} \lambda^k}{k!} = \lambda, \tag{2.114}$$

i.e., $E[X] = \lambda$ for the Poisson distribution.

2.7 Derive the formula for the variance of t in the exponential distribution with the mean rate λ:

$$\int_0^{\infty} \left(t - \frac{1}{\lambda} \right)^2 \lambda e^{-\lambda t} \, \mathrm{d}t = \frac{1}{\lambda^2}. \tag{2.115}$$

2.8 Given the PDF

$$f(x) = \frac{1}{\sqrt{2\pi}\sigma} e^{-\frac{1}{2}(x-\mu)^2/\sigma^2}, \quad -\infty < x < \infty, \tag{2.116}$$

use calculus to show that

$$E[X] = \int_{-\infty}^{\infty} x f(x) \, \mathrm{d}x = \mu \tag{2.117}$$

and

$$\mathrm{Var}[X] = \int_{-\infty}^{\infty} (x - \mu)^2 f(x) \, \mathrm{d}x = \sigma^2. \tag{2.118}$$

2.9 Following Figure 2.6, use R or Python to simulate and plot the joint and marginal distributions under the assumption of normal distribution in both x and y.

2.10 (a) Use R or Python to fit the January monthly precipitation data of Omaha, Nebraska, USA, from 1950 to 2019 to the Gamma distribution. You may find the data from the NOAA Climate Data Online (CDO) or another source.

(b) Plot the fitted distribution and the histogram of data on the same figure. You may put your the density scale of the fitted distribution on the right using the blue color, as shown in Figure 2.9.

2.11 Repeat the Gamma distribution problem for the June monthly precipitation data of a location of your interest. You may find the monthly data from your data source or from the NOAA Climate Data Online (CDO).

2.12 (a) Use R or Python to fit the January precipitation data of Madison, Wisconsin, USA, from 1950 to 2019 to the lognormal distribution. You may find the data from the NOAA Climate Data Online (CDO) or another source.

(b) Plot the fitted distribution and the histogram of data on the same figure. See Figure 2.9 as a reference.

2.13 Repeat the lognormal distribution problem for the June precipitation data of a location of your interest. You may find the monthly data from your data source or from the NOAA Climate Data Online (CDO).

2.14 Use R or Python to plot the Poisson distributions with five different rates λ on the same figure. Explain why the PDF peaks at $k = \lambda$.

2.15 Use R or Python to plot the exponential distributions with five different rates λ on the same figure.

2.16 Use R or Python to plot the lognormal distributions with five different sets of μ and σ values on the same figure.

2.17 Use R or Python to plot the Gamma distributions with five different sets of a, b, and c values on the same figure.

2.18 Use R or Python to plot chi-square distributions with different degrees of freedom on the same figure.

2.19 (a) Identify a GHCN station of your interest that has at least 30 years of high-quality monthly precipitation records, and download the data.

(b) Use R or Python to plot the May precipitation data against the time [year].

(c) Compute the standard precipitation index (SPI) from the May precipitation data for the driest year.

2.20 Compute the summer SPI indices based on the June-July-Aug mean precipitation at Omaha, Nebraska, USA, from 1950 to 2019. You may find the data from the NOAA Climate Data Online (CDO) or another source.

2.21 Repeat the previous problem for the June precipitation data.

2.22 (a) Compute the standardized anomalies of the NOAAGlobalTemp monthly temperature data from January 1880 to December 2015 over the 5-degree grid box centered at (47.5° N, 107.5° W). This data stream has 1,632 months and can be broken into 272 6-month periods. Then, compute

$$X = \sum_{n=1}^{6} T_n^2 \tag{2.119}$$

for every six months.

(b)   Plot the histogram of these 272 X data.

(c)   Using the mean and SD computed from the data, plot the distribution χ_6^2, and compare the shapes of the distribution and the histogram. Use Figure 2.9 as a reference for your figure appearance.

2.23   Repeat the previous problem for the grid box of your interest, such as the grid box that contains your hometown.

## 3

Estimation and Decision-Making

A goal of climate statistics is to make estimates from climate data and use the estimates to make quantitative decisions with a given probability of success, described by confidence interval. For example, based on the estimate from the NOAAGlobalTemp data and given 95% probability, what is the interval in which the true value of the global average decade mean of the 2010s lies? Was the 1980–2009 global temperature significantly different from the 1950 to 1979 temperature, given the 5% probability of being wrong? To answer questions like these is to make a conclusive decision based on the estimate of a parameter, such as the global average annual mean temperature, from data. With this in mind, we introduce the basic methods of parameter estimation from data, and then introduce decision-making using confidence intervals and hypothesis testing. The procedures of estimation and decision-making are constrained by sample size. Climate data are usually serially correlated, and the individual data entries may not be independent. The actual sample size, n_{eff}, may be much smaller than the number of data records n. The uncertainty of the estimates from the climate data is much larger when taking into account of n_{eff}. Examples are provided to explain the correct use of n_{eff}. The incorrect use of sample size by climate scientists can lead to erroneous decisions. We provide a way to test serial correlation and compute actual sample size.

Fundamentally, three elements are involved: data, estimate, and decision-making. Then we may ask our first questions: What are climate data? How are they defined? Can we clearly describe the main differences between the definition and observation of climate data?

3.1 From Data to Estimate

3.1.1 Sample Mean and Its Standard Error

In statistics, a population refers to the set of all members in a specific group about which information is sought, for example, El Niño events. A sample drawn from a population in the course of investigation is a subset of observations, such as the El Niño events observed between 1950 and 2000. The sample data are the quantitative values of the samples, such as the maximum sea surface temperature anomalies of El Niños over a specific region over the

Eastern Tropical Pacific. Our goal here is to use the sample data $(x_1, x_2, \ldots, x_n)$ to estimate a statistical parameter for the population, such as the sample mean:

$$\bar{x} = \frac{1}{n} \sum_{i=1}^{n} x_i. \tag{3.1}$$

The true population mean μ is never known, but in our imagination it exists. We wish to use the sample mean $\bar{x}$ as an estimate for the true mean. We assume that the samples are random variables. Then $\bar{x}$ is also a random variable. Some textbooks would use uppercase letters to denote the random variables to explain statistical concepts and lowercase letters to denote data for computational convenience. The two notations often get confused in those books. In this chapter, we use lowercase letters for both random variables and data. The lowercase letters $(x_1, x_2, \ldots, x_n)$ and $\bar{x}$ can be interpreted as random variables in statistical interpretation and also treated as data in computing.

The central limit theorem (CLT) states that the sample mean $\bar{x}$ is normally distributed when the sample size n is sufficiently large. Some textbooks consider 30 to be "large" and others 50. Climate scientists often use 30. The CLT needs an assumption that each sample x_i has finite variance. Climate data usually satisfy this assumption.

We do not expect that the true mean μ is equal to the sample mean, rather we ask a question: what is the probability that the true mean lies in an interval around the sample mean $\bar{x}$? Apparently, if the interval is large, it has a better chance for the true mean to be inside the interval. However, a large interval is less accurate and less useful, because in this case the true mean μ may be far away from the sample mean $\bar{x}$.

If the population has small variability, then it is easier to obtain accurate observations. The sample data will have a smaller standard deviation. Hence even a small interval may have a big chance to contain the true mean.

Therefore, the interval depends on two factors: variability of the population, measured by the standard deviation σ, and number of the observations, given by the sample size n. Our intuition is that we would get a more accurate estimate for the population mean when we have a large sample size n. These two factors lead us to define the *sample standard deviation* and the *standard deviation of the sample mean*. The unbiased estimate of the sample standard deviation is defined by

$$s = \sqrt{\frac{1}{n-1} \sum_{i=1}^{n} (x_i - \bar{x})^2}. \tag{3.2}$$

Further, when the random sample data $(x_1, x_2, \ldots, x_n)$ are independent from each other and are from the same population with variance equal to σ^2, the standard deviation of the sample mean $\bar{x}$ is

$$\mathrm{SD}[\bar{x}] = \frac{s}{\sqrt{n}}. \tag{3.3}$$

This is also called the *standard error* of the sample mean, denoted by $\mathrm{SE}[\bar{x}]$ or SE. The standard error quantifies the size of the "error bar" $\bar{x} \pm \mathrm{SE}$ when approximating the true mean using the sample mean.

Formula (3.3) can be easily derived as follows:

$$
\begin{aligned}
(\mathrm{SD}[\bar{x}])^2 &= \mathrm{Var}[\bar{x}] \\
&= \mathrm{Var}\left[\frac{x_1 + x_2 + \cdots + x_n}{n}\right] \\
&= \mathrm{Var}\left[\frac{x_1}{n}\right] + \mathrm{Var}\left[\frac{x_2}{n}\right] + \cdots + \mathrm{Var}\left[\frac{x_n}{n}\right] \\
&= \frac{\mathrm{Var}[x_1]}{n^2} + \frac{\mathrm{Var}[x_2]}{n^2} + \cdots + \frac{\mathrm{Var}[x_n]}{n^2} \\
&= \frac{\sigma^2}{n^2} + \frac{\sigma^2}{n^2} + \cdots + \frac{\sigma^2}{n^2} \\
&= \frac{\sigma^2}{n}.
\end{aligned}
\tag{3.4}
$$

This description and derivation imply that sample standard deviation s estimates the spread of the population, while standard deviation of the mean $\mathrm{SD}(\bar{x})$ is an indicator of the accuracy of using the sample mean $\bar{x}$ to approximate the population mean μ. This is why $\mathrm{SD}[\bar{x}]$ is also called the standard error. To further understand this concept, we use numerical simulation results shown in Figure 3.1. We generate 1,000,000 normally distributed and independent data according to $N(0, 10^2)$. So the population mean is $\mu = 0$ and standard deviation is $\sigma = 10$. These are given for the simulation. Figure 3.1(a) shows the histogram of the 1,000,000 data. This histogram has a standard deviation equal to 10. Then, we take sample mean of size $n = 100$. The 1,000,000 data allow us to take the sample mean 10,000 times, which leads to 10,000 sample means $\bar{x}$. Figure 3.1(b) shows the histogram of these 10,000 sample means $\bar{x}$, whose standard deviation is $\mathrm{SD}(\bar{x}) = \sigma/\sqrt{n} = 10/\sqrt{100} = 1$.

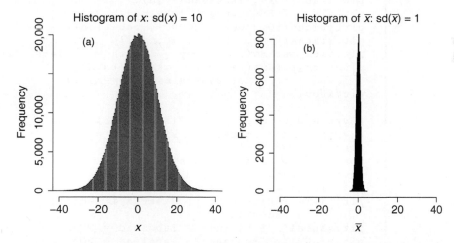

Figure 3.1 (a) Histogram of a population x, and (b) histogram of sample mean $\bar{x}$ generated from the numerical simulation according to the normal distribution $N(0, 10^2)$.

Figure 3.1 can be generated by the following computer code.

```
#R plot Fig. 3.1: Histograms
#Simulation of the standard error of mean
```

```
setwd("/Users/sshen/climstats")
setEPS() # save the .eps figure file to the working directory
postscript("fig0301.eps", height = 5, width = 10)
mu = 0; sig = 10; m=10000; n = 100; k = m*n
x = rnorm(k, mean = mu, sd = sig)
par(mar = c(4, 4.5, 2, 1.5))
par(mfrow=c(1,2))
hist(x, breaks = 201, xlim = c(-40,40),
 main = expression('Histogram␣of␣x:␣sd(x)␣=␣10'),
     cex.lab = 1.5, cex.axis = 1.5,
     cex.main = 1.5)
text(-25, 19000, '(a)', cex =1.3)
xmat = matrix(x, ncol = n)
xbar = rowMeans(xmat)
hist(xbar, breaks = 31, xlim = c(-40,40),
     xlab = expression(bar(x)),
     main = expression('Histogram␣of␣'*bar(x)*
        '␣:␣sd('*bar(x)*')␣=␣1'),
     cex.lab = 1.5, cex.axis = 1.5,
     cex.main = 1.5)
text(-25, 750, '(b)', cex =1.3)
dev.off()
```

```
#Python plot Fig. 3.1a: Histogram
mu = 0; sig = 10; m=10000; n = 100; k = m*n
x = np.random.normal(mu, sig, size = k)
plt.hist(x, bins = 201)
plt.title("Histogram␣of␣x:␣sd(x)␣=␣10",
          pad = 15)
plt.xlabel("x", labelpad = 20)
plt.ylabel("Frequency", labelpad = 20)
plt.yticks([0, 10000, 20000], rotation=90)
plt.xticks([-40, -20, 0, 20, 40])
plt.tick_params(axis = 'y', length = 7, labelsize=20)
plt.tick_params(axis='x', which='both',
                bottom=False, top=False)
plt.show()

#Python plot Fig. 3.1b: Histogram of means
x = np.random.normal(mu, sig, size = k)
xmat = x.reshape(m,n)
xbar = np.mean(xmat, axis - 1)
np.size(xbar)
plt.hist(xbar, bins = 31)
plt.title(r'Histogram␣of␣$\bar{x}$:␣sd($\bar{x}$)␣=␣1',
          pad = 15)
plt.xlabel(r"$\bar{x}$", labelpad = 20)
plt.ylabel(r"Frequency", labelpad = 20)
plt.yticks([0, 200, 400, 600, 800], rotation=90)
plt.xticks([-40, -20, 0, 20, 40])
plt.tick_params(axis = 'y', length = 7, labelsize=20)
plt.tick_params(axis='x', which='both',
                bottom=False, top=False)
plt.show()
```

However, most climate data $x_1, x_2, \ldots, x_n$ are not independent, and are correlated with $\text{cor}(x_i, x_j) \neq 0$. Then, the n sample data have some redundancy. The extreme case is that all $x_1, x_2, \ldots, x_n$ are equal. Thus, the n sample data really mean only one datum. Thus, $x_1, x_2, \ldots, x_n$ do not serve as n independent pieces of information. The mean $\bar{x}$ will not achieve the purpose of shrinking the standard error $s/\sqrt{n}$. Then, how many independent pieces of information are in the data $x_1, x_2, \ldots, x_n$? We use n_{eff} to denote the number, called the effective sample size, which is less than n. Thus, a more accurate estimate of the standard error is

$$\text{SE} = \frac{s}{\sqrt{n_{\text{eff}}}}. \tag{3.5}$$

More details about the effective sample size can be found in Section 3.3.

3.1.2 Confidence Interval for the True Mean

The error margin at a 95% confidence level is

$$\text{EM} = 1.96 \frac{s}{\sqrt{n}}, \tag{3.6}$$

where 1.96 comes from the 95% probability in $(\mu - 1.96\sigma, \mu + 1.96\sigma)$ for a normal distribution $N(\mu, \sigma^2)$. When the confidence level is raised from 0.95 to a larger value, the number 1.96 will be increased to a larger number accordingly.

The confidence interval (CI) for a true mean μ is then defined as

$$(\bar{x} - \text{EM}, \bar{x} + \text{EM}) \text{ or } (\bar{x} - 1.96\frac{s}{\sqrt{n}}, \bar{x} + 1.96\frac{s}{\sqrt{n}}). \tag{3.7}$$

This means that the probability for the true mean to be inside the confidence interval $(\bar{x} - \text{EM}, \bar{x} + \text{EM})$ is 0.95:

$$P(\bar{x} - \text{EM} < \mu < \bar{x} + \text{EM}) = 0.95. \tag{3.8}$$

In this expression, we treat both $\bar{x}$ and EM as random variables, and the unknown true mean μ as a fixed value. When the CI $(\bar{x} - \text{EM}, \bar{x} + \text{EM})$ is estimated with numerical values, we have then made a single realization of the CI. A particular realization may not cover the true mean, although we are 95% confident that it does because about 95% of realizations do.

One can use computers to simulate this statement by repeatedly generating normally distributed numbers $(x_1, x_2, \ldots, x_n)$ m times with given mean and standard deviation. We then compute the intervals $(\bar{x} - \text{EM}, \bar{x} + \text{EM})$ using $\text{EM} = 1.96\frac{s}{\sqrt{n}}$ and verify that among the m number of experiments, 95% of them have the true mean inside the interval $(\bar{x} - \text{EM}, \bar{x} + \text{EM})$. The computer code is below.

```
#Verification of 95% CI by numerical simulation
m=10000 #10,000 experiments
x = 1:m
n = 30 #sample size n
truemean = 8
da = matrix(rnorm(n*m, mean = truemean, sd = 1), nrow = m)
esmean = rowMeans(da) #sample mean
```

```
library(matrixStats)
essd = rowSds(da) #sample SD
upperci = esmean + 1.96*essd/sqrt(n) #interval upper limit
lowerci = esmean - 1.96*essd/sqrt(n) #interval lower limit
l=0
for(k in 1:m){
  if(upperci[k] >= truemean & lowerci[k] <= truemean )
    l=l+1
} #Determine if the true mean is inside the interval
l/m #Percentage of truth
#[1] 0.9425 #which is approximately 0.95
```

```
#Verify 95% CI by numerical simulation: Python code
m = 10000 #10,000 experiments
n = 30 #sample size n
truemean = 8
da = np.random.normal(truemean, 1, size = (m,n))
esmean = da.mean(1)
essd = da.std(1)
upperci = esmean + 1.96*essd/np.sqrt(n)
lowerci = esmean - 1.96*essd/np.sqrt(n)
l=0
for k in np.arange(0, m):
    if upperci[k] >= truemean and lowerci[k] <= truemean :
        l = l + 1
l/m
#[1] 0.9398 #which is approximately 0.95
```

You can also plot and visualize these simulations. See the exercise problems of this chapter.

Similarly, the probability for the true mean to be inside the error bar $\bar{x} \pm \text{SE}$ is 0.68. See Figure 3.2 for the confidence intervals at 95% and 68% confidence levels.

When the sample size n goes to infinity, the error margin (EM) goes to zero. Accordingly, the sample mean is equal to the true mean. This conclusion is correct with 95% probability, and wrong with 5% probability.

One can also understand the sample confidence interval for a new variable:

$$z = \frac{\bar{x} - \mu}{s/\sqrt{n}}, \tag{3.9}$$

which is a normally distributed variable with mean equal to zero and standard deviation equal to one, i.e., it is a standard normal distribution. The variable $y = -z$ also satisfies the standard normal distribution. So, the probability of $-1 < z < 1$ is 0.68, and that of $-1.96 < z < 1.96$ is 0.95. The set $-1.96 < z < 1.96$ is equivalent to $\bar{x} - 1.96s/\sqrt{n} < \mu < \bar{x} + 1.96s/\sqrt{n}$. Thus, the probability of the true mean being in the confidence interval of the sample mean $\bar{x} - 1.96s/\sqrt{n} < \mu < \bar{x} + 1.96s/\sqrt{n}$ is 0.95. This is visually displayed in Figure 3.2.

In addition, the formulation $\bar{x} = \mu + zs/\sqrt{n}$ corresponds to a standard statistics problem for an instrument with observational errors:

$$y = x \pm \varepsilon, \tag{3.10}$$

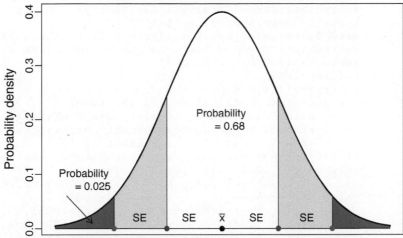

True mean as a normally distributed random variable

Figure 3.2 Schematic illustration of confidence intervals and confidence levels of a sample mean for a large sample size. The confidence interval at the 95% confidence level is between the two red points, and that at 68% is between the two blue points. SE stands for the standard error, and 1.96 SE is approximately regarded as 2 SE in this figure.

where ε stands for error, x is the true but never-known value to be observed, and y is the observational data. Thus, data are equal to the truth plus random errors. The *expected value* of the error is zero and the standard deviation of the error is $s/\sqrt{n}$, also called the standard error.

The 95% confidence level comes into the equation when we require that the observed value must lie in the interval $(\mu - EM, \mu + EM)$ with a probability equal to 0.95. This corresponds to the requirement that the standard normal random variable z is found in a symmetric interval (z_-, z_+) with a probability equal to 0.95, which implies that $z_- = -1.96$ and $z_+ = 1.96$. Thus, the confidence interval of the sample mean at the 95% confidence level is

$$(\bar{x} - 1.96s/\sqrt{n}, \bar{x} + 1.96s/\sqrt{n}), \qquad (3.11)$$

or

$$(\bar{x} - z_{\alpha/2}s/\sqrt{n}, \bar{x} + z_{\alpha/2}s/\sqrt{n}), \qquad (3.12)$$

where $z_{\alpha/2} = z_{0.05/2} = 1.96$. So, $1 - \alpha = 0.95$ is used to represent the probability of the variate lying inside the confidence interval, while $\alpha = 0.05$ is the "tail probability" outside the confidence interval. Outside the confidence interval means occurring on either the left side or the right side of the distribution indicated by the red area of Figure 3.2. Each side represents $\alpha/2 = 0.025$ tail probability.

Figure 3.2 can be plotted by the following computer code.

```
#R plot Fig. 3.2: Confidence intervals and confidence levels
#Plot confidence intervals and tail probabilities
setwd("/Users/sshen/climstats")
setEPS() # save the .eps figure file to the working directory
postscript("fig0302.eps", height = 4.5, width = 6.5)
par(mar=c(2.5,3.5,2.0,0.5))
rm(list=ls())
par(mgp=c(1.4,0.5,0))
curve(dnorm(x,0,1), xlim=c(-3,3), lwd=3,
      main='Confidence Intervals and Confidence Levels',
      xlab="True mean as a normally distributed random variable",
      ylab="", xaxt="n", cex.lab=1.2)
title(ylab='Probability density', line=2, cex.lab=1.2)
polygon(c(-1.96, seq(-1.96,1.96,len=100), 1.96),
        c(0,dnorm(seq(-1.96,1.96,len=100)),0),col='skyblue')
polygon(c(-1.0,seq(-1.0, 1, length=100), 1),
        c(0, dnorm(seq(-1.0, 1, length=100)), 0.0),col='white')
polygon(c(-3.0,seq(-3.0, -1.96, length=100), -1.96),
        c(0, dnorm(seq(-3.0, -1.96, length=100)), 0.0),col='red')
polygon(c(1.96,seq(1.96, 3.0, length=100), 3.0),
        c(0, dnorm(seq(1.96, 3.0, length=100)), 0.0),col='red')
points(c(-1,1), c(0,0), pch=19, col="blue")
points(0,0, pch=19)
points(c(-1.96,1.96),c(0,0),pch=19, col="red")
text(0,0.02, expression(bar(x)), cex=1.0)
text(-1.50,0.02, "SE", cex=1.0)
text(-0.60,0.02, "SE", cex=1.0)
text(1.50,0.02, "SE", cex=1.0)
text(0.60,0.02, "SE", cex=1.0)
text(0,0.2, "Probability
     = 0.68")
arrows(-2.8,0.06,-2.35,0.01, length=0.1)
text(-2.5,0.09, "Probability
     =0.025")
dev.off()
```

```
#Python plot Fig. 3.2: Confidence intervals and levels
# Create normal probability function
def intg(x):
    return 1/( np.sqrt(2 * np.pi))*np.exp( - x**2 /2)
a_ , b_ = -3,3 # Set desired quantities
a, b = -1.96, 1.96
a1, b1 = -1.0, 1.0 # integral limits

x = np.linspace(-3, 3)
y = intg(x)
# Set up figure
fig, ax = plt.subplots(figsize=(12,10))
ax.plot(x, y, 'black', linewidth=1)
ax.set_ylim(bottom=0)

# Make the shaded region
ix_ = np.linspace(a_,b_)
```

```
ix = np.linspace(a, b)
ix1 = np.linspace(a1, b1)

# Using the intg function
iy_ = intg(ix_)
iy = intg(ix)
iy1 = intg(ix1)

# Creating desired vertical separation
verts_ = [(a_, 0), *zip(ix_, iy_), (b_, 0)]
verts = [(a, 0), *zip(ix, iy), (b, 0)]
verts1 = [(a1, 0), *zip(ix1, iy1), (b1, 0)]

# Creating desired polygons
from matplotlib.patches import Polygon
poly_ = Polygon(verts_, facecolor='red', edgecolor='black',
                linewidth=0.5)
poly =Polygon(verts,facecolor='lightblue',edgecolor='black')
poly1 =Polygon(verts1, facecolor='white', edgecolor='black')

# Adding polygon
ax.add_patch(poly_)
ax.add_patch(poly)
ax.add_patch(poly1)

# Adding appropriate text
ax.text(0.5 * (a + b), 0.24, "Probability =_0.68",
    horizontalalignment='center', verticalalignment="center",
    fontsize=15, fontweight="bold")
ax.text(-1.55, .02, "SE", fontsize=15)
ax.text(1.40, .02, "SE", fontsize=15)
ax.text(-0.9, .02, "SE", fontsize=15)
ax.text(0.65, .02, "SE", fontsize=15)
ax.text(-0.05, .02, r'$\bar{x}$', fontsize=15)

# Creating desired arrows
plt.arrow(-2.2,.02,-0.5,0.05)
ax.text(-3,0.08,"Probability_=_\n_0.025", fontsize=15,
        fontweight="bold")

# Label the figure
plt.ylabel("Probability density")
plt.xlabel("True mean as a random variable")
plt.title("Confidence Intervals and Confidence Levels")

ax.set_xticks((a,a1,0,b1,b))
ax.set_xticklabels(('','','','',''))

plt.show()
```

In practice, we often regard 1.96 as 2.0, and the 2σ-error bar as the 95% confidence interval.

Example 3.1 Figure 3.3 shows the January surface air temperature anomalies of Edmonton, Canada, based on the 1971–2000 monthly climatology. We compute the mean and its confidence interval for the first 50 years from 1880 to 1929 at the 95% confidence level. The result is as follows:

(i)   Sample mean: $\bar{x} = -2.47°C$ (with sample size $n = 50$).

(ii) Sample standard deviation: $s = 4.95°C$.

(iii) Error margin at the 95% confidence level without the consideration of serial correlation: $EM = 1.96 \times s/\sqrt{n} = 1.37°C$.

(iv) Confidence interval of the mean at the 95% confidence level: $\bar{x} \pm EM$, i.e., $[-3.84, -1.10]°C$.

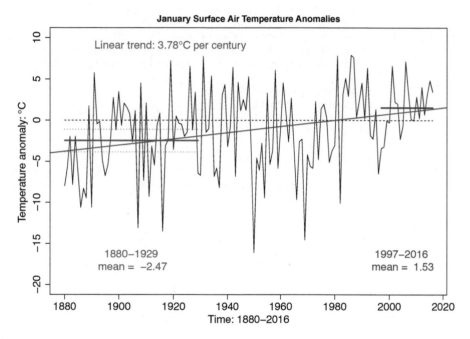

January Surface Air Temperature Anomalies

Figure 3.3   The January mean surface air temperature anomalies from 1850 to 2016 based on the 1971–2000 monthly climatology for the $5° \times 5°$ grid box centered around ($52.5°$ N, $112.5°$ W), that covers Edmonton, Canada (data source: NOAAGlobalTemp Version 3.0).

See Figure 3.3 for the mean $\bar{x} = -2.47°C$ and the confidence interval $[-3.84, -1.10]°C$. These statistics can be computed using the following computer code.

```
#R code for Edmonton data statistics NOAAGlobalTemp
setwd('/Users/sshen/climstats')
da1 =read.csv("data/Lat52.5_Lon-112.5.csv", header=TRUE)
dim(da1)
#[1] 1642    3 #1642 months: Jan 1880 - Oct 2016
da1[1:2,]
#       Date    Value
#1 1880-01-01 -7.9609
#2 1880-02-01 -4.2510
```

```
jan = seq(1, 1642, by=12)
Tjan = da1[jan,]
TJ50 = Tjan[1:50, 3]
xbar = mean(TJ50)
sdEdm = sd(TJ50)
EM = 1.96*s/sqrt(50)
CIupper = xbar + EM
CIlower = xbar - EM
round(c(xbar, sdEdm, EM, CIlower, CIupper), digits =2)
#[1] -2.47  4.95  1.37 -3.84 -1.10
```

```
#Python code for Edmonton data statistics from NOAAGlobalTemp
import pandas as pd
from statistics import stdev
da1 = pd.read_csv("data/Lat52.5_Lon-112.5.csv");
da1.shape
#(1642, 3) #1642 months: Jan 1880 - Oct 2016
da1.iloc[0:1,:]
#        Level Name      Date      Value
#0       Surface Temp    1880-01-01      -7.9609
jan = np.arange(0, 1641, 12)
Tjan = np.array(da1.iloc[jan,:])
TJ50 = Tjan[0:50,2]
#print(TJ50)
xbar = np.mean(TJ50)
sdEdm = stdev(TJ50)
EM = 1.96*sdEdm/np.sqrt(50)
CIupper = xbar + EM
CIlower = xbar - EM
np.round(np.array([xbar, sdEdm, EM, CIlower, CIupper]), 2)
#array([-2.47,  4.95,  1.38, -3.85, -1.1 ])
```

When the serial correlation is considered, the error margin will be larger. We will discuss this later in this chapter.

Figure 3.3 can be produced by the following computer code.

```
#R plot Fig. 3.3: Edmonton January temp
#52.5N, 112.5W, Edmonton, NOAAGlobalTemp Version 3.0
setwd("/Users/sshen/climstats")
setEPS() # save the .eps figure file to the working directory
postscript("fig0303.eps", height = 7, width = 10)
da1 =read.csv("data/Lat52.5_Lon-112.5.csv", header=TRUE)
jan =seq(1, 1642, by=12)
Tjan = da1[jan,]
t=1880:2016
regJan = lm(Tjan[,3] ~ t)
par(mar=c(3.6,4.5,2,1.3), mgp = c(2.3, 0.9, 0))
plot(t, Tjan[,3], type="l",
     main = "Edmonton January Surface Air Temperature Anomalies",
     xlab = "Time: 1880-2016",
```

```
    ylab = expression("Temperature␣Anomaly:"~degree~C),
    ylim=c(-20,10), cex.lab=1.4, cex.axis=1.4)
m19972006 = mean(Tjan[118:137, 3]) #1997--2016 Jan mean
m18801929 = mean(Tjan[1:50, 3]) #1880--1929 Jan mean
EM = 1.96*sd(Tjan[1:50, 3])/sqrt(50)
lines(t, rep(0,length(t)), lty = 2)
lines(t[118:137], rep(m19972006, 20),
    col = 'blue', lwd = 3, lty = 1)
lines(t[1:50], rep(m18801929, 50),
    col = 'darkgreen', lwd = 3, lty = 1)
lines(t[1:50], rep(m18801929 - EM, 50),
    col = 'darkgreen', lwd = 1.5, lty = 3)
lines(t[1:50], rep(m18801929 + EM, 50),
    col = 'darkgreen', lwd = 1.5, lty = 3)
abline(regJan, col='red', lwd = 2)
text(1920, 9,
    expression("Linear␣trend:␣3.78"~degree~C~"per␣century"),
    col="red", cex=1.4)
text(2005, -17,
    paste('1997-2016', '\nmean␣=␣',
        round({m19972006}, digits = 2)),
    col = 'blue', cex = 1.4)
text(1905, -17,
    paste('1880-1929', '\nmean␣=␣',
        round({m18801929}, digits = 2)),
    col = 'darkgreen', cex = 1.4)
dev.off()
```

```
# Python plot Fig. 3.3: Edmonton January temp
#52.5N, 112.5W, Edmonton, NOAAGlobalTemp Version 3.0
import pandas as pd
from statistics import stdev
from sklearn.linear_model import LinearRegression
da1 = pd.read_csv("data/Lat52.5_Lon-112.5.csv");
jan = np.arange(0, 1641, 12)
Tjan = np.array(da1.iloc[jan,:])
Tda = Tjan[:,2]
t = np.arange(1880,2017)
t1 = pd.DataFrame(t)
Tda1 = pd.DataFrame(Tda)
reg = LinearRegression().fit(t1, Tda1)
predictions = reg.predict(t.reshape(-1, 1))
plt.plot(t1, Tda1, color = 'k')
plt.ylim(-18, 12)
plt.grid()
plt.title('Edmonton␣Jan␣Surface␣Air␣Temperature␣Anomalies')
plt.xlabel('Time:␣1880-2016')
plt.ylabel('Temperature␣Anomaly␣[$\degree$C]')
plt.plot(t, predictions, '-', color="red")
plt.text(1885,10, 'Linear␣trend:␣3.78$\degree$C␣per␣Century',
        fontsize = 20, color ='red')
m19972006 = np.mean(Tda1[117:137]) #1997--2016 Jan mean
m18801929 = np.mean(Tda1[0:50]) #1880--1929 Jan mean
```

```
EM = 1.96*stdev(TJ50)/np.sqrt(50)
plt.plot(t[117:137], np.repeat(m19972006, 20),
         color = 'blue')
plt.plot(t[0:50], np.repeat(m18801929, 50),
         color = 'darkgreen')
plt.text(1985, -15,
         '1997-2016' '\n' 'mean␣=␣1.53$\degree$C',
     color = 'blue', fontsize = 20)
plt.text(1880, -15,
         '1880-1929' '\n' 'mean␣=␣-2.47$\degree$C',
     color = 'blue', fontsize = 20)
plt.show()
```

3.2 Decision-Making by Statistical Inference

One type of decision-making process is based on deterministic logic. If it rains, you cover yourself when walking outside, perhaps with an umbrella or a rain coat. If the temperature is hot in the room, you turn on the air conditioning.

Another type of decision-making involves incomplete information and uncertainties. This process is not deterministic but probabilistic. For example, Figure 3.3 seems to suggest that in the last 20 years (1997–2016), Edmonton January temperature anomalies are greater than zero. However, the temperature goes up and down. It is unclear whether the suggested "greater than zero" claim is true. We need data and procedures to make a decision: accept or reject the claim. This decision can be correct or wrong with four possible outcomes, as shown in Table 3.1, the decision contingency table.

		DECISION	
		Accept H_0	Accept H_1
ACTUAL	H_0 True	Correct Decision Significance level: α Confidence level: $1 - \alpha$	Type I Error False alarm Error probability: α Risk of false cases
	H_1 True	Type II Error Missed detection Error probability: β Risk of missed cases	Correct Decision Power: $1 - \beta$

Table 3.1 Contingency table for H_0 and H_1

H_0: Null hypothesis H_1: Alternative hypothesis

3.2.1 Contingency Table for Decision-Making with Uncertainty

The true and false determination of an event, such as the positive average of anomalies, and the correct and wrong decisions on the event, form four cases: TT (true event correctly detected as true), FF (false event correctly detected as false), TF (true event, but wrongly detected as false), and FT (false event, but wrongly detected as true). TF is a missed detection, and FT is a false alarm. These four cases form a 2×2 table, called the *contingency table* for decision-making. The correct detections (TT and FF) are diagonal elements, and the wrong detections (TF and FT) are off diagonal elements. Table 3.1 describes the contingency table in formal statistical terms.

We wish to make a decision based on the data: can we statistically justify that a suggestion from data makes sense, with the significance level 0.05, or 5%? Figure 3.4 shows the significance level as the tail probability on the right of the critical value x_c.

Making a decision with probability considerations goes through the following five steps. The process is called *hypothesis testing*, which has to consider uncertainties.

3.2.2 Steps of Hypothesis Testing

The first step is to assume that the situation under consideration is the same as what was before: business as usual. This is called the null hypothesis, denoted by H_0. Here, the word "null" may be understood as a synonym for nullified, abandoned, nonexistent, or unchanged. However, the business-as-usual is not deterministic and has uncertainty. Thus, H_0 has a probabilistic distribution, illustrated by the solid bell-shape curve shown in Figure 3.4.

The second step is to state the hypothesis of a changed scenario. In general, this occurs with a small probability, such as a fire accident in a building. This hypothesis is called the alternative hypothesis, denoted by H_1, whose distribution is shifted to the right in Figure 3.4 by the "Difference" marked in the figure. When the difference is large, the two distributions have little overlap. Because H_1 occurs with a small probability, accepting H_1 requires strong evidence for H_1 to be established. For example, strong evidence must be present in the diagnostic process in order to conclude that a patient has cancer; a minor symptom is insufficient. However, the required strength of evidence depends on the specific problem. The fire alarm does not need strong evidence, while the cancer conclusion requires very strong evidence. We see many false fire alarms, but very few false cancer diagnoses. Global warming is in the nature of a fire alarm because the stakes are too high.

The third step is directly related to the aforementioned decision: what is the small-tail probability α, indicated by the pink area in Figure 3.4, that the alternative scenario is true? When the difference is large, we can be more certain that the alternative hypothesis should be true. The chance for being wrong to accept H_1 is α, because the incorrect acceptance of H_1 implies that the reality is H_0, but we have incorrectly rejected H_0. We thus should use the tail distribution of H_0, i.e., the pink region, which is determined by the H_0 distribution

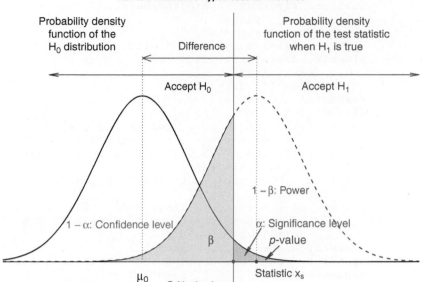

Figure 3.4 The distributions of null and alternative hypotheses: Critical value, p-value, and significance level α.

and the critical value x_c. Hence, when α is chosen and H_0 distribution is predetermined by the nature of the problem, then x_c can be calculated. For example, if H_0 is normally distributed $N(0,1)$, and $\alpha = 0.025$, then $x_c = 1.96$.

The fourth step is to calculate an index from the data, such as mean and variance. The index is called the test statistic. The distributions of H_0 and H_1 depend on the definition of the test statistic. For example, the distributions are normal if the test statistic x_s is the mean with a large sample size; or the distributions are student-t if the sample size is small and the standard deviation of the population is unknown and is estimated as the sample standard deviation.

The fifth step is to use the x_s value to make a decision based on the distributions of H_0 and H_1. If x_s is in the small-tail probability region of the H_0 distribution (i.e., $x_s > x_c$ in Figure 3.4), then the alternative hypothesis H_1 is accepted. The probability of this decision being wrong is equal to α. This is the chance of a false alarm: claiming a significant difference but the reality is business as usual. Namely, a fire alarm is triggered when there is no fire. This is called the Type I error, or α-error, in the statistical decision process. See Table 3.1 and next section for further explanation of the decision errors.

Figure 3.4 may be plotted by the following computer code.

```
#R plot Fig. 3.4: Null and alternative hypotheses
setwd("/Users/sshen/climstats")
setEPS() # save the .eps figure file to the working directory
postscript("fig0304.eps", height = 7, width = 10)
x = seq(-3,6, len=1000)
par(mar=c(0.5,0,2.5,0.0))
plot(x, dnorm(x,0,1), type="l",
    lwd=2, bty='n',
```

```
        main='Probability␣Density␣Functions␣of
Null␣and␣Alternative␣Hypotheses␣for␣Inference',
        xlab='', ylab='',
        xaxt="n",yaxt="n",
        cex.lab=1.2, ylim=c(-0.05,0.6))
lines(x, dnorm(x,2.5,1),
        lwd=2, lty=2,
        type="l", col='blue')
segments(-3,0, 6,0)
#lines(c(-3,6),c(0,0))
polygon(c(2.0,seq(2.0, 4, length=100), 4),
        c(0, dnorm(seq(2, 4, length=100)), 0.0),col='pink')
arrows( 2.6,0.08, 2.25,0.015, length=0.2, angle=8,
        lwd=2, col='red')
text(3.5,0.09,expression(alpha*":␣Significance␣level"),
        col='red', cex=1.4)
polygon(c(-1,seq(-1, 2, length=100), 2),
        c(0, dnorm(seq(-1, 2, length=100), 2.5,1), 0),
        col='lightblue')
lines(x, dnorm(x,0,1), type="l")
text(1.5,0.05,expression(beta), col='blue', cex=1.5)
segments(2,-0.05,2,0.6, col='red')
points(2,0, col='red', pch=16)
text(1.3,-0.06, expression("Critical␣value␣"*x[c]),
        cex=1.4, col='red')
segments(0,0, 0,0.5, lty=3, lwd=1.5)
segments(2.5,0, 2.5,0.5, lty=3, lwd=1.5, col='blue')
#polygon(c(2.5,seq(2.5, 4, length=100), 4),
#        c(0, dnorm(seq(2.5, 4, length=100)), 0.0),
#        col='green')
lines(x, dnorm(x,0,1), type="l")
arrows( 3.0,0.038, 2.7,0.005, length=0.2, angle=8,
        lwd=2, col='blue')
text(3.2,0.05,"p-value", col='blue', cex=1.5)
points(2.5,0, col='blue', pch=16)
text(-1.3,0.55, "Probability␣density
function␣of␣the
H0␣distribution", cex=1.4)
text(4,0.55, "Probability␣density
function␣of␣the␣test␣statistic
when␣H1␣is␣true", cex=1.4, col='blue')
text(3,-0.03, expression("Statistic␣"*x[s]),
        cex=1.4, col='blue')
arrows( 2.0,0.45, 6,0.45, length=0.2, angle=10,
        lwd=1.5, col='blue', code=3)
text(4,0.42, "Accept␣H1", cex=1.4, col='blue')
arrows( -2,0.45,2.0,0.45,  length=0.2, angle=10,
        lwd=1.5, code=3)
text(1,0.42, "Accept␣H0", cex=1.4)
text(-0.5,0.09, expression(1- alpha *":␣Confidence␣level"),
        col='grey40', cex=1.4)
text(3,0.17, expression(1-beta*':␣Power:'),
        col='darkgreen', cex=1.4)
arrows(0,0.49, 2.5,0.49, length=0.2, angle=20,
        lwd=1.5, col='maroon', code=3)
text(1.3,0.52, "Difference",  col='maroon', cex=1.4)
```

```
text(0,-0.04,expression(mu[0]),  cex=1.6)
dev.off()
```

```python
#Python plot Fig. 3.4: Null and alternative hypotheses
from scipy.stats import norm
x = np.linspace(-3,6,1000)
fig = plt.figure(figsize = (20,10))
#the black left curve
left_curve = norm.pdf(x, loc = 0, scale = 1)
plt.plot(x, left_curve, 'k')
right_curve = norm.pdf(x, loc = 2.5, scale = 1)
plt.plot(x, right_curve, 'b--') #'b--' for blue dashed lines
plt.title("Probability Density Functions of \n\
          Null and Alternative Hypotheses for Inference",
          fontsize = 25)
#Alpha region
alpha_region = np.linspace(2,4,222)
plt.fill_between(alpha_region, left_curve[555:777],
                 0, color = 'pink' )
plt.text(2.3, 0.05, r"$\alpha$: Significance level",
         size = 20, color = 'r') #alpha label
plt.annotate('', xy=(2.05,0.005), xytext=(2.4,0.04),
             arrowprops=dict(arrowstyle='->', color = 'r'))
#Beta region
beta_region = np.linspace(-1,2,333)
plt.fill_between(beta_region, right_curve[222:555],0,
                 color = 'cyan')
plt.text(1,0.03, r"$\beta$", size = 20, color = 'b')
#Critical Value
plt.axvline(x=2, color = 'r') #vertical line
plt.text(1, -0.05, r"Critical Value $x_C$", size = 20,
         color = 'r')
plt.plot(2, 0, 'ro', markersize = 10)
#mu 0 line
plt.axvline(x = 0, color = 'k', linestyle = '--') #for mu_0
plt.text(0, -0.05, r"$\mu_0$", size = 20)
#p value
plt.axvline(x = 2.5, color = 'b', linestyle = '--')
plt.text(2.6, 0.03, "p-value", size = 20, color = 'b')
plt.plot(2.5, 0, 'bo', markersize = 10)
plt.annotate('', xy=(2.6,0.005), xytext=(2.8,0.03),
             arrowprops=dict(arrowstyle='->', color = 'b'))
#comment on the left side
plt.text(-3, 0.45, 'Probability density', size = 20)
plt.text(-2.8, 0.43, 'function of the', size = 20)
plt.text(-2.8, 0.41, '$H_0$ distribution', size = 20)
#comment on the right side
plt.text(4, 0.45, "Probability density" ,
         size = 20, color = 'b')
plt.text(3.6, 0.43, "function of the test statistic",
         size = 20, color = 'b')
plt.text(4.1, 0.41, "when $H_1$ is true",
         size = 20, color = 'b')
#Accept H0
```

```
plt.annotate('', xy=(2,0.4), xytext=(-3,0.4),
             arrowprops=dict(arrowstyle='<->'))
#Accept H1
plt.annotate('',xy=(2,0.4), xytext=(6,0.4),
             arrowprops=dict(arrowstyle='<->', color = 'b'))
plt.annotate('', xy = (0, 0.45), xytext = (2.5, 0.45),
    arrowprops = dict(arrowstyle= '<->', color = 'maroon'))
#Python plot for Fig. 3.4  continued
plt.text(0.75, 0.46, "Difference",
         size = 20, color = 'maroon')
plt.text(-1.3, 0.1, r'$1-_\alpha_$:_Confidence_level',
         size = 20)
plt.text(2.5, 0.1, r"$1-\beta$:_Power_level",
         size = 20, color = 'g')
plt.text(4, 0.35, "Accept_$H_1$", size = 20, color = 'b')
plt.text(0.75, 0.35, "Accept_$H_0$", size = 20)
plt.ylim(-0.02,0.5) #limits the y axis to include more space
plt.axis('off') #turns off x and y axis
plt.show()
```

3.2.3 Interpretations of Significance Level, Power, and *p*-Value

When the distributions of H_0 and H_1 are determined, the values of α and β are determined by the critical value x_c. If x_c is large, it requires a larger difference in order to reject H_0 and accept H_1. Accepting H_1 when H_0 is true, is a false alarm, i.e., the fire siren is triggered, but there is no fire. This is called a Type I error of decision. The probability of Type I error is equal to α: the tail probability of H_0 because H_0 is true. A smaller α means a less sensitive detector, and less chance of a false alarm, because a smaller α means a larger x_c, and hence a larger difference is needed for accepting H_1, i.e., much stronger evidence is required to claim a fire event.

Of course, a smaller α means a large β: the tail probability of H_1. A small α for an insensitive detector means a larger chance of missing detection: a real fire not detected, a patient's real cancer not diagnosed, or a real climate change not identified. This kind of missed detection is called a Type II error: H_1 is true but is wrongly rejected. The probability of the wrong rejection of H_1 is β. To make β smaller, we have to make α larger, to be a more sensitive detector.

Then how can we minimize our probability of wrong decision? Figure 3.4 suggests two scenarios. The first is wide separation of the two distributions H_0 and H_1. If the difference is large, then the H_1 is evident, and accepting H_1 has a very small chance of being wrong. The second is that two distributions are slim and tall, namely, having a very small standard deviation s, which means accurate sensors, or a large sample size n. This is the common sense of seeing different doctors to seek separate opinions, if a patient feels seriously sick but has not been diagnosed with causes.

If the instruments and sample size are determined, then a more sensitive detector (i.e., a larger α) implies a smaller probability of Type II error (i.e., a smaller β). One cannot require both α and β to be small at the same time. Thus, you may choose a predetermined

α appropriate to your problem. If, for example, you design a fall detection sensor to go with smart rollator walkers for seniors, this could be a matter of life or death, and thus must be very sensitive so a larger α, say 10% or higher, should be used. Our Earth's climate change detection must also be very sensitive, since we cannot afford to allow a missed detection to ruin our only Earth. When testing the quality of shoes, on the other hand, the detection does not need to be that sensitive. A smaller α, say 5%, would be appropriate.

The significance of difference to be checked in hypothesis testing is measured by the significance level α and the *power of statistical inference*

$$P = 1 - \beta \tag{3.13}$$

in Figure 3.4. The power is between α and 1. Two extreme cases corresponding to $P = 1$ and $P = \alpha$ are as follows:

(i) $P = 1$: In this case, the difference is infinite. Then β is zero, hence $P = 1 - \beta = 1$. When $P = 1$, it means the maximum power of inference. It is certain that the alternative hypothesis H_1 should be accepted since H_1 is well separated from H_0. The probability of Type II error is zero: $\alpha = 0$, and there is no missed detection.

(ii) $P = \alpha$: In this case, the distributions of the two hypotheses H_0 and H_1 fully overlap. Thus, the difference is zero. The power is α, which is the minimum power to accept H_1 and implies that it is certain to accept the null hypothesis, because there is no difference, and it is business as usual. There is a very small chance to make a Type I error, because the alarm is very likely not triggered. If alarm is triggered, it is likely a false alarm. Thus, the probability of Type I error is large.

All the other cases are in between the these two extremes. The tail probability on the right of the test statistic is called the *p-value*. As discussed earlier, if the *p*-value is less than α, then we can conclude that the alternative hypothesis is accepted at the α significance level, and that the statistical inference power is $1 - \beta$. The *p*-value is determined by x_s and the H_0 distribution can be interpreted as the probability due to sampling data for the extremes. For example, a medical study produced a *p*-value 0.01. This means that if the medication has no effect (the H_0 scenario) as a whole, one would conclude a valid medical effect in only 1% of studies due to random data error. A large *p*-value thus means that there is a larger chance for the sample data to support the null hypothesis. Thus, the *p*-value measures the compatibility of the sample data and null hypothesis.

3.2.4 Hypothesis Testing for the 1997–2016 Mean of the Edmonton January Temperature Anomalies

Figure 3.3 seems to suggest that the mean of temperature anomalies in 1997–2016 for the January Edmonton is positive. We wish to substantiate this conclusion by a hypothesis testing procedure. We can then formally conclude that the mean of temperature anomalies of Edmonton from 1977 to 2016 is significantly greater than zero at a given significance level, say 5%.

The sample size is $n = 20$. We will show later that the samples are independent and have no serial correlations. Since $n < 30$, the sample size is treated as small. The mean of the samples should follow a t-distribution. The null and alternative hypotheses are $H_0: \bar{x} = 0°C$ and $H_1: \bar{x} > 0°C$, respectively.

The numerical results are as follows:

  (i) Sample mean: $\bar{x} = 1.53°C$ (with sample size $n = 20$)
 (ii) Sample standard deviation: $s = 2.94°C$
(iii) Degrees of freedom: 19
 (iv) The t-statistic with $t_s = \bar{x}/(s/\sqrt{n}) = 2.33°C$
  (v) The critical value of for the t-distribution at the 5% significance level: $t_c = 1.73°C$.

The test statistic $t_s > t_c$ is in the rejection domain of H_0. Thus, H_1 is accepted. The mean temperature anomalies of Edmonton from 1977 to 2016 are significantly greater than zero at the 5% significance level. The p-value at $t_s = 2.33°C$ is 0.0154, which is less than 2%. This substantiates the acceptance of H_1.

The computer code for calculating this result is as follows.

```
#R code for the Edmonton 1997-2016 hypothesis testing example
setwd("/Users/sshen/climstats")
da1 =read.csv("data/Lat52.5_Lon-112.5.csv", header=TRUE)
jan =seq(1, dim(da1)[1], by=12) #Jan indices
Tjan = da1[jan,] #Jan data matrix
da3=Tjan[118:137, 3] #1997- 2016 data string
round(c(mean(da3), sd(da3)), digits = 2)
#[1] 1.53 2.94
mean(da3)/(sd(da3)/sqrt(20))
#[1] 2.331112 #t-statistic ts
1 - pt(2.331112, 19)
#[1] 0.01545624  # p value < the significance level 0.05
```

```
#Python code for the Edmonton 1997-2016 hypothesis testing
from scipy.stats import t
da1 = pd.read_csv("data/Lat52.5_Lon-112.5.csv")
jan = np.arange(0, 1641, 12)
Tjan = np.array(da1.iloc[jan,:])
Tda = Tjan[:,2]
da3 = Tda[117:137]#1997- 2016 data string
xbar = np.mean(da3)
sdEdm = np.std(da3, ddof=1)
np.round(np.array([xbar, sdEdm]), 2)
#1.53 2.94
xbar/(sdEdm/np.sqrt(20))
#2.331112 #t-statistic ts
#calculate p-value
1 - t.cdf(2.331112, 19)
#0.01545624  # p value < the significance level 0.05
```

3.3 Effective Sample Size

Randomly flipping coins, tossing dice, or drawing playing cards can generate independent samples. A sample is independent from the other samples. As derived earlier, if n samples $x_1, x_2, \ldots, x_n$ are independent from the same population, and each sample has the same standard deviation σ estimated by

$$s = \sqrt{\frac{\sum_{i=1}^{n}(x_i - \bar{x})^2}{n-1}} \qquad (3.14)$$

where

$$\bar{x} = \frac{\sum_{i=1}^{n} x_i}{n} \qquad (3.15)$$

is the sample mean, then the standard deviation of the mean $\bar{x}$, also known as standard error of the mean, is

$$SE = \frac{s}{\sqrt{n}}. \qquad (3.16)$$

However, many (if not most) climate datasets are recorded from a sequence of times and violate this independence assumption. For example, a string of n monthly temperature anomaly records at a location may be serially correlated, i.e., not independent. Then, the standard error of the mean of the n records is larger than $s/\sqrt{n}$,

There is often a *serial correlation* and this could reduce the sample size n to a smaller value, the *effective sample size* n_{eff}. This is important in estimating standard deviation $\hat{\sigma}$ of a mean $\bar{x}$, because $n_{\text{eff}} < n$ and hence $s/\sqrt{n}$ leads to an erroneous underestimate of the spread of expected values in the mean estimate $\bar{x}$. The correct estimate of the standard error for the mean should be

$$SE = \frac{s}{\sqrt{n_{\text{eff}}}}, \qquad (3.17)$$

which is larger than the erroneous

$$\frac{s}{\sqrt{n}}. \qquad (3.18)$$

Thus, if we overcount the number of independent pieces of information, we will be led into the trap of claiming more confidence in a parameter's value than we are entitled to.

Use ρ to denote the lag-1 autocorrelation coefficient. Figure 3.5 shows the AR(1) time series[1] and their lagged autocorrelations for two different values of ρ: 0.9 and 0.6. The figure shows that AR(1) processes have an initial memory depending on ρ and become random after the memory. With the ρ value, the n_{eff} value can be computed from

$$n_{\text{eff}} \approx \frac{1-\rho}{1+\rho} n. \qquad (3.19)$$

---

[1] Time series will be discussed in Chapters 7 and 8.

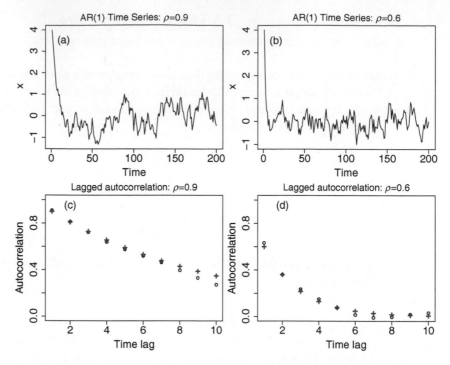

Figure 3.5   Realizations of AR(1) time series (a) and (b) for $n = 1,000$, but only the first 200 are shown, and their autocorrelation (c) and (d). The blue dots in Panels (c) and (d) are the exact correlations, and the black circles indicate the correlations computed from the simulated data.

The parameter ρ is the lag-1 correlation coefficient and can be estimated from the data:

$$\hat{\rho} = \frac{\sum_{t=2}^{n}(x_t - \bar{x}_+)(x_{t-1} - \bar{x}_-)}{\sqrt{\sum_{t=2}^{n}(x_t - \bar{x}_+)^2}\sqrt{\sum_{t=2}^{n}(x_{t-1} - \bar{x}_-)^2}}, \tag{3.20}$$

where

$$\bar{x}_+ = \frac{\sum_{t=2}^{n} x_t}{n-1}, \tag{3.21}$$

$$\bar{x}_- = \frac{\sum_{t=2}^{n} x_{t-1}}{n-1}. \tag{3.22}$$

If there is no serial correlation, then $\hat{\rho} = 0$ and $n_{\text{eff}} = n$. White noise is such an example. If there is a perfect correlation, then $\hat{\rho} = 1$ and $n_{\text{eff}} = 0$, for example if the data are all the same, which is a trivial case and does not need further analysis.

The approximation formula for effective degrees of freedom can have a large error for a large ρ close to 1.0, and a short time series (i.e., a small n value). The estimation formula (3.20) for ρ also has a large error for a short time series data string. In practice, you may use your own experience and the nature of the problem to determine when the approximation formulas (3.19) and (3.20) are acceptable. More discussions on the simulations and applications about independence and serial correlations of time series are included in Chapters 6 and 7, where sampling errors and forecasting are described.

Figure 3.5 can be generated by the following computer code.

```
#R plot Fig. 3.5: AR(1) time series
#AR(1) process and its simulation
setwd("/Users/sshen/climstats")
setEPS() # save the figure to an .eps file
postscript("fig0305.eps", height = 8, width = 10)
par(mar=c(4.0,4.5,2,1.0))
par(mfcol = c(2, 2))

lam=0.9; alf=0.25; y0=4 #AR1 process parameters
library(forecast) #Load the forecast package
#set.seed(1234)
n <- 1000 #Generate an AR1 process of length n
x <- rep(0,n)
w <- rnorm(n)
x[1] <- y0
# loop to create time series x
for (t in 2:n) x[t] <- lam * x[t-1] + alf * w[t]
par(mar=c(4.5,4.5,2,0.5))
plot(x[1:200],type='l',
     xlab="Time", ylab="x",
     main=expression("AR(1)␣Time␣Series:␣" * rho *"=0.9"),
     cex.lab=1.8, cex.axis=1.8, cex.main=1.6)
text(15,3.5, "(a)", cex=1.8)

#Calculate the auto-correlation
M <- 10
rho <- rep(0,M)
for (m in 1:M){
   rho[m]=cor(x[1:(n-m)],x[(1+m):n])
}
par(mar=c(4.5,4.5,2,0.5))
plot(rho, type='p', ylim=c(0,1),
     xlab="Time␣lag",
     ylab="Auto-correlation",
main=expression("Lagged␣auto-correlation:␣" * rho *"=0.9"),
     lwd=1.8, cex.lab=1.8, cex.axis=1.8, cex.main=1.6)
text(2,0.95, "(c)", cex=1.8)

rhotheory <- rep(0,M) #Theoretical auto-correlation
for (m in 1:M){rhotheory[m]=lam^m}
points(rhotheory, col="blue", pch=3, lwd=3)
#AR1 process for rho = 0.6
lam=0.6; alf=0.25; y0=4
n <- 1000; x <- rep(0,n); w <- rnorm(n); x[1] <- y0

# loop to create time series x
for (t in 2:n) x[t] <- lam * x[t-1] + alf * w[t]
par(mar=c(4.5,4.5,2,0.5))
plot(x[1:200],type='l',
     xlab="Time", ylab="x",
main=expression("AR(1)␣Time␣Series:␣" * rho *"=0.6"),
     cex.lab=1.8, cex.axis=1.8, cex.main=1.6)
text(15,3.5, "(b)", cex=1.8)
```

```
#Calculate the auto-correlation
M <- 10
rho <- rep(0,M)
for (m in 1:M){
   rho[m]=cor(x[1:(n-m)], x[(1+m):n])
}
par(mar=c(4.5,4.5,2,0.5))
plot(rho, type='p', ylim=c(0,1),
     xlab="Time␣lag",
     ylab="Auto-correlation",
main=expression("Lagged␣auto-correlation:␣" * rho *"=0.6"),
     lwd=1.8, cex.lab=1.8, cex.axis=1.8, cex.main=1.6)
text(2,0.95, "(d)", cex=1.8)

rhotheory <- rep(0,M) #Theoretical auto-correlation
for (m in 1:M){rhotheory[m]=lam^m}
points(rhotheory, col="blue", pch=3, lwd=3)
dev.off()
# Back to the original graphics device
par(mfrow = c(1, 1))
```

```
#Python plot Fig. 3.5: AR(1) time series
#establishing subplots
fig, axs = plt.subplots(2,2, figsize = (20,10))
fig.tight_layout(pad = 7.0)
#establishing constants
lamb = 0.9; alpha = 0.25; y0 = 4
n = 1000 #length of array
x = np.zeros(n)
w = np.random.normal(size = 1000)
x[0] = y0
for t in range(1,n):
    x[t] = x[t-1]* lamb + alpha*w[t]
#time values from 200-400
t200_400 = np.linspace(201, 400, 200)
axs[0,0].plot(t200_400, x[200:400], color = 'k')
axs[0,0].set_title(r"AR(1)␣Time␣Series:␣$\rho␣=␣0.9$",
                     size = 25)
axs[0,0].set_xlabel('Time', size = 20);
axs[0,0].set_ylabel('x', size = 20)
axs[0,0].tick_params(axis='both', which='major',
                     labelsize=15)
#correlation
M = 10
rho = np.zeros(M)
for m in range(M):
    rho[m] = np.corrcoef(x[0:n-m], x[m:n+1])[0,1]
axs[1,0].scatter(np.linspace(1,10,10), rho,
                 color = 'none', edgecolor = 'k')
axs[1,0].set_ylim(0,1.2)
axs[1,0].set_xlabel('Time␣lag', size = 20);
axs[1,0].set_ylabel('Auto-correlation', size = 20)
axs[1,0].set_title(r'Lagged␣auto-correlation␣:␣$\rho␣=␣0.9$'
```

```
                          size = 25)
#theoretical
rho_theory = np.zeros(M)
for m in range(M):
    rho_theory[m] = lamb**m
axs[1,0].scatter(np.linspace(1,10,10), rho_theory,
                 color = 'b', marker = '+')
axs[1,0].tick_params(axis='both', which='major',
                     labelsize=15)
# For lambda = 0.6
lamb = 0.6; alpha = 0.25; y0 = 4
n = 1000 #length of array
x = np.zeros(n)
w = np.random.normal(size = 1000)
x[0] = y0
for t in range(1,n):
    x[t] = x[t-1]* lamb + alpha*w[t]
#time values from 200-400
t200_400 = np.linspace(201, 400, 200)
axs[0,1].plot(t200_400, x[200:400], color = 'k')
axs[0,1].set_title(r"AR(1)_Time_Series:_$\rho_=_0.6$",
                   size = 25)
axs[0,1].set_xlabel('Time', size = 20);
axs[0,1].set_ylabel('x', size = 20)
axs[0,1].tick_params(axis='both', which='major',
                     labelsize=15)
 #Python plot Fig. 3.5 continued
#correlation
M = 10
rho = np.zeros(M)
for m in range(M):
    rho[m] = np.corrcoef(x[0:n-m], x[m:n+1])[0,1]
 axs[1,1].scatter(np.linspace(1,10,10), rho,
                 color = 'none', edgecolor = 'k')
axs[1,1].set_ylim(0,1.2)
axs[1,1].set_xlabel('Time_lag', size = 20);
axs[1,1].set_ylabel('Auto-correlation', size = 20)
axs[1,1].set_title(r'Lagged_auto-correlation_:_$\rho_=_0.6$'
                   size = 25)
#theoretical
rho_theory = np.zeros(M)
for m in range(M):
    rho_theory[m] = lamb**m
axs[1,1].scatter(np.linspace(1,10,10),
                 rho_theory, color = 'b', marker = '+')
axs[1,1].tick_params(axis='both', which='major',
                     labelsize=15)
```

Example 3.2   Figure 3.6 shows the mean NOAAGlobalTemp data from 1880 to 2019, indicated by the left most thick solid green horizontal line. We calculate the confidence interval of temperature data in the period of 1920–1949. The results are as follows: $\bar{x} = -0.38°C, s = 0.16°C, \hat{\rho} = 0.78, n = 30, n_{\text{eff}} = 4, t_c = t_{0.975,4} = 2.78$.

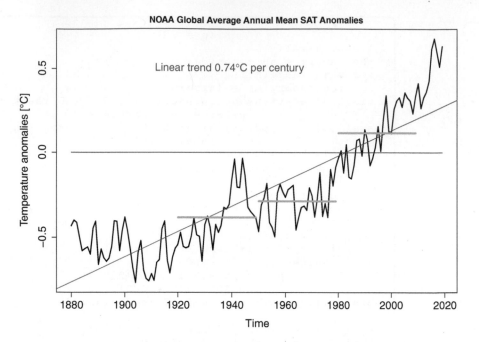

NOAA Global Average Annual Mean SAT Anomalies

Figure 3.6 NOAA global average annual mean temperature from 1880 to 2019 (data source: NOAAGlobalTemp V5, 2020).

The confidence interval is computed by substituting the above into the following formula:

$$\bar{x} \pm t_c \times \frac{s}{\sqrt{n_{\text{eff}}}}. \tag{3.23}$$

The numerical result is $[-0.63, -0.13]°C$.

If the serial correlation were not considered, the confidence interval would be $[-0.44, -0.32]°C$. The length of this CI is $0.12°C$ and is incorrect. The correct length is $0.50°C$. The conclusion of the smaller uncertainty is incorrect, because the 30 samples are not random and independent. The effective sample size is only 4.

Figure 3.6 may be plotted by the following computer code.

```
#R plot Fig. 3.6: NOAAGlobalTemp V5 1880-2019
setwd("/Users/sshen/climstats")
da1 =read.table("data/NOAAGlobalTempAnn2019.txt",
                header=FALSE) #read data
dim(da1)
#[1] 140   6 #140 years of anomalies data
da1[1:2,1:5] #column 1: year; column 2: data
#    V1          V2        V3        V4        V5
#1 1880 -0.432972 0.009199 0.000856 0.000850
#2 1881 -0.399448 0.009231 0.000856 0.000877
da1[139:140,1:5]
#      V1          V2        V3        V4      V5
#139 2018 0.511763 0.005803 8.6e-05 2e-06
#140 2019 0.633489 0.005803 8.6e-05 2e-06
t=da1[,1]; Tann = da1[,2]
regAnn = lm(Tann ~ t)
```

```
setEPS() # save the figure to an .eps file
postscript("fig0306.eps", height = 7, width = 10)
par(mar=c(4.0,4.5,2,1.0))
plot(t, Tann, type="l", lwd=2.5,
main = "NOAA␣Global␣Average␣Annual␣Mean␣SAT␣Anomalies",
     xlab = "Time", ylab = "Temperature␣Anomalies␣[deg␣C]",
     cex.lab=1.4, cex.axis=1.4)
abline(lm(Tann ~t), col='red', lwd=1.3)
lines(t, rep(0, 140), col='blue')
lines(t[41:70], rep(mean(Tann[41:70]), 30),
     col='green', lwd=4) # 1920-1949 mean
lines(t[71:100], rep(mean(Tann[71:100]), 30),
     col='green', lwd=4) #1950-1979 mean
lines(t[101:130], rep(mean(Tann[101:130]), 30),
     col='green', lwd=4) #1980-2010 mean
lm(Tann ~t)
#(Intercept)               t
#-14.741366      0.007435
text(1940, 0.5,
     expression("Linear␣trend␣0.74"~degree~C~"per␣century"),
     col='red', cex=1.4)
#0.007435 #0.7 deg C per century
dev.off()
```

```
#Python plot Fig. 3.6: NOAAGlobalTemp V5 1880-2019
import pandas as pd
da1 = pd.read_csv("data/NOAAGlobalTempAnn2019.csv",
               header = None)
print(da1.shape) # 140 years of anomalies data
print("")#leave a line of space
print(da1.iloc[0:2, 0:5]) # column 1: year; column 2: data
print("")#leave a line of space
print(da1.iloc[138:142, 0:5])
t = np.array(da1.iloc[:, 0])
Tann = np.array(da1.iloc[:,1])
lm = LinearRegression()# create linear regression
lm.fit(t.reshape(-1,1), Tann)
regAnn = lm.predict(t.reshape(-1,1))
print("")#leave a line of space
print(lm.intercept_, lm.coef_)
# plot time series
plt.plot(t, Tann, 'k', linewidth = 2)
plt.title("NOAA␣Global␣Average␣Annual␣Mean␣SAT␣Anomalies",
         pad = 10)
plt.xlabel("Time", labelpad = 10)
plt.ylabel(r"Temperature␣Anomalies␣[$\degree$C]",
         labelpad = 10)
plt.yticks([-0.5,0,0.5])
plt.plot(t, regAnn, 'r')# plot regression line
plt.axhline(y=0, color = 'b')# plot y = 0
# plot 1920-1949 mean
plt.axhline(y = np.mean(Tann[40:70]),
             xmin = 0.3, xmax = 0.5,
```

```
                       color = 'limegreen', linewidth = 2)
# plot 1950-1979 mean
plt.axhline(y = np.mean(Tann[70:100]),
            xmin = 0.5, xmax = 0.69,
            color = 'limegreen', linewidth = 2)
# plot 1980-2010 mean
plt.axhline(y = np.mean(Tann[100:130]),
            xmin = 0.69, xmax = 0.9,
            color = 'limegreen', linewidth = 2)
# plot text
plt.text(1910,0.5,
         r"Linear␣trend␣0.74␣$\degree$C␣per␣century",
         size = 20, color = 'r')
plt.show()
```

You may use the following computer code to compute the statistics of this example.

```
#R code for the 1920-1949 NOAAGlobalTemp statistics
setwd("/Users/sshen/climstats")
da1 =read.table("data/NOAAGlobalTempAnn2019.txt",
                header=FALSE) #read data
Ta = da1[41:70,2]
n = 30
xbar = mean(Ta)
s = sd(Ta)
r1 = cor(Ta[1:29], Ta[2:30])
neff = n*(1 - r1)/(1 + r1)
neff = 4 #[1] 3.677746 approximately equal to 4
tc0 = qt(0.975, 29, lower.tail=TRUE)
tc = qt(0.975, 3, lower.tail=TRUE)
CI1 = xbar - tc0*s/sqrt(n); CI2 = xbar + tc0*s/sqrt(n)
CI3 = xbar - tc*s/sqrt(neff); CI4 = xbar + tc*s/sqrt(neff)
print(paste('rho␣=', round(r1, digits =2),
            'neff␣=', round(neff, digits =2)))
#[1] "rho = 0.78 neff = 4"
round(c(xbar, s, r1, CI1, CI2, CI3, CI4, tc), digits = 2)
#[1] -0.38  0.16  0.78 -0.44 -0.32 -0.63 -0.13  3.18
```

```
#Python code for the 1920-1949 NOAAGlobalTemp statistics
import pandas as pd
import scipy.stats
from scipy.stats import pearsonr
da = pd.read_csv("data/NOAAGlobalTempAnn2019.csv",
                 header = None)
da1 = np.array(da.iloc[:,1])
Ta = da1[40:70]
n = 30
xbar = np.mean(Ta)
s = np.std(Ta, ddof =1)
r1, _ = pearsonr(Ta[0:29], Ta[1:30])
```

```
neff = n*(1 - r1)/(1 + r1)
print('rho1=', np.round(r1,2), 'neff=', np.round(neff, 2))
neff = 4 #rho1= 0.78 neff= 4 #3.68 rounded to 4
tc0 = scipy.stats.t.ppf(q=0.975, df=29)
tc = scipy.stats.t.ppf(q=0.975, df=4)
CI1=xbar - tc0*s/np.sqrt(n); CI2=xbar + tc0*s/np.sqrt(n)
CI3=xbar - tc*s/np.sqrt(neff); CI4=xbar + tc*s/np.sqrt(neff)
result = np.array([xbar, s, r1, CI1, CI2, CI3, CI4, tc])
print(np.round(result, 2))
#[-0.38  0.16  0.78 -0.44 -0.32 -0.63 -0.13   2.78]
```

Example 3.3 Is there a significant difference between the 1950–1979 and 1980–2009 means, indicated by the two thick horizontal green lines on the right in Figure 3.6, of the global average temperature anomalies based on the NOAAGlobalTemp data V5?

The relevant statistics are as follows:

1950–1979: $\bar{x}_1 = -0.29°C, s_1 = 0.11°C, n = 30, \rho_1 = 0.22, n_{1\text{eff}} = 19$
1980–2009: $\bar{x}_2 = 0.12°C, s_2 = 0.16°C, n = 30, \rho_1 = 0.75, n_{2\text{eff}} = 4$

The t-statistic is

$$t_s = \frac{\bar{x}_1 - \bar{x}_2}{\sqrt{s_1^2/n_{1\text{eff}} + s_2^2/n_{2\text{eff}}}} = -4.89. \tag{3.24}$$

The degrees of freedom is $n_{1\text{eff}} + n_{2\text{eff}} - 2 = 21$. The critical t value for $\bar{x}_1 \neq \bar{x}_2$ at the two-sided tail 5% significance level is -2.08 or 2.08. Thus, $t_s = -4.89$ is in the H_0 rejection region and we conclude that there is a significant difference between the 1950–1979 and the 1980–2009 temperature anomalies.

If the serial correlation is not considered, then $t_s = -9.53$, and H_0 is also rejected. This is a stronger conclusion of rejection and has a smaller p-value.

These statistics may be computed using the following computer code.

```
#R code for the 1950-1979 and 1980-2009 statistics
#  The 1950-1979 NOAAGlobalTemp statistics
Ta = da1[71:100,2]
n = 30
xbar = mean(Ta)
s = sd(Ta)
r1 = cor(Ta[1:29], Ta[2:30])
neff = n*(1 - r1)/(1 + r1)
neff #[1] 19.02543 approximately equal to 19
neff = 19
round(c(xbar, s, r1, neff), digits = 2)
#[1] -0.29  0.11  0.22 19.00

# The 1980-2009 NOAAGlobalTemp statistics
```

```
Ta = da1[101:130,2]
n = 30
xbar = mean(Ta)
s = sd(Ta)
r1 = cor(Ta[1:29], Ta[2:30])
neff = n*(1 - r1)/(1 + r1)
neff #[1] 4.322418 approximately equal to 4
neff = 4
round(c(xbar, s, r1, neff), digits = 2)
#[1] 0.12 0.16 0.75 4.00

#t-statistic for 1950-1979 and 1980-2009 difference
ts = (-0.29 - 0.12)/sqrt(0.11^2/19 + 0.16^2/4)
ts  #[1] -4.887592
```

```
#Python code for the 1950-1979 and 1980-2009 statistics
#Python code for the 1950-1979 and 1980-2009 statistics
import pandas as pd
import scipy.stats
from scipy.stats import pearsonr
da = pd.read_csv("data/NOAAGlobalTempAnn2019.csv",
                header = None)
da1 = np.array(da.iloc[:,1])
Ta = da1[70:100] #1950-1979
n = 30
xbar = np.mean(Ta)
s = np.std(Ta)
r1, _ = pearsonr(Ta[0:29], Ta[1:30])
neff = n*(1 - r1)/(1 + r1)
print(neff) #[1] 19.02543 approximately equal to 19
neff = 19
res1 = np.array([xbar, s, r1, neff])
print(np.round(res1,2))
#-0.29  0.11  0.22 19.00

Ta = da1[100:130]#1980-2009
n = 30
xbar = np.mean(Ta)
s = np.std(Ta)
r1, _ = pearsonr(Ta[0:29], Ta[1:30])
neff = n*(1 - r1)/(1 + r1)
print(neff) #4.3224 approximately equal to 4
neff = 4
res2 = np.array([xbar, s, r1, neff])
print(np.round(res2,2))
#0.12 0.16 0.75 4.00

#t-statistic for 1950-1979 and 1980-2009 difference
ts = (-0.29 - 0.12)/np.sqrt(0.11**2/19 + 0.16**2/4)
print(ts)  #The result is -4.887592
```

3.4 Test Goodness of Fit

This section uses examples to answer two types of questions using the chi-square distribution. First, given the weather distribution of a location based on the long historical record, is the observed weather of this month significantly different from the given distribution? Second, after fitting climate data to a certain distribution, how good is the fit? The chi-square statistic measures the difference between the histogram and the proposed PDF, discrete or continuous. A small chi-square statistic means a good fit.

3.4.1 The Number of Clear, Partly Cloudy, and Cloudy Days

Based on past observations, the mean number of clear, partly cloudy, cloudy days of Dodge City, Kansas, USA, in October are 15, 7, and 9, respectively (data source: National Weather Service, NOAA, www.weather.gov/ddc/cloudydays).

If a future October will observe 9 clear days, 6 partly cloudy days, and 16 cloudy days, is this October weather significantly different from the historical record?

The expected data are $E_1 = 15, E_2 = 7, E_3 = 9$, respectively. The observed data are $O_1 = 9, O_2 = 6, O_3 = 16$. We use the chi-square statistic to measure the differences between the observed and expected data as follows:

$$\chi^2 = \sum_{i=1}^{3} \frac{(O_i - E_i)^2}{E_i} = 7.9873. \tag{3.25}$$

If this χ^2-statistic is small, then there is no significant difference. Then, how small should the χ^2-statistic be for a given significance level? This decision question can be answered by the following χ^2 calculations.

This chi-square distribution has 3 degrees of freedom, equal to the number of categories minus the number of constraints. We have three categories, and one constraint. The constraint is that the total number of days of the three categories has to be equal to $n = 31$ for October. The tail probability of the $\chi^2(2)$ distribution in the region $x > x_s = 7.9873$ is 0.0184, which is the p-value and is less than 5%. Thus, this future observed October weather is significantly different from the past record at the 5% significance level.

Another argument is that the statistic $x_s = 7.9873 > x_c = 5.99$ hence is in the rejection region. Thus, the null hypothesis of no significant difference should be rejected.

The computer code for computing the above statistics is as follows.

```
#R code: Chi-square test for Dodge City, Kansas, USA
((9-15)^2)/15+ ((6-7)^2)/7 + ((16-9)^2)/9
#7.987302 #This is the chi-statistic
1 - pchisq(7.987302, df=2)  #Compute the tail probability
#[1] 0.01843229 #p-value
1 - pchisq(5.99, df=2) #Compute the tail probability
#0.05003663 # Thus, xc = 5.99
```

```
#Python code: Chi-square test for Dodge City, Kansas, USA
from scipy.stats import chi2
((9-15)**2)/15 + ((6-7)**2)/7 + ((16-9)**2)/9
#7.987302 #This is the chi-statistic
1 - chi2.cdf(7.987302, df=2)  #Compute the tail probability
#[1] 0.01843229 #p-value
1 - chi2.cdf(5.99, df=2) #Compute the tail probability
#[1] 0.05003663 # Thus, xc = 5.99
```

3.4.2 Fit the Monthly Rainfall Data to a Gamma Distribution

In Chapter 2, the Omaha, Nebraska, June precipitation from 1948 to 2019 was fitted to a Gamma distribution with shape equal to 1.5176, and rate equal to 0.0189. See Figure 3.7 for the histogram and its fit to the Gamma distribution.

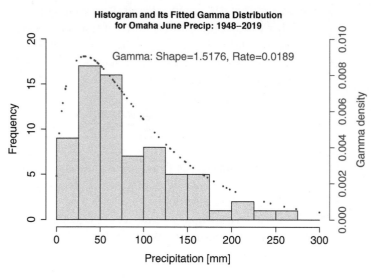

Figure 3.7 Histogram and its fit to a Gamma distribution for the June precipitation data from 1948 to 2019 for Omaha, Nebraska, USA (data source: Climate Data Online (CDO), NOAA, www.ncdc.noaa.gov/cdo-web).

The following computer code can generate Figure 3.7.

```
#R plot Fig. 3.7: June precipitation and Gamma distribution
#Read the Omaha data
setwd("/Users/sshen/climstats")
Omaha=read.csv("data/OmahaP.csv", header=TRUE)
dim(Omaha)
#[1] 864   7  : Jan 1948-Dec 2019: 864 months, 72 years*12
daP = matrix(Omaha[,7], ncol=12, byrow=TRUE)
y = daP[,6] #Omaha June precipitation data
#Fit the data
#install.packages('fitdistrplus')
```

```
library(fitdistrplus)
fitdist(y, distr = "gamma", method = "mle")
#          estimate  Std. Error
#shape 1.51760921 0.229382819
#rate  0.01889428 0.003365757

#Plot the figure
setEPS() # save the .eps figure
postscript("fig0307.eps", height = 5.6, width = 8)
par(mar=c(4.2,4.2,2.5,4.5))
hist(y, breaks= seq(0,300,by=25),
    xlim=c(0,300), ylim=c(0,20),
    main="Histogram and Its Fitted Gamma Distribution
      for Omaha June Precip: 1948-2019",
    xlab="Precipitation [mm]", cex.lab=1.4, cex.axis=1.4)
par(new=TRUE)
density = function(y) dgamma(y, shape=1.5176, rate=0.0189)
plot(y, density(y), col="blue",
    pch=20, cex=0.5, ylim=c(0,0.01),
    axes=FALSE, bty="n", xlab="", ylab="")
axis(4, cex.axis=1.4, col='blue', col.axis='blue')
mtext("Gamma Density", cex=1.4,
      side = 4, line = 3, col="blue")
text(140, 0.009, cex=1.4, col="blue",
    "Gamma: Shape=1.5176, Rate=0.0189")
dev.off()
```

```
#Python plot Fig. 3.7: Precipitation and Gamma distribution
# read the data file from the folder named "data"
Omaha = pd.read_csv("data/OmahaP.csv")
print(Omaha.shape) #  data dimensions
#(864, 7)
# change dataset into a matrix
daP = pd.DataFrame(Omaha['PRCP'])
daP = np.array(daP)
shape = (72,12)
daP = np.reshape(daP, shape)
y = daP[:, 5]
# plot the figure
fig, ax = plt.subplots(figsize=(10,8))
plt.hist(y, bins = 11, color = 'white', edgecolor = 'black')
ax.set_xlim(-10,300)
ax.set_ylim(0,20)
ax.set_title('Histogram and its Fitted Gamma Distribution\n\
for Omaha June Precip: 1948-2019', pad = 20)
ax.set_xlabel('Precipitation [mm]', labelpad = 20)
ax.set_ylabel('Frequency', labelpad = 20)
ax.set_yticks(np.linspace(0,20,5))
ax.spines['top'].set_visible(False)
ax.spines['right'].set_visible(False)
ax.set_xticks([0, 50, 100, 150, 200, 250])
yl = stats.gamma.rvs(a=1.5176,scale=1/0.0189,size=1000)
xl = np.linspace(y.min(),y.max(),35)
```

```
plt.plot(xl,stats.gamma.pdf(xl,a=1.5176,scale=1/0.0189))
ax1 = ax.twinx()
# plot gamma distribution
ax1.plot(xl,stats.gamma.pdf(xl,a=1.5176,scale=1/0.0189),
         'o',color = "b")
ax1.spines['right'].set_color('blue')
ax1.spines['top'].set_visible(False)
ax1.tick_params(length=6, width=2,
                labelsize=21, color = 'b', labelcolor = 'b')
ax1.set_yticks([0.000, 0.002, 0.004, 0.006, 0.008, 0.010])
ax1.set_xticks([0, 50, 100, 150, 200, 250])
plt.show()
```

Next we examine goodness of fit based on the chi-square test. The chi-square statistic can be computed by the following formula:

$$\chi^2 = \sum_{i=1}^{12} \frac{(O_i - E_i)^2}{E_i} = 6.3508. \tag{3.26}$$

This can be computed using the following computer code.

```
#R code: Chi-square test for the goodness of fit: Omaha prcp
setwd("/Users/sshen/climstats")
Omaha=read.csv("data/OmahaP.csv", header=TRUE)
dim(Omaha)
#[1] 864   7  : Jan 1948-Dec 2019: 864 months, 72 years*12
daP = matrix(Omaha[,7], ncol=12, byrow=TRUE)
y = daP[,6] #Omaha June precipitation data
n = 72 #Total number of observations
m = 12 #12 bins for the histogram in [0, 300] mm
p1 = pgamma(seq(0,300, by=300/m),
            shape=1.5176, rate=0.0189)
p1[m+1] = 1
p2 = p1[2:(m+1)]-p1[1:m]
y # The 72 years of Omaha June precipitation
cuts = cut(y, breaks = seq(0,300,by=300/m))
#The cut function in R assigns values into bins
Oi <- c(t(table(cuts))) #Extract the cut results
Ei = round(p2*n, digits=1) #Theoretical results
rbind(Oi, Ei)
# [,1] [,2] [,3] [,4] [,5] [,6] [,7] [,8] [,9] [,10] [,11] [,12]
#Oi 9 17.0 16.0  7.0  8.0  5.0  5.0  1.0  2.0  1.0   1.0   0.0
#Ei13 15.6 12.8  9.5  6.8  4.7  3.2  2.1  1.4  0.9   0.6   1.2
sum(((Oi - Ei)^2)/Ei)
#[1] 6.350832 #Chi-square statistic
1 - pchisq(19.68, df=11) #Compute the tail probability
#[1] 0.04992718 # Thus, xc = 19.68
1 - pchisq(6.3508, df=11) #Compute the tail probability
#[1] 0.8489629 #p-value
```

```
#Python code: Chi-square test for the goodness of fit
from scipy.stats import gamma
Omaha = pd.read_csv("data/OmahaP.csv")
daP = pd.DataFrame(Omaha['PRCP'])
daP = np.array(daP)
shape = (72,12)
daP = np.reshape(daP, shape)
y = daP[:, 5] #June precipitation
n = 72 #Total number of observations
m = 12 #12 bins for the histogram in [0, 300] mm
x = np.arange(0, 301, 300/m)
p1 = gamma.cdf(x, a = 1.5176, scale = 1/0.0189)
p1[m] = 1
p2 = p1[1:(m+1)]-p1[0:m]
freq, bins = np.histogram(y, x)
Oi = freq
Ei = np.round(p2*n, 1)
count1 = np.stack((Oi, Ei), axis=0)
print(count1)
np.sum((Oi - Ei)**2/Ei)
#6.350832032876924
1 - chi2.cdf(19.68, df=11) #Compute the tail probability
#0.049927176237663295 #approximately 0.05
1 - chi2.cdf(6.3508, df=11) #=0.84896
```

The critical value for the 5% significance level of the $\chi^2(11)$ is $x_c = 19.68$, since $P(\chi^2(11) > 19.68) = 0.04993$. Our statistic $x_s = 6.3508$ is much smaller than x_c, which implies that there is no significant difference between the histogram and the fitted Gamma distribution. Therefore, we conclude that the Gamma distribution is a very good fit to the data.

Of course, you can use the computed p-value $p = 0.8490 > 0.05$ to conclude the good fit.

3.5 Kolmogorov–Smirnov Test Using Cumulative Distributions

The chi-square test for goodness of fit examines the sum of the squared differences between the histogram of the observed data and the expected PDF. The Kolmogorov–Smirnov (K-S) test for goodness of fit examines the maximum difference between the cumulative percentage of the observed data and the expected CDF. The K-S test can be applied to check differences between two cumulative distributions, such as if a dataset satisfies a given distribution, or if two datasets are from the same distribution. However, the K-S test assumes that the parameters of the expected CDF are independent of the sample data (i.e., the observed data), while the chi-square test does not have such an assumption.

The test statistic is

$$D_n = \max_x |F_{\text{obs}}(x) - F_{\text{exp}}(x)|, \tag{3.27}$$

where $F_{\text{exp}}(x)$ is the expected CDF, and $F_{\text{obs}}(x)$ is the cumulative distribution from n independent and identically distributed samples and can be computed by ranking the observed data. The R command for computing $F_{\text{obs}}(x)$ is ecdf(obs.data).

If n is sufficiently large, $\sqrt{n}D_n$ is approximately the Kolmogorov distribution, which can help determine the critical value for D_n. The Kolmogorov distribution is quite complex and is not discussed in this book. In applications, computer software packages for the K-S test give the p-value, from which the decision of accepting or rejecting the null hypothesis can be made: accept the null hypothesis at the 5% significance level if the p-value is greater than 0.05.

We use R command x = rnorm(60) to generate data x. Then check if x is from the standard normal population $N(0,1)$ by using the K-S test. The answer is of course, yes. The computer code for the K-S test is as follows.

```
#R code for K-S test for N(0,1) data
x = rnorm(60)
ks.test(x, "pnorm", mean=0, sd=1)
#D = 0.10421, p-value = 0.4997
#The large p-value implies no significant difference

#K-S test of uniformly distributed data vs N(0,1) population
u = runif(60)
ks.test(u, "pnorm", mean=0, sd=1)
#D = 0.50368, p-value = 1.366e-14
#The small p-value implies significant difference
```

```
#Python code for the K-S test
from scipy.stats import kstest
#K-S test of normal data vs N(0,1)
x1 = np.random.normal(0,1,60)#generate 60 normal data
kstest(x1, 'norm')#perform Kolmogorov-Smirnov test
#statistic=0.07706, pvalue=0.86830
#The large p-value implies no significant difference
#K-S test of uniform data vs N(0,1)
x2 = np.random.uniform(0, 1, 60)#generate 60 uniform data
kstest(x2, 'norm')#perform Kolmogorov-Smirnov test
#statistic=.52023, pvalue=1.2440e-15
#The smallp-value implies significant difference
```

The large p-value 0.4997 implies that the null hypothesis is accepted and the data are indeed from the standard normal population.

If data u is from a uniform distribution, then the K-S test will give a very small p-value, approximately zero p-value = 1.366e-14, as shown in the K-S test computer code. Thus, there is a significant difference between the data distribution and $N(0,1^2)$. Of course, this is expected, since these sample data u are from a uniform distribution and are not from the $N(0,1^2)$ population.

The K-S test is applicable to any distribution, and is often used in climate science. However, the K-S test should not be used when the parameters of the expected CDF are estimated from the sample data. If the sample data are used to estimate the model parameter and to compute the K-S statistic D_n, then the D_n value may be erroneously reduced. The consequence is the exaggeration of Type II error: the null hypothesis should be rejected, but is not. Thus, it may lead to a false negative and miss detecting a signal.

The following computer code tests that the standard anomalies of the monthly Edmonton temperature data from January 1880 to December 2015 are from the standard normal population. The almost-zero p-value implies the significant difference between the Edmonton data and the standard normal distribution. Figure 3.8 shows the difference between the cumulative distribution from data and that from $N(0, 1)$.

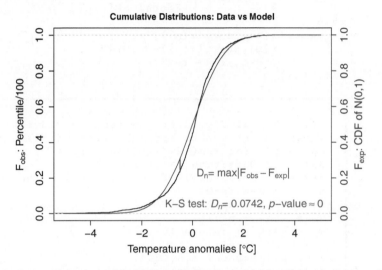

Figure 3.8   Cumulative distributions of the observed data and a model. The observed data are the standardized monthly surface air temperature anomalies of Edmonton from January 1880 to December 2015. The model is $N(0, 1^2)$, where D_n is equal to the maximum difference between the two CDF curves.

Figure 3.8 may be generated by the following computer code.

```
#R plot Fig. 3.8: K-S test for Edmonton temp
setEPS() # save the .eps figure
postscript("fig0308.eps", height = 5.6, width = 8)
par(mar=c(4.2,4.5,2.5,4.5))
setwd("/Users/sshen/climstats")
da1 =read.csv("data/EdmontonT.csv", header=TRUE)
x=da1[,3]
m1 = mean(x)
s1 = sd(x)
xa = (x- m1)/s1
ks.test(xa, "pnorm", mean=0, sd=1)
#Dn = 0.074186, p-value = 2.829e-08 => significant difference
plot(ecdf(xa), pch =3,
    xlim=c(-5,5),ylim=c(0,1),
  main="Cumulative␣Distributions:␣Data␣vs␣Model",
  xlab=expression(paste('Temperature␣Anomalies␣[',degree,'C]')),
  ylab=expression(paste(F[obs],':␣Percentile/100')),
  cex.lab=1.4, cex.axis=1.4)
```

```
par(new=TRUE)
x=seq(-5,5, by=0.1)
lines(x,pnorm(x), col='blue')
axis(4, cex.axis=1.4, col='blue',
     col.axis='blue')
mtext(expression(paste(F[exp],':␣CDF␣of␣N(0,1)')),
      cex=1.4, side = 4, line = 3, col="blue")
text(2, 0.05, cex=1.4, col="red",
     expression(paste('K-S␣test:␣', D[n],
                    '=␣0.0742,␣p-value' %~~% '0' )))
text(2, 0.22, cex=1.4, col="red",
     expression(paste(D[n], '=␣max(',
                    F[obs] -F[exp], ')')))
m=46
segments(x[m],pnorm(x[m]),
         x[m],pnorm(x[m])- 0.0742,
         lwd=2, col='red')
dev.off()
```

```
#Python plot Fig. 3.8: K-S test for Edmonton temp
#Read Edmonton data from the gridded NOAAGlobalTemp
da1 = pd.read_csv('data/EdmontonT.csv')
da1 = pd.DataFrame(da1['Value'])
da2 = np.array(da1)# turn dataset into an array
m1 = np.mean(da2)
s1 = np.std(da2)
xa = (da2 - m1)/s1
# convert xa into a 1D array
xa1 = xa.ravel()
x = np.linspace(-5,5,1642)
# function of ecdf made from xa
ecdf = ECDF(xa1)
# percentiles, length of xa is 1642
vals = ecdf(np.linspace(min(x), max(x),1642))
fig, ax1 = plt.subplots(figsize=(12,8))
# plot cdf
ax1.plot(x, vals, 'k')
ax1.set_title("Cumulative␣Distributions:␣Data␣vs␣Model",
              pad = 20)
ax1.set_xlabel("Temperature␣Anomalies␣[$\degree$C]",
              labelpad = 20)
ax1.set_ylabel("$F_{obs}$:␣Percentile/100",
              labelpad = 20)
ax2 = ax1.twinx()
# plot ecdf
x = np.linspace(-5,5,101)
ax2.plot(x,scistats.norm.cdf(x), color = 'b')
ax2.set_xlim((-5,5))
ax2.tick_params(length=6, width=2,
                labelsize=20, color = 'b',
                labelcolor = 'b')
ax2.spines['right'].set_color('b')
ax2.text(0.5,0.3, '$D_n␣=␣max(F_{obs}␣-␣F_{exp})$',
         size = 20, color = 'r')
```

```
ax2.text(-0.3,0.2, 'K-S test: Dn = 0.0742, p-value = 0',
         size = 20, color = 'r')\

ax2.text(6,0.2, "$F_{exp}$: CDF of N(0,1)",
         rotation = 'vertical', color = 'b',
         size = 22)
# add arrow
kwargs = {'color' : 'r'}
plt.arrow(-0.4,0.35, 0., -0.07, **kwargs)
# plot horizontal lines
plt.axhline(y=1, ls = '--', color = 'grey')
plt.axhline(y=0, ls = '--', color = 'grey')
plt.show()
```

You can use the K-S test to check if the Omaha June precipitation data from 1948 to 2019 follows a Gamma distribution. The *p*-value of the K-S test is 0.8893. The null hypothesis is retained, and the Gamma distribution is a good fit to the Omaha June precipitation data. The K-S test leads to the same conclusion as the χ^2-test. The difference between the tests is that the K-S test is based on a continuous distribution, while the χ^2-test is applied to a discrete histogram in this Omaha June precipitation example. You may use the following computer code to compute the K-S statistic and its corresponding *p*-value.

```
#R: K-S test for Omaha June precip vs Gamma
Omaha=read.csv("data/OmahaP.csv", header=TRUE)
daP = matrix(Omaha[,7], ncol=12, byrow=TRUE)
y = daP[,6] #June precipitation 1948-2019
#install.packages('fitdistrplus')
library(fitdistrplus)
omahaPfit = fitdist(y, distr = "gamma", method = "mle")
ks.test(y, "pgamma", shape = 1.51760921, rate= 0.01889428)
#D = 0.066249, p-value = 0.8893 #y is the Omaha data string
#You may verify the K-S statistic using another command
#install.packages('fitdistrplus')
library(fitdistrplus)
gofstat(omahaPfit)
#Kolmogorov-Smirnov statistic  0.06624946
```

```
#Python: K-S test for Omaha June precip vs Gamma
from scipy.stats import gamma
Omaha = pd.read_csv("data/OmahaP.csv")
daP = pd.DataFrame(Omaha['PRCP'])
daP = np.array(daP)
shape = (72,12)
daP = np.reshape(daP, shape)
y = daP[:, 5] #June precipitation
shape, loc, scale = gamma.fit(y, floc=0)
print('shape=', shape, 'scale =', scale)
#shape= 1.5178 scale = 52.9263
stats.kstest(y, 'gamma', (shape, scale))
#statistic=0.066249, pvalue=0.8893
```

The K-S test can also check if two sample datasets x and y are from the same population. The R command is `ks.test(x, y)`. The Python command is

```
stats.ks_2samp(x, y)
```

3.6 Determine the Existence of a Significant Relationship

Climate scientists often need to determine whether there is a significant relationship between two variables using two datasets. The statistical demonstration of the relationship further motivates scientists to explore physical relationships or causality. Or this process can be reversed: when a physical relationship is determined by the first principles, scientists need to observe and analyze data to support the relationship quantitatively.

3.6.1 Correlation and t-Test

The simplest relationship is a linear one between two random variables X and Y:

$$Y = aX + b. \tag{3.28}$$

The strength of this relationship is determined by the Pearson correlation coefficient, computed from the sample data x and y. If X and Y are normally distributed, then the nonzero correlation coefficient r determines the existence of a linear relationship:

$$r = \frac{\sum_{i=1}^{n}(x_i - \bar{x})(y_i - \bar{y})}{\sqrt{\sum_{i=1}^{n}(x_i - \bar{x})^2}\sqrt{\sum_{i=1}^{n}(y_i - \bar{y})^2}}, \tag{3.29}$$

where $\bar{x}$ is the mean of data $x = (x_1, x_2, \ldots, x_n)$, and $\bar{y}$ is the mean of data $y = (y_1, y_2, \ldots, y_n)$.

The null hypothesis is $r = 0$ for the nonexistence of the linear relationship. The alternative hypothesis is, of course, $r \neq 0$.

For given sample data (x, y), the t-statistic for the hypothesis testing is

$$t = r\sqrt{\frac{n-2}{1-r^2}}, \tag{3.30}$$

which satisfies a t-distribution with degrees of freedom equal to $n - 2$.

To avoid erroneous conclusions, we should use caution when applying this test. First, this test is based on a linear relationship. A very good nonlinear relationship between x and y data may still lead to mistaken acceptance of the no-relationship null hypothesis. For example, if 100 pairs of data (x, y) lie uniformly on a unit circle centered around zero, then the nonlinear relationship should exist. However, the correlation coefficient is zero, and the existence of a linear relationship is rejected. Therefore, rejection of a linear relationship by the correlation coefficient does not exclude the existence of a nonlinear relationship. It is important that statistical analysis should incorporate physics, and statistical results should have physical interpretations.

Second, the t-statistic assumes that the sample data are from normal distributions. When the sample size is large, say greater than 50, this assumption can be regarded as approximately true even if x and y are from nonnormal distributions.

Third, the correlation coefficient is sensitive to the data and can be greatly altered by some outliers. If the outliers are erroneous, then the test conclusion can be wrong too. For example, Figure 3.9 shows such a case. The random data points around the origin apparently show the nonexistence of any relationship. Of course, these random points do not imply a linear relationship either. The correlation coefficient is only 0.19. However, when a pair of outliers $(1,1)$ is added to the random data, the correlation coefficient becomes 0.83, which supports a strong linear relationship. It is clear if the data pair $(1,1)$ is an erroneous outlier, then the rejection of the null hypothesis is a wrong decision.

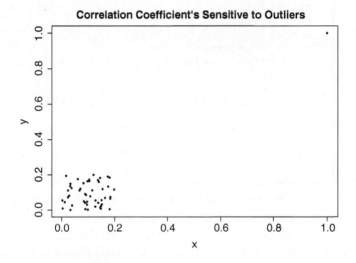

Figure 3.9   Correlation coefficient is sensitive to outliers.

Figure 3.9 and its related statistics can be calculated by the following computer code.

```
#R plot Fig. 3.9: Correlation coefficient sensitive to outliers
setEPS() # save the .eps figure
postscript("fig0309.eps", height = 5.6, width = 8)
par(mar=c(4.2,4.5,2.5,4.5))
setwd("/Users/sshen/climstats")
par(mar=c(4,4,0.5,0.5))
x = c(0.2*runif(50),1)
y = c(0.2*runif(50),1)
plot(x,y, pch=19, cex=0.5,
     cex.axis =1.4, cex.lab = 1.4)
dev.off()

#t-statistic
n=51
r = cor(x, y)
t=r*(sqrt(n-2))/sqrt(1-r^2)
t
#[1] 10.19999
```

```
qt(0.975, df=49)
#[1] 2.009575 #This is the critical t value
1 - pt(10.19999, df=49)
#[1] 5.195844e-14 # a very small p-value
```

```
#Python plot Fig. 3.9: Correlation sensitive to outliers
import numpy as np
from scipy.stats import pearsonr
# generate 50 random floats in [0,1)
# multiply each float by 0.2
x = np.append(0.2*np.random.uniform(0,1,50), 1)
y = np.append(0.2*np.random.uniform(0,1,50), 1)
plt.plot(x, y, 'o', color = 'black')
plt.show()
#t-statistic
n = 51
r, _  = pearsonr(x, y)
ts = r*(np.sqrt(n-2))/np.sqrt(1-r**2)
print(tc)
#11.1269
tc = scipy.stats.t.ppf(q=.975,df=49)
print(tc)
#2.009575 #This is the critical t value
1 - scipy.stats.t.cdf(ts, df=49)
#3.05311e-14 # a very small p-value
```

The critical t value at the two-sides 5% significance level is approximately 2.01 in the R code calculation, and the p-value is almost zero. Thus, the null hypothesis is rejected at the 5% significance level. However, this rejection is all caused by the pair of outlier data (1,1), and is most likely wrong. To avoid this kind of error, one may try a Kendall tau test to make a decision.

3.6.2 Kendall Tau Test for the Existence of a Relationship

For the same data, the Kendall tau test accepts the null hypothesis, because the p-value is 0.2422 and is larger than 0.025. Thus, we conclude that the linear relationship does not exist. The computer code for the Kendall tau test is as follows.

```
#R code for a Kendall tau test
#install.packages("Kendall")
library(Kendall)
x = c(0.2*runif(50), 1)
y = c(0.2*runif(50), 1)
Kendall(x,y)
#tau = 0.114, 2-sided pvalue =0.24216
```

```
#Python code for a Kendall tau test
import numpy as np
from scipy.stats import pearsonr
```

```
x = np.append(0.2*np.random.uniform(0,1,50), 1)
y = np.append(0.2*np.random.uniform(0,1,50), 1)
tau, p_value = stats.kendalltau(x, y)
print('tau␣=', tau, 'pvalue␣=', p_value)
#tau = -0.09647, 2-sided pvalue = 0.3178
```

The Kendall tau score here is defined according to the positive or negative slope of the line segment determined by two pairs of data $(x_i, y_i), (x_j, y_j)$:

$$m_{ij} = \frac{y_j - y_i}{x_j - x_i}. \tag{3.31}$$

If $m_{ij} > 0$, the two pairs are said to be concordant, else discordant. If $m_{ij} = 0$, then the two pairs are called a tie. When $x_j = x_i$, the data pair are excluded from the Kendall tau score calculation. Finally the Kendall tau score is defined as

$$\tau = \frac{N_c - N_d}{N_c + N_d}, \tag{3.32}$$

where N_c is the number of concordant pairs and N_d is the number of discordant pairs. A tie adds 1/2 to both N_c and N_d. In this way, the tau score is insensitive to outliers because the tau score depends on the ratios of data (i.e., the slope), not the data themselves.

The Kendall tau test has the advantage of being insensitive to outliers. It also does not assume the distributions of data. However, it still requires data independence. When the data are serially correlated, the Kendall tau score needs to be revised.

Although our Kendall tau test result has successfully accepted the null hypothesis, once in a while the randomly generated data (x, y) may still have a small p-value in the Kendall tau test, and hence reject the null hypothesis and support the existence of a linear relationship. This implies the importance of interpreting the statistical results in terms of climate science. There is no definite way to statistically determine whether the data $(1, 1)$ are erroneous and hence should be excluded, but there might be a physical way to determine whether the data $(1, 1)$ are unlikely to occur and to justify whether a linear relationship is reasonable. Statistical methods help identify signals in climate data, and climate science helps interpret the statistical results. This science-data-statistics-science cycle can go on for multiple rounds, in order to obtain correct and useful conclusions.

3.6.3 Mann–Kendall Test for Trend

In the Kendall tau case, if x is monotonic time, then the slope signs depend on y data only. We can define the so called Mann–Kendall score for the y data:

$$S = \frac{\sum_{j=1}^{n-1} \sum_{i=j+1}^{n} \text{sgn}(y_j - y_i)}{2}, \tag{3.33}$$

where the sgn function is defined as

$$\text{sgn}(x) = \begin{cases} 1 & x > 0 \\ 0 & x = 0 \\ -1 & x < 0 \end{cases}. \tag{3.34}$$

The R command for the Mann–Kendall test is `MannKendall(data)`. For the monthly Edmonton standardized temperature data from January 1880 to December 2015, the Mann–Kendall trend test result is $S = 206254$ and the p-value approximately zero. Thus, there is a linear trend in the Edmonton temperature data.

The computer code is as follows.

```
#R code for Mann-Kendall test: Edmonton data
setwd("/Users/sshen/climstats")
#Read Edmonton data from the gridded NOAAGlobalTemp
da1 =read.csv("data/EdmontonT.csv", header=TRUE)
x=da1[,3]
m1 = mean(x)
s1 = sd(x)
xa = (x- m1)/s1 #standardized anomalies
#install.packages('Kendall')
library(Kendall)
summary(MannKendall(xa))
#Score =  206254 , Var(Score) = 492349056
#tau = 0.153, 2-sided pvalue =< 2.22e-16
```

```
#Python code for Mann-Kendall test: Edmonton data
#pip install pymannkendall
#pymannkendall is not a standard Python package
import pymannkendall as mk
#Read Edmonton data from the gridded NOAAGlobalTemp
da1 = pd.read_csv('data/EdmontonT.csv')
da1 = pd.DataFrame(da1['Value'])
da2 = np.array(da1)# turn dataset into an array
m1 = np.mean(da2)
s1 = np.std(da2)
xa = (da2 - m1)/s1
mk.original_test(xa)
#tau = 0.153, p = 0.0 #yes, a significant trend exists
#s =  206254 , var_s = 492349056
```

3.7 Chapter Summary

This chapter has provided the basics of estimation and decision-making based on data. These methods are commonly used in simple statistical analyses of scientific data, and are helpful in understanding the statistical results in most literature. However, when data involve both time and space, they often require spatial statistics methods, such as empirical orthogonal functions for spatial patterns and principal components for temporal patterns. The space-time data involve considerable matrix theory and will be discussed in Chapters 5 and 6.

This chapter began by discussing the fundamental concept of estimating the standard deviation for a population and estimating the standard error of the mean from the point of view of the accuracy assessment. A numerical simulation was provided to illustrate the concept of standard error (see Fig. 3.1). From this concept, many statistics questions arise. Among them, we have discussed the following topics:

(i) Search for the true mean by a confidence interval;
(ii) Contingency table for decision-making with probabilities of false alarm and missed detection;
(iii) Steps of hypothesis testing as the standard procedure of decision-making based on data;
(iv) Serial correlation of data which lead to a reduced effective sample size;
(v) A suite of typical climate statistical decision examples: Determine the goodness of fit to a model by χ^2-distribution, determine change of a probability distribution by the Kolmogorov–Smirnov test, determine a significant correlation by a t-test or a non-parametric Kendall tau test, and determine a significant trend by the nonparametric Mann–Kendall test.

Real climate data examples were provided to illustrate the applications of the statistical methods. You can directly apply these methods to the climate data of your choice. You can also use these methods to verify the statistical conclusions in your climate science literature.

References and Further Reading

[1] L. Chihara and T. Hesterberg, 2011: *Mathematical Statistics with Resampling and R.* Wiley.

> Chapters 1–3 and 6–8 of the book are a good reference to the materials presented in our chapter and the book has many R codes and numerical examples.

[2] R. A. Johnson and C. K. Bhattacharyya, 1996: *Statistics: Principles and Methods.* 3rd ed., John Wiley & Sons.

> This is a popular textbook for a basic statistics course. It emphasizes clear concepts, and provides accurate descriptions of statistical methods and their assumptions, such as the four assumptions of a linear regression. Many real-world data are analyzed in the book. The 8th edition was published in 2019.

[3] S. S. P. Shen and R. C. J. Somerville, 2019: *Climate Mathematics: Theory and Applications.* Cambridge University Press.

> This book uses real climate data from both observations and numerical models, and has computer codes of R and Python to read the data and reproduce the figures and statistics in the book.

[4] H. Von Storch and F. W. Zwiers, 2001: *Statistical Analysis in Climate Research.* Cambridge University Press.

> This is a comprehensive climate statistics book that helps climate scientists apply statistical methods correctly. The book has both rigorous mathematical theory and practical applications. Numerous published research results are discussed.

[5] D. S. Wilks, 2011: *Statistical Methods in the Atmospheric Sciences.* 3rd ed., Academic Press.

This excellent textbook is easy to read and contains many simple examples of analyzing real climate data. It is not only a good reference manual for climate scientists, but also a guide tool that helps scientists in other fields make sense of the data analysis results in climate literature.

Exercises

3.1 Visualize a confidence interval.

(a) Use computer to generate 30 normally distributed numbers $x_1, x_2, \ldots, x_n$ $(n = 30)$ with mean equal to 8 and standard deviation equal to 2.

(b) Compute the mean $\bar{x}$ and standard deviation s from the 30 numbers.

(c) On the x-axis, plot the dots of $x_1, x_2, \ldots, x_n$ $(n = 30)$.

(d) On same the x-axis, plot the confidence interval based on the formula

$$(\bar{x} - 1.96\frac{s}{\sqrt{n}}, \bar{x} + 1.96\frac{s}{\sqrt{n}}). \tag{3.35}$$

(e) Plot the point for the true mean 8 and check if the true mean is inside the confidence interval.

3.2 Visualize 20 confidence intervals on a single figure by repeating the numerical experiment of the previous problem 20 times. *Hint: One way to display the result is to use the horizontal axis as the order of experiment from 1 to 20 and the vertical axis for the CI, and to plot a blue horizontal line for the true mean. Then visually check whether the blue horizontal line intersects with all the CIs.*

3.3 (a) Repeat the numerical simulation of histograms of Figure 3.1 with a total of 2,000,000 normally distributed data according to $N(6, 9^2)$, and $n = 100$.

(b) Discuss the two histograms and explain the meaning of standard error.

3.4 Find the confidence interval of the NOAAGlobalTemp data of the global average annual mean for the period of 1961–1990 when the serial correlation is taken into account. What would be the confidence interval if the serial correlation is ignored?

3.5 Use the NOAAGlobalTemp data of the global average annual mean and go through the t-test procedures to find out whether there is a significant difference between the 1920–1949 and 1950–1979 temperature anomalies, when the serial correlation is taken into account.

3.6 Use the t-test to determine if there a significant nonzero trend in the NOAAGlobalTemp data of the global average annual mean during the period of 1971–2010 at the 5% significance level. What is the p-value?

3.7 Repeat the previous problem for the monthly data of Januaries.

3.8 Repeat the previous problem for the monthly data of Julies.

3.9 Use the chi-square test to check if the January monthly precipitation data of Omaha, Nebraska, USA, from 1948 to 2019 fits well to a Gamma distribution. What are the shape and rate of the fitting? You can find the data on the NOAA Climate Data Online website www.ncdc.noaa.gov/cdo-web.

3.10 Do the same for the July monthly precipitation data of Omaha from 1948 to 2019.

3.11 Do the same for the annual mean precipitation data of Omaha from 1948 to 2019.

3.12 Use the chi-square test to check if the June monthly precipitation data of Omaha, Nebraska, USA, from 1948 to 2019 fits well to a lognormal distribution. Find μ and σ in the lognormal distribution.

3.13 Fit a long-term monthly precipitation data of a station of your choice to the Gamma distribution. Choose a station and a month, such as Paris and June. The record should be at least 50 years long. Use the chi-square test to check if the monthly precipitation data fits well to a Gamma distribution. What are the shape and rate of the fitting?

3.14 Examine the long-term monthly surface air **temperature** data of a station of your choice. Explore if the standardized anomaly data satisfy the standard normal distribution using the K-S test.

3.15 Examine the long-term monthly **precipitation** data of a station of your choice. Explore if the standardized anomaly data satisfy the standard normal distribution using the K-S test.

3.16 For the standardized January monthly mean precipitation and temperature data of Madison, Wisconsin, USA, from 1961 to 2010, use a t-test to examine whether there exists a significant correlation between temperature and precipitation. You can find the data on the NOAA Climate Data Online website www.ncdc.noaa.gov/cdo-web.

3.17 Examine the same correlation using the Kendall tau test.

3.18 (a) Use the Mann–Kendall trend test to examine whether there is a significant positive temporal trend in the Madison precipitation data at the 5% significance level. What is the p-value?

(b) Do the same for the temperature data.

3.19 Use the Mann–Kendall trend test to examine whether there is a significant positive temporal trend in the NOAAGlobalTemp annual mean data from 1880 to 2019 at the 5% significance level. What is the p-value? See Figure 3.6 for the linear trend line.

3.20 Use the Mann–Kendall trend test to examine whether there is a significant positive temporal trend in the NOAAGlobalTemp September monthly mean data from 1880 to 2019 at the 2.5% significance level. What is the p-value?

3.21 Use the Mann–Kendall trend test to examine whether there is a temporal trend (positive or negative) in the global annual mean precipitation from 1979 to 2018 at the 5% significance level. What is the p-value? You may use the NASA Global Precipitation Climatology Project (GPCP data).

## 4

Regression Models and Methods

The word "regression" means "a return to a previous and less advanced or worse form, state, condition, or way of behaving," according to the Cambridge dictionary. The first part of the word – "regress" – originates from the Latin *regressus*, past participle of *regredi* ("to go back"), from re- ("back") + gradi ("to go"). Thus, "regress" means "return, to go back" and is in contrast to the commonly used word "progress." The *regression* in statistical data analysis refers to a process of returning from irregular and complex data to a simpler and less perfect state, which is called a model and can be expressed as a curve, a surface, or a function. The function or curve, less complex or less advanced than the irregular data pattern, describes a way of behaving or a relationship. This chapter covers linear models in both uni- and multivariate regressions, least square estimations of parameters, confidence intervals and inference of the parameters, and fittings of polynomials and other nonlinear curves. By running diagnostic studies on residuals, we explain the assumptions of a linear regression model: linearity, homogeneity, independence, and normality. As usual, we use examples of real climate data and provide both R and Python codes.

4.1 Simple Linear Regression

The simplest model of regression is a straight line, i.e., a linear model. This involves only two variables x and y. Therefore, a *simple linear regression* means going from the data pairs $(x_i, y_i)(i = 1, 2, \ldots, n)$ on the xy-plane back to a simple straight line model:

$$y = a + bx. \tag{4.1}$$

The data will determine the values of a and b by the criterion of the best fit. The model can be used for prediction: predict y when a value of x is given.

4.1.1 Temperature Lapse Rate and an Approximately Linear Model

It is known that, as an approximation, in a mountain terrain temperature decreases linearly according to elevation. The decrease rate is called the temperature *lapse rate* in meteorology. The linear relationship and the lapse rate are shown in Figure 4.1 for Colorado, which is a Rocky Mountain state of the United States and has a large elevation range from 1010 to 4401 meters, according to Colorado Tourism www.colorado.com. The dots in the figure

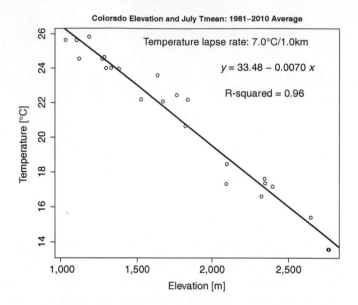

Figure 4.1   The dots are the scatter plot of the data of elevation and 1981–2010 average July surface air temperature Tmean taken from the 24 USHCN stations. The thick straight line is the linear regression model computed from the data.

are the scatter plot of the data of elevation and 1981–2010 average July surface air temperature Tmean of the 24 Colorado stations in the United States Historical Climatological Network (USHCN) (Menne et al. 2009). The straight line of the figure corresponds to the linear model:

$$y = 33.48 - 0.0070x. \tag{4.2}$$

This equation is a result from regression of data, referred to as observations in climate science, or sample data in statistics.

Figure 4.1 shows a scattered and complex data pattern regressed to a simpler form, in this case, a straight-line model. This is called the simple linear regression process. The regression result has climate science interpretations, such as the lapse rate of temperature.

The horizontal axis of Figure 4.1 is for elevation. The elevation data in units of meters for the 24 stations are as follows:

```
1671.5, 1635.6, 2097.0, 1295.4, 1822.7, 2396.9, 2763.0, 1284.7,
1525.2, 1328.6, 1378.9, 2323.8, 2757.8, 1033.3, 1105.5, 1185.7,
2343.9, 1764.5, 1271.0, 2347.3, 2094.0, 2643.2, 1837.9, 1121.7
```

The vertical axis is for temperature, which is the 30-year average of the July daily mean temperature (Tmean) from 1981 to 2010, computed from the USHCN monthly data, which have been adjusted for the time of observation bias (TOB). Some stations had missing data denoted by −9,999. When computing the 30-year average, the entries of −9,999 were omitted. Thus, some averages were computed from fewer than 30 years. The resulting average temperature data in the unit of °C for the 24 stations are as follows:

```
22.064, 23.591, 18.464, 23.995, 20.645, 17.175, 13.582, 24.635,
22.178, 24.002, 23.952, 16.613, 13.588, 25.645, 25.625, 25.828,
17.626, 22.433, 24.539, 17.364, 17.327, 15.413, 22.174, 24.549
```

With these data, Figure 4.1 can be generated by the following computer code.

```
#R plot Fig. 4.1: Colorado temperature lapse rate
x= c(
1671.5, 1635.6, 2097.0, 1295.4, 1822.7, 2396.9, 2763.0, 1284.7,
1525.2, 1328.6, 1378.9, 2323.8, 2757.8, 1033.3, 1105.5, 1185.7,
2343.9, 1764.5, 1271.0, 2347.3, 2094.0, 2643.2, 1837.9, 1121.7)
y= c(
22.064, 23.591, 18.464, 23.995, 20.645, 17.175, 13.582, 24.635,
22.178, 24.002, 23.952, 16.613, 13.588, 25.645, 25.625, 25.828,
17.626, 22.433, 24.539, 17.364, 17.327, 15.413, 22.174, 24.549)
setEPS() # save the .eps figure
postscript("fig0401.eps", width = 8)
par(mar=c(4.5,4.5,2.5,0.5))
plot(x,y,
     xlab="Elevation␣[m]",
     ylab=expression("Temperature␣["~degree~"C]"),
  main="Colorado␣Elevation␣and␣July␣Tmean:␣1981-2010␣Average",
     cex.lab=1.5, cex.axis=1.5, cex.main =1.2)
reg=lm(y~x)
reg
#(Intercept)              x
# 33.476216    -0.006982    #-7.0 degC/1km
summary(reg)
#R-squared:  0.9631
abline(reg,lwd=3)
text(2100, 25.5,
expression("Temperature␣lapse␣rate:␣7.0"~degree~"C/1.0km"),
     cex=1.5)
text(2350, 24, "y=␣33.48␣-␣0.0070␣x", cex=1.5)
text(2350, 22.5,"R-squared␣=␣0.96", cex=1.5)
dev.off()
```

```
#Python plot Fig. 4.1: Colorado temperature lapse rate
x = np.array([
1671.5, 1635.6, 2097.0, 1295.4, 1822.7,2396.9,2763.0,1284.7,
1525.2, 1328.6, 1378.9, 2323.8, 2757.8,1033.3,1105.5,1185.7,
2343.9, 1764.5, 1271.0, 2347.3, 2094.0,2643.2,1837.9,1121.7
])
y = np.array([
22.064, 23.591, 18.464, 23.995, 20.645,17.175,13.582,24.635,
22.178, 24.002, 23.952, 16.613, 13.588,25.645,25.625,25.828,
17.626, 22.433, 24.539, 17.364, 17.327,15.413,22.174,24.549
])
# calculate correlation coefficients
corrMatr = np.corrcoef(x,y)
# R-squared
Rsqu = corrMatr[0,1]**2
# trend line
reg1 = np.array(np.polyfit(x, y, 1))
abline = reg1[1] + x*reg1[0]
```

```
fig, ax = plt.subplots(figsize=(12,12))
ax.plot(x,y, 'ko');
ax.plot(x,abline, 'k-');
ax.set_title("Colorado␣Elevation␣vs.␣July␣Tmean:␣\n␣\
1981␣-␣2010␣Average",
           size = 25, fontweight = 'bold', pad = 20)
ax.set_xlabel("Elevation␣[$m$]", size = 25, labelpad = 20)
ax.set_ylabel("Temperature␣[$\degree$C]",
           size = 25, labelpad = 20)
ax.tick_params(length=6, width=2, labelsize=20);
ax.set_xticks(np.round(np.linspace(1000, 3000, 5), 2))
ax.set_yticks(np.round(np.linspace(12, 26, 8), 2))
ax.text(1750, 25.5,
       r"Temperature␣lapse␣rate:␣7.0␣$\degree␣$C/km",
       color= 'k', size = 20)
ax.text(2250, 24.5, r"$y␣=␣33.48␣-␣0.0070~␣x$" % Rsqu,
       color= 'k', size = 20)
ax.text(2250, 23.5, r"$R-squared␣=␣%.2f$" % Rsqu,
       color= 'k', size = 20)
plt.show()
```

The Colorado July temperature lapse rate (TLR) 7.0°C/1.0 km is consistent with earlier studies using 104 stations on the west slope of Colorado for an elevation range of 1,500–3,600 meters: 6.9°C/1.0 km (Fall 1997). For the annual mean temperature, the LTR is 6.0°C/1.0 km. These two results are comparable to an LTR study for the Southern Ecuadorian Andes in the elevation range of 2,610–4,200 meters (Cordova et al. 2016). The annual mean temperature TLR is 6.88°C/1.0 km.

The TLR may be applied to approximately predict the temperature of a mountain region at a given location. This is useful in cases when it is hard to maintain a weather station due to high elevation or complex terrain, while it is relatively easy to obtain elevation data based on the Geographical Information System (GIS) or a digital elevation model (DEM) dataset.

4.1.2 Assumptions and Formula Derivations of the Single Variate Linear Regression

This subsection describes the statistical concepts and mathematical theory of the aforementioned linear regression procedure. We pay special attention to the assumptions about the linear regression model.

4.1.2.1 Linear Model and Its Assumptions

The linear model for TLR may be written as

$$Y = a + bx + \varepsilon. \tag{4.3}$$

Here,

(i) x is an independent variable, also called explanatory variable, and is deterministic (not random);

(ii) Y is the dependent variable, also called the response variable, and is a random variable with a constant variance $\mathrm{Var}[Y|x] = \sigma^2$ for any given x;

(iii) ε is the random error term, which is assumed to have zero mean $\mathrm{E}[\varepsilon] = 0$, and constant variance σ^2: $\mathrm{Var}[\varepsilon|x] = \sigma^2$, and is also assumed to be uncorrelated with each other: $\mathrm{Cor}[\varepsilon|x_1, \varepsilon|x_2] = 0$ if $x_1 \neq x_2$;

(iv) a, b, and σ are called parameters and are to be estimated from data $(x_i, y_i)(i = 1, 2, \ldots, n)$, also called samples, or sample data; and

(v) The expected value of Y for a given x is $a + bx$, i.e.,

$$\mathrm{E}[Y|x] = a + bx. \tag{4.4}$$

The linear model Eq. (4.3) at the given points x_i is

$$Y_i = a + bx_i + \varepsilon_i, \quad i = 1, 2, \ldots, n, \tag{4.5}$$

where both ε_i and Y_i are random variables with

$$\mathrm{E}[Y_i] = a + bx_i, \tag{4.6}$$

$$\mathrm{E}[\varepsilon_i] = 0, \tag{4.7}$$

$$\mathrm{Var}[Y_i] = \mathrm{E}[(Y_i - (a + bx_i))^2] = \sigma^2, \tag{4.8}$$

$$\mathrm{Var}[\varepsilon_i] = \mathrm{E}[(\varepsilon_i - 0)^2] = \sigma^2, \tag{4.9}$$

$$\mathrm{Cor}[\varepsilon_i, \varepsilon_j] = \delta_{ij}, \tag{4.10}$$

where δ_{ij} is the Kronecker delta

$$\delta_{ij} = \begin{cases} 1, & \text{if } i = j, \\ 0, & \text{if } i \neq j. \end{cases} \tag{4.11}$$

The last equation means that the error terms ε_i at the observational points x_i are uncorrelated with each other.

The observations at the given points x_i are y_i. Both are fixed values, such as temperature $y_1 = 22.064°C$ at the given elevation $x_1 = 1671.5$ meters in the Colorado July temperature example. Here, y_1 is a sample value for the random variable Y_1. The corresponding error datum is e_1, which is called a residual and is regarded to be a sample value of ε_1 with respect to the linear model.

4.1.2.2 Estimating a and b from Data

Given a set of sample data $(x_i, y_i), i = 1, 2, \ldots, n$, the parameters a and b in the regression model can be estimated and denoted with $\hat{a}, \hat{b}$. Thus, $\hat{a}, \hat{b}$ correspond to a particular dataset, i.e., another dataset $(u_i, v_i), i = 1, 2, \ldots, m$ will yield a pair of different estimators that may be denoted by $\hat{a}_{uv}, \hat{b}_{uv}$. In contrast, a, b are the general notations of constant coefficients for a linear regression model.

The estimated linear model may be written in the following way:

$$y_i = \hat{a} + \hat{b} x_i + e_i, \quad i = 1, 2, \ldots, n, \tag{4.12}$$

where

$$e_i = y_i - (\hat{a} + \hat{b} x_i) \tag{4.13}$$

are called *residuals* of the linear model, and are equal to the observed values y_i minus the predicted values

$$\hat{y}_i = \hat{a} + \hat{b} x_i. \tag{4.14}$$

The mean of the n equations (4.12) yields

$$\bar{y} = \hat{a} + \hat{b}\bar{x}, \tag{4.15}$$

where

$$\bar{x} = \frac{x_1 + x_2 + \cdots + x_n}{n}, \tag{4.16}$$

$$\bar{y} = \frac{y_1 + y_2 + \cdots + y_n}{n}. \tag{4.17}$$

Because of the unbiased model assumption $E[\varepsilon] = 0$, we require that

$$\bar{e} = \frac{e_1 + e_2 + \cdots + e_n}{n} = 0, \tag{4.18}$$

equivalently,

$$\sum_{i=1}^{n} e_i = 0. \tag{4.19}$$

Equations (4.15)–(4.17) give an estimate for a from data:

$$\hat{a} = \bar{y} - \hat{b}\bar{x}, \tag{4.20}$$

but $\hat{b}$ is to be estimated by

$$\hat{b} = \frac{\mathbf{y}_a \cdot \mathbf{x}_a}{\mathbf{x}_a \cdot \mathbf{x}_a}. \tag{4.21}$$

Here,

$$\mathbf{x}_a = \begin{bmatrix} x_{a,1} \\ x_{a,2} \\ \vdots \\ x_{a,n} \end{bmatrix} \quad \mathbf{y}_a = \begin{bmatrix} y_{a,1} \\ y_{a,2} \\ \vdots \\ y_{a,n} \end{bmatrix} \quad \mathbf{e} = \begin{bmatrix} e_1 \\ e_2 \\ \vdots \\ e_n \end{bmatrix}, \tag{4.22}$$

with

$$x_{a,i} = x_i - \bar{x}, \tag{4.23}$$

$$y_{a,i} = y_i - \bar{y}. \tag{4.24}$$

In the last two formulas, $x_{a,i}$ and $y_{a,i}$ may be regarded as anomaly data, hence with the subscript "a." Climate scientists often use the concept of anomaly data, departures from a climate normal.

Equivalence of Orthogonality to Minimization

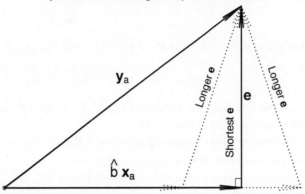

Figure 4.2 The condition of the least sum of the residual squares SSE is equivalent to the orthogonality condition $\mathbf{e} \cdot \mathbf{x}_a = 0$.

The $\hat{b}$ estimate formula (4.21) can be derived from the best fit condition which minimizes the mean square errors (MSE)

$$\text{MSE} = \frac{|\mathbf{e}|^2}{n} = \frac{\sum_{i=1}^{n} e_i^2}{n} = \frac{\sum_{i=1}^{n}(y_i - \hat{y}_i)^2}{n}, \tag{4.25}$$

equivalently minimizing the sum of the square errors (SSE)

$$\text{SSE} = |\mathbf{e}|^2 = \sum_{i=1}^{n} e_i^2 = \sum_{i=1}^{n}(y_i - \hat{y}_i)^2 = n \times \text{MSE}, \tag{4.26}$$

where $|\mathbf{e}|$ denotes the Euclidean length of vector $\mathbf{e}$. Thus, this method is also called the least square estimate. The least square condition is equivalent to the following orthogonality condition (see Fig. 4.2)[1]:

$$\mathbf{e} \cdot \mathbf{x}_a = 0. \tag{4.27}$$

Inserting Eq. (4.20) into the linear model with data (4.12) leads to

$$y_i - \bar{y} = \hat{b}(x_i - \bar{x}) + e_i, \quad i = 1, 2, \ldots, n. \tag{4.28}$$

or

$$y_{a,i} = \hat{b}x_{a,i} + e_i, \quad i = 1, 2, \ldots, n, \tag{4.29}$$

---

[1] SSE can be written as

$$\text{SSE} = |\mathbf{e}|^2 = |\mathbf{y_a} - \hat{b}\mathbf{x_a}|^2.$$

When $\hat{b}$ is optimized to minimize SSE, the derivative of this equation with respect to $\hat{b}$ is zero.

$$\frac{\text{d}}{\text{d}\hat{b}} \text{SSE} = (\mathbf{y_a} - \hat{b}\mathbf{x_a}) \cdot \mathbf{x_a} = \mathbf{e} \cdot \mathbf{x_a} = 0.$$

The orthogonality condition, $\mathbf{e} \cdot \mathbf{x}_a = 0$, can also be derived from a geometric point of view. The three vectors $\hat{b}\mathbf{x}_a, \mathbf{e}$ and $\mathbf{y}_a$ in $\mathbf{y}_a = \hat{b}\mathbf{x}_a + \mathbf{e}$ form a triangle. The side $\mathbf{e}$, depending on $\hat{b}$, is the shortest when $\hat{b}\mathbf{x}_a$ is perpendicular to $\mathbf{e}$ by adjusting parameter $\hat{b}$.

or

$$\mathbf{y}_a = \hat{b}\mathbf{x}_a + \mathbf{e}. \tag{4.30}$$

The dot product of both sides of Eq. (4.29) with $\mathbf{x}_a$ yields

$$\hat{b} = \frac{\mathbf{y}_a \cdot \mathbf{x}_a - \mathbf{e} \cdot \mathbf{x}_a}{\mathbf{x}_a \cdot \mathbf{x}_a}. \tag{4.31}$$

With the orthogonality condition Eq. (4.27) $\mathbf{e} \cdot \mathbf{x}_a = 0$, Eq. (4.31) is reduced to the least square estimate of b: Eq. (4.21).

In summary, the estimates $\hat{a}$ and $\hat{b}$ are derived by requiring (i) zero mean of the residuals, and (ii) minimum sum of the residual squares.

As shown in the computer code for Figure 4.1, the R command for the simple linear regression is $\text{lm}(y \sim x)$. The TLR dataset yields the estimate of the intercept $\hat{a} = 33.476216$ and slope $\hat{b} = -0.006982$:

```
lm(y ~ x)
#(Intercept)              x
#33.476216        -0.006982
```

The estimated linear model is thus

$$y = 33.476216 - 0.006982\, x, \tag{4.32}$$

or

$$\text{Temp } [^{\circ}\text{C}] = 33.476216 - 0.006982 \times \text{Elevation [m]}. \tag{4.33}$$

The corresponding Python code is as follows.

```
#Python linear regression as the first-order polynomial fit
reg = np.polyfit(x, y, 1)
print(reg)
#[-6.98188456e-03   3.34763004e+01]
```

The 24 residuals $e_i, i = 1, 2, \ldots, 24$ are

```
reg = lm(y ~ x)
round(reg$residuals, digits = 5)
#0.25792   1.53427  -0.37129  -0.43697  -0.10542   0.43358
#-0.60335   0.12833  -0.64953  -0.19817   0.10302  -0.63880
#-0.63366  -0.61692  -0.13283   0.63012   0.51454   1.27623
#-0.06333   0.27628  -1.52923   0.39122   1.52971  -1.09572
```

The mean of the 24 residuals is zero, demonstrated by the following R code.

```
mean(reg$residuals)
#[1] 1.62043e-17
```

The least square condition $\mathbf{e} \cdot \mathbf{x}_a = 0$ is demonstrated by the following R code:

```
xa = x - mean(x)
sum(xa*reg$residuals)
#[1] -2.83773e-13
```

The unbiased MSE is

$$s^2 = \frac{1}{n-2} \sum_{i=1}^{n} (y_i - \hat{y}_i)^2. \tag{4.34}$$

Here, we subtract 2 due to the two constraints of estimating a and b. The unbiased MSE s^2 for the Colorado TLR regression can be computed using the following R code.

```
sum((reg$residuals)^2)/(length(y) -2)
#[1] 0.6096193
```

Thus, one can use the data to estimate $\hat{b}$ first using (4.21), then $\hat{a}$ using (4.20), and finally $\hat{\sigma}^2$, also denoted by s^2, using (4.34).

The preceding overall computer code for the Colorado TLR regression is summarized as follows.

```
#R code for the Colorado TLR regression analysis
lm(y ~ x)
#(Intercept)             x
#33.476216      -0.006982

reg = lm(y ~ x)
round(reg$residuals, digits = 5)
mean(reg$residuals)
#[1] 1.62043e-17

xa = x - mean(x)
sum(xa*reg$residuals)
#[1] -2.83773e-13
```

```
#Python code for the Colorado TLR regression analysis
reg = np.polyfit(x, y, 1)
print(reg) #slope and intercept
#[-6.98188456e-03  3.34763004e+01]

regfit = np.polyval(reg,x)
regresiduals = y - regfit
print(np.round(regresiduals,5))
print(np.mean(regresiduals))
#-1.850371707708594e-17 #should be almost zero

xa = np.array(x) - np.mean(x)
np.sum(xa*regresiduals)
print(np.dot(xa, regresiduals)) #orthogonality
#-1.48929757415317e-11 #should be almost zero

np.sum((regresiduals)**2)/(y.size -2)
#0.6096193452251238 #unbiased MSE
```

Once again, we emphasize that two assumptions are used in the derivation of the formulas to estimate $\hat{a}$ and $\hat{b}$:

(i) The unbiased model assumption: The mean residual is zero $\bar{e} = 0$, and

(ii) The optimization assumption: The residual vector **e** is perpendicular to the x-anomaly vector $\mathbf{x}_a = x - \bar{x}$.

Under these two assumptions, the estimators $\hat{a}$ and $\hat{b}$ are highly sensitive to the Y outliers, particularly the outlier data corresponding to the smallest and largest x values. One outlier can completely change the $\hat{a}$ and $\hat{b}$ values, in agreement with our intuition. This is an endpoint problem often encountered in data analysis. To suppress the sensitivity, many robust regression methods have been developed and R packages are available, such as the Robust Regression, by the UCLA Institute for Digital Research & Education:

https://stats.idre.ucla.edu/r/dae/robust-regression/

Figure 4.2 can be generated by the following computer code.

```
#R plot Fig. 4.2: Geometric derivation of the least squares
par(mar=c(0.0,0.5,0.0,0.5))
plot(0,0, xlim=c(0,5.2), ylim=c(0,2.2),
     axes = FALSE, xlab="", ylab="")
arrows(0,0,4,0, angle=5, code=2, lwd=3, length=0.5)
arrows(4,0,4,2, angle=5, code=2, lwd=3, length=0.5)
arrows(0,0,4,2, angle=5, code=2, lwd=3, length=0.5)
arrows(5,0,4,2, angle=7, code=2, lwd=2, lty=3, length=0.5)
arrows(0,0,5,0, angle=7, code=2, lwd=2, lty=3, length=0.5)
arrows(3,0,4,2, angle=7, code=2, lwd=2, lty=3, length=0.5)
arrows(0,0,3,0, angle=7, code=2, lwd=2, lty=3, length=0.5)
segments(3.9,0, 3.9, 0.1)
segments(3.9, 0.1, 4.0, 0.1)
text(2,0.2, expression(hat(b)~bold(x)[a]), cex=2)
text(2,1.2, expression(bold(y)[a]), cex=2)
text(4.1,1, expression(bold(e)), cex=2)
text(3.8,0.6, expression(paste("Shortest ",bold(e))),
     cex=1.5, srt=90)
text(3.4,1.1, expression(paste("Longer ",bold(e))),
     cex=1.5, srt=71)
text(4.6,1.1, expression(paste("Longer ",bold(e))),
     cex=1.5, srt=-71)
```

```
#Python plot Fig. 4.2: Geometry of the least squares
plt.figure(figsize=(10,8))# Define figure size
plt.xlim([0, 5.2]); plt.ylim([0, 2.2])
plt.axis('off')
plt.annotate('', ha = 'center', va = 'bottom',
              xytext = (0, 0),xy = (4,  0),
              arrowprops = {'facecolor' : 'black'})
plt.annotate('', ha = 'center', va = 'bottom',
              xytext = (4, 0),xy = (4,  2),
              arrowprops = {'facecolor' : 'black'})
plt.annotate('', ha = 'center', va = 'bottom',
              xytext = (0, 0),xy = (4,  2),
              arrowprops = {'facecolor' : 'black'})
plt.plot([4, 5], [0,  0], color ='k', linewidth = 12,
         dashes = (3, 1))
```

```
plt.plot([3,4], [0,  2], color ='k', linewidth = 3,
         dashes = (3, 1))
plt.plot([5,4], [0,  2], color ='k', linewidth = 3,
         dashes = (4,1))
plt.plot([0, 3], [0,  0], color ='k', linewidth = 12,
         dashes = (4,1))
plt.plot([3.9, 3.9], [0,  0.1], color ='k')
plt.plot([3.9, 4.0], [0.1,  0.1], color ='k')
plt.text(2, 0.1, r'$\hat{b}~\mathbf{x}_a$', fontsize = 30)
plt.text(2, 1.2, r'$\mathbf{y}_a$', fontsize = 30)
plt.text(4.08, 0.8, r'$\mathbf{e}$', fontsize = 30)
plt.text(3.8, 0.6, r'Shortest_$\mathbf{e}$',
         rotation=90, fontsize = 20)
plt.text(3.3, 1.1, r'Longer_$\mathbf{e}$',
         rotation=71, fontsize = 20)
plt.text(4.3, 1.1, r'Longer_$\mathbf{e}$',
         rotation= -71, fontsize = 20)
plt.show()
```

4.1.2.3 Relationship between Slope and Correlation

Slope $\hat{b}$ and correlation r_{xy} are related in the following way:

$$\hat{b}|\mathbf{x}_a| = r_{xy}|\mathbf{y}_a|. \tag{4.35}$$

This has two extreme cases: $r_{xy} = 1$ and $r_{xy} = 0$.

Case (a): Perfect correlation, $r_{xy} = 1$, when the points on the scatter plot of the x, y data lie exactly on a straight line, which implies perfect prediction:

$$\hat{b}|\mathbf{x}_a| = |\mathbf{y}_a|. \tag{4.36}$$

Case (b): Perfect noise with zero correlation, $r_{xy} = 0$, which implies

$$\hat{b} = 0, \tag{4.37}$$

i.e., the best prediction is the mean.

Formula (4.35) can be derived as follows. The slope estimate formula (4.21) can be rewritten to provide geometric and science interpretations.

(i) Geometric projection interpretation:

$$\hat{b} = \frac{\mathbf{y}_a \cdot (\mathbf{x}_a/|\mathbf{x}_a|)}{|\mathbf{x}_a|}. \tag{4.38}$$

Since $\mathbf{x}_a/|\mathbf{x}_a|$ is a unit vector in the direction of $\mathbf{x}_a$, the slope is the projection of the y anomaly data vector on the x anomaly data vector and then normalized by the x anomaly data vector.

Or we write the $\hat{b}$ formula in the following way:

$$\hat{b} = \frac{\mathbf{y}_a}{|\mathbf{x}_a|} \cdot \frac{\mathbf{x}_a}{|\mathbf{x}_a|}. \tag{4.39}$$

The first fraction on the right-hand side is the y anomaly data vector normalized by $|\mathbf{x}_a|$. Thus, the slope is the projection of this normalized y anomaly data vector onto the unit x anomaly data vector.

(ii) Correlation interpretation:

$$\hat{b} = r_{xy}\frac{|\mathbf{y}_a|}{|\mathbf{x}_a|}, \tag{4.40}$$

where r_{xy} is the correlation coefficient, or simply called the correlation, defined by

$$r_{xy} = \frac{\mathbf{y}_a \cdot \mathbf{x}_a}{|\mathbf{x}_a||\mathbf{y}_a|}. \tag{4.41}$$

The slope depends directly on the correlation coefficient between the x and y anomaly data. The correlation is scaled by the ratio of the length of the y anomaly data vector to the length of the x anomaly data vector.

The following is the computer code corresponding to the two interpretations for the Colorado July TLR example.

```r
#R code for estimating regression slope b

#Method 1: Using vector projection
xa = x - mean(x)  #Anomaly the x data vector
nxa = sqrt(sum(xa^2)) #Norm of the anomaly data vector
ya = y - mean(y)
nya=sqrt(sum(ya^2))
sum(ya*(xa/nxa))/nxa #Compute b
#[1] -0.006981885  #This is an estimate for b

#Method 2:  Using correlation
corxy=cor(xa, ya) #Compute the correlation between xa and ya
corxy
#[1] -0.9813858 #Very high correlation
corxy*nya/nxa #Compute b
#[1] -0.006981885 #This is an estimate for b
```

```python
#Python code for estimating regression slope b

#Method 1: Using vector projection
xa = x - np.mean(x)  #Anomaly the x data vector
nxa = np.sqrt(np.sum(xa**2)) #Norm of the anomaly vector
ya = y - np.mean(y)
nya = np.sqrt(sum(ya**2))
print(np.sum(ya*(xa/nxa))/nxa) #Compute b
#[1] -0.006981885  #This is an estimate for b

#Method 2:  Using correlation
from scipy.stats import pearsonr
```

```
corxy, _ = pearsonr(xa, ya) #Correlation between xa and ya
print(corxy)
#[1] -0.9813858 #Very high correlation
print(corxy*nya/nxa) #Compute b
#[1] -0.006981885 #This is an estimate for b
```

4.1.2.4 Percentage of Variance Explained: $R^2 \times 100\%$

The unbiased variance of the model-predicted data $\hat{y}_i(i = 1, 2, \ldots, n)$ is defined as

$$\mathrm{MV} = \frac{\sum_{i=1}^{n} \left[(\hat{y}_i - \bar{\hat{y}}) \right]^2}{n-1}, \tag{4.42}$$

where $\bar{\hat{y}}$ is the mean of $\hat{y}_i(i = 1, 2, \ldots, n)$.

The unbiased variance of the Y data is

$$\mathrm{YV} = \frac{\sum_{i=1}^{n} \left[(y_i - \bar{y}) \right]^2}{n-1}, \tag{4.43}$$

where $\bar{y}$ is the mean of the original station temperature data $y_i(i = 1, 2, \ldots, n)$.

We wish to measure how good the model is in terms of variance, and thus define the ratio of MV to YV which is named R-squared, i.e., R^2:

$$R^2 = \frac{\mathrm{MV}}{\mathrm{YV}}. \tag{4.44}$$

The value $R^2 \times 100\%$ indicates the percentage of variance explained by the linear model. A larger R^2 value suggests a better model.

The variance of the model-predicted data can be re-written as

$$\begin{aligned}
\mathrm{MV} &= \frac{\sum_{i=1}^{n} \left[(\hat{a} + \hat{b}x_i) - (\hat{a} + \hat{a}\bar{x}) \right]^2}{n-1} \\
&= \hat{b}^2 \frac{\sum_{i=1}^{n} [x_i - \bar{x}]^2}{n-1} \\
&= \left[r_{xy} \frac{|\mathbf{y}_a|}{|\mathbf{x}_a|} \right]^2 \frac{|\mathbf{x}_a|^2}{n-1} \\
&= r_{xy}^2 \frac{|\mathbf{y}_a|^2}{n-1} \\
&= r_{xy}^2 \times \mathrm{YV}. \tag{4.45}
\end{aligned}$$

Therefore,

$$R^2 = r_{xy}^2. \tag{4.46}$$

The R^2 value is equal to the square of the correlation coefficient computed from the data $(x_i, y_i)(i = 1, 2, \ldots, n)$ with $0 \le R^2 \le 1$.

For the Colorado TLR data used earlier, we have $r_{xy} = -0.9813858$, and $R^2 = 0.9631181$.

The variances MV, YV, and R^2 can be computed by the following computer code in multiple ways.

```
#R code for computing MV
var(reg$fitted.values)
#[1] 15.22721
#Or another way
yhat = reg$fitted.values
var(yhat)
#[1] 15.22721
#Or still another way
n = 24
sum((yhat - mean(yhat))^2)/(n-1)
#[1] 15.22721

#R code for computing YV
sum((y - mean(y))^2)/(n-1)
# [1] 15.81033
#Or another way
var(y)
#[1] 15.81033

#R code for computing R-squared value
var(reg$fitted.values)/var(y)
#[1] 0.9631181  #This is the R-squared value

cor(x,y)
#[1] -0.9813858
(cor(x,y))^2
#[1] 0.9631181 #This is the R-squared value
```

```
#Python code for Compute MV, YV, and R^2
from statistics import variance
reg = np.polyfit(x, y, 1)
yhat = np.polyval(reg,x)
#computing MV
print('MV␣=␣', variance(yhat))
#MV =  15.227212262175959
#Or another way
n = 24
print('MV␣=␣',
      np.sum((yhat - np.mean(yhat))**2)/(n-1))
#MV =  15.227212262175959

#computing YV
print('YV␣=', np.sum((y - np.mean(y))**2)/(n-1))
# [1] 15.81033
#Or another way
print('YV␣=', variance(y))
#YV = 15.81032641847826

#computing R-squared value
print('R-squared␣=␣', variance(yhat)/variance(y))
#R-squared =  0.9631181456430407
```

```
#computing correlation and R-squared
from scipy.stats import pearsonr
corxy, _ = pearsonr(x, y)
print('Correlation␣r_xy␣=␣', corxy)
#Correlation r_xy =  -0.9813858291431772
print('R-squared␣=␣', corxy**2)
#R-squared =  0.9631181456430413
```

4.1.2.5 The Regression Estimates by the Least Square or the Maximum Likelihood

A more complicated and commonly used derivation is based on the minimization of the sum of squared errors (SSE):

$$SSE = \sum_{i=1}^{n} e_i^2 = \sum_{i=1}^{n} [y_i - (\hat{a} + \hat{b}x_i)]^2. \tag{4.47}$$

This approach can be found in most statistics books and the Internet. The terminology of "least square" regression comes from this minimization principle.

This minimization is with respect to $\hat{a}$ and $\hat{b}$. The minimization condition leads to two linear equations that determine $\hat{a}$ and $\hat{b}$. The "least square" minimization condition for $\hat{b}$ is equivalent to the orthogonality condition (4.27). Geometrically, the minimum distance of a point to a line is defined as the length of the line segment that is orthogonal to the line and connects the point to the line, i.e., it is the minimum distance between the point and the line (see Fig. 4.2). Thus, the orthogonality condition and minimization condition are equivalent.

Another commonly used method to estimate the parameters is to maximize a likelihood function defined as

$$L(a,b,\sigma) = \prod_{i=1}^{n} \frac{1}{\sqrt{2\pi}\sigma} \exp(-(y_i - (a + bx_i))^2 / (2\sigma^2)). \tag{4.48}$$

The solution for the maximum $L(a,b,\sigma)$ yields the estimate of $\hat{a}$, $\hat{b}$, and $\hat{\sigma}^2$. The result is exactly the same as the ones obtained by the method of least squares or perpendicular projection.

However, the maximum likelihood approach shown here explicitly assumes the normal distribution of the error term ε_i and the response variable Y_i. For the least squares approach, the assumption of normal distribution is needed only when for the inference of the parameters $\hat{a}$, $\hat{b}$, and $\hat{\sigma}^2$, but is not needed to estimate them.

4.1.3 Statistics of Slope and Intercept: Distributions, Confidence Intervals, and Inference

4.1.3.1 Mean and Variance of the Slope

The normal distribution of a and b is now assumed in this section. This assumption was not needed earlier to estimate $\hat{a}$ and $\hat{b}$ and to compute R-squared.

Different datasets will yield different estimates of $\hat{a}$, $\hat{b}$, and $\hat{\sigma}^2$, which form a distribution corresponding to the random datasets. Thus, we may regard $\hat{a}$, $\hat{b}$, and $\hat{\sigma}^2$ as random variables. Their expected values are a, b and σ^2 if the random datasets satisfy the linear model assumptions (linearity, constant variance, independent errors, and normal distribution). However, in practical applications, we just have one dataset, estimate $\hat{a}$, $\hat{b}$, and $\hat{\sigma}^2$ once, and then interpret the results.

Following Eq. (4.21), instead of using data y_i to estimate the slope, we use the corresponding random variables Y_i to define the slope B as a random variable:

$$
\begin{aligned}
B &= \frac{\mathbf{Y}_a \cdot \mathbf{x}_a}{\mathbf{x}_a \cdot \mathbf{x}_a} \\
&= \frac{\sum_{i=1}^n (Y_i - \bar{Y})(x_i - \bar{x})}{\sum_{i=1}^n (x_i - \bar{x})^2} \\
&= \sum_{i=1}^n \left(\frac{x_i - \bar{x}}{\sum_{i=1}^n (x_i - \bar{x})^2} \right) Y_i,
\end{aligned}
\tag{4.49}
$$

because

$$
\sum_{i=1}^n (x_i - \bar{x}) = 0.
\tag{4.50}
$$

Equation (4.49) shows that B is a linear combination of n terms Y_i. If $Y_i \sim N(a+bx, \sigma^2)$, then B is also normally distributed when n is large, because of the central limit theorem.

The expected value of B is b, as shown:

$$
\begin{aligned}
\mathrm{E}[B] &= \sum_{i=1}^n \left(\frac{x_i - \bar{x}}{\sum_{i=1}^n (x_i - \bar{x})^2} \right) E[Y_i] \\
&= \sum_{i=1}^n \left(\frac{x_i - \bar{x}}{\sum_{i=1}^n (x_i - \bar{x})^2} \right) (a+bx_i) \\
&= b \times \sum_{i=1}^n \left(\frac{x_i - \bar{x}}{\sum_{i=1}^n (x_i - \bar{x})^2} \right) x_i \\
&= b \times \left(\frac{\sum_{i=1}^n (x_i - \bar{x})(x_i - \bar{x})}{\sum_{i=1}^n (x_i - \bar{x})^2} \right) \\
&= b.
\end{aligned}
\tag{4.51}
$$

The variance of B is

$$
\mathrm{Var}(B) = \frac{\sigma^2}{\|\mathbf{x}\|_a^2}.
\tag{4.52}
$$

This can be shown as follows:

$$\mathrm{Var}[B] = \sum_{i=1}^{n} \left(\frac{x_i - \bar{x}}{\sum_{i=1}^{n}(x_i - \bar{x})^2} \right)^2 \mathrm{Var}(Y_i)$$

$$= \frac{\sigma^2}{\sum_{i=1}^{n}(x_i - \bar{x})^2}$$

$$= \frac{\sigma^2}{\|\mathbf{x}\|_a^2}. \tag{4.53}$$

Statistics literature often denotes

$$S_{xx} = \|\mathbf{x}\|_a^2. \tag{4.54}$$

Hence,

$$\mathrm{Var}(B) = \frac{\sigma^2}{S_{xx}}. \tag{4.55}$$

When n goes to infinity, S_{xx} also goes to infinity. Thus,

$$\lim_{n \to \infty} \mathrm{Var}(B) = 0. \tag{4.56}$$

This means that more data points yield a better estimate for the slope, supporting our intuition.

4.1.3.2 Mean and Variance of the Intercept

Instead of estimating the intercept using data y_i following Eq. (4.20), we define the intercept as a random variable using the same formula but with random variable Y:

$$A = \bar{Y} - B\bar{x}. \tag{4.57}$$

Here we show that $E[A] = a$:

$$\mathrm{E}[A] = E[\bar{Y}] - E[B]\bar{x}$$

$$= \frac{\sum_{i=1}^{n} Y_i}{n} - b\frac{\sum_{i=1}^{n} x_i}{n}$$

$$= \frac{\sum_{i=1}^{n}(Y_i - bx_i)}{n}$$

$$= \frac{\sum_{i=1}^{n} a}{n}$$

$$= a. \tag{4.58}$$

The variance of A is equal to

$$\mathrm{Var}[A] = \frac{\sigma^2}{n}\left(1 + \frac{\bar{x}^2}{\sigma_x^2}\right), \tag{4.59}$$

where

$$\sigma_x^2 = \frac{S_{xx}}{n} \tag{4.60}$$

may be considered the estimated variance of the x_i data. This result can be derived as follows:

$$\begin{aligned} \text{Var}[A] &= \text{Var}[\bar{Y}] + \text{Var}[B]\bar{x}^2 \\ &= \frac{\sigma^2}{n} + \frac{\sigma^2}{S_{xx}}\bar{x}^2 \\ &= \frac{\sigma^2}{n}\left(1 + \frac{\bar{x}^2}{S_{xx}/n}\right). \end{aligned} \tag{4.61}$$

This expression implies that as n goes to infinity, the variance $\text{Var}[A]$ also goes to zero. Thus, the intercept estimate is better when there are more data points.

4.1.3.3 Confidence Intervals of Slope and Intercept

The $(1-\alpha) \times 100\%$ confidence interval of the slope is

$$\left(\hat{b} - t_{\alpha/2,n-2}\frac{s}{\sqrt{S_{xx}}}, \quad \hat{b} + t_{\alpha/2,n-2}\frac{s}{\sqrt{S_{xx}}}\right), \tag{4.62}$$

where

$$s = \sqrt{\frac{SSE}{n-2}} \tag{4.63}$$

is the unbiased estimator of σ, and the degrees of freedom (dof) of the t-distribution is $n-2$.

If the confidence interval does not include zero, then the slope is significantly different from zero at the $\alpha \times 100\%$ significance level. The Colorado TLR example has $\hat{b} = -0.006982$. For the 95% confidence interval, the R's quantile function qt of the t-distribution can produce the quantile interval corresponding to probabilities 0.025 and 0.975.

```
qt(c(.025, .975), df=22)
#[1] -2.073873  2.073873
```

The estimated standard error for $\hat{b}$ is

$$\text{SE}[\hat{b}] = \frac{s}{\sqrt{S_{xx}}} = 0.0002913, \tag{4.64}$$

based on the R output from summary(reg). Thus, the confidence interval for $\hat{b}$ is

$$(-0.0069818 - 2.073873 \times 0.0002913, -0.0069818 + 2.073873 \times 0.0002913)$$
$$= (-0.0076, -0.0064). \tag{4.65}$$

This confidence interval does not include zero, and hence the slope is significantly less than zero at the 5% confidence level. The 95% confidence interval for TLR is $(6.4, 7.6)°\text{C/km}$ based on the Colorado data used in this chapter.

The $(1-\alpha) \times 100\%$ confidence interval of the intercept is

$$\left(\hat{a} - t_{\alpha/2,n-2}\,s\sqrt{\frac{1}{n} + \frac{\bar{x}^2}{S_{xx}}}, \quad \hat{a} + t_{\alpha/2,n-2}\,s\sqrt{\frac{1}{n} + \frac{\bar{x}^2}{S_{xx}}}\right). \tag{4.66}$$

The R output from `summary(reg)` gives the standard error of $\hat{a}$

$$\text{SE}[\hat{a}] = s\sqrt{\frac{1}{n} + \frac{\bar{x}^2}{S_{xx}}} = 0.5460279 \tag{4.67}$$

Thus, the confidence interval for $\hat{a}$ is

$$(33.4762157 - 2.073873 \times 0.5460279, 33.4762157 + 2.073873 \times 0.5460279)$$
$$= (32.3438, 34.6086). \tag{4.68}$$

The mathematical expressions of the confidence interval can be derived as follows. Formulas (4.51), (4.55), (4.58), and (4.61) imply that

$$A \sim N\left(a, \frac{\sigma^2}{n}\left(1 + \frac{\bar{x}^2}{\sigma_x^2}\right)\right), \tag{4.69}$$

$$B \sim N\left(b, \frac{\sigma^2}{S_{xx}}\right) \tag{4.70}$$

$$\tag{4.71}$$

for a given σ^2.

When σ^2 is unknown and is to be estimated by s^2, then A and B follow a t-distribution. To be exact, the t-statistics with $n - 2$ dof can be defined as

$$T_a = \frac{\hat{a} - a}{s\sqrt{\frac{1}{n} + \frac{\bar{x}^2}{S_{xx}}}}, \tag{4.72}$$

$$T_b = \frac{\hat{b} - b}{s/\sqrt{S_{xx}}}, \tag{4.73}$$

where a and b are given, and $\hat{a}$ and $\hat{b}$ are estimated from data. The confidence interval $(-t_{\alpha/2,n-2}, t_{\alpha/2,n-2})$ for T_a and T_b leads to the confidence intervals for $\hat{a}$ and $\hat{b}$. Thus, these two statistics T_a and T_b can be used for a hypothesis test.

4.1.3.4 Hypothesis Test for the Slope Using the t-Test

We wish to test the hypothesis that the slope is equal to a given constant β. The null and alternative hypotheses are

$$H_0: \quad b = \beta, \tag{4.74}$$

$$H_1: \quad b \neq \beta. \tag{4.75}$$

We define a t-statistic

$$T_b = \frac{\hat{b} - \beta}{s/\sqrt{S_{xx}}}. \tag{4.76}$$

At the $\alpha \times 100\%$ significance level, we reject H_0 if

$$|T_b| > t_{1-\alpha/2,n-2}. \tag{4.77}$$

For the Colorado July TLR example, if we want to test whether the TLR is equal to 7.3°C/km, we calculate

$$T_b = \frac{-0.0069818 - (-0.0073)}{0.0002913} = 1.092345. \tag{4.78}$$

The quantile

$$t_{0.975,22} = 2.073873 \tag{4.79}$$

Thus, $|T_b| < t_{0.975,22}$, and hence the null hypothesis is not rejected at the 5% significance level. In other words, the obtained TLR 7.0°C/km is not significantly different from the given value 7.3°C/km.

4.1.3.5 Confidence Intervals of Prediction: The Fitted Model and the Response Variable

For an assigned point x^*, which is not on the data points x_i, the fitted model prediction is

$$\hat{Y} = \hat{a} + \hat{b}x^*. \tag{4.80}$$

The variance of this model prediction is

$$\mathrm{Var}[\hat{a} + \hat{b}x^*] = \sigma^2 \left[\frac{1}{n} + \frac{(x^* - \bar{x})^2}{S_{xx}} \right]. \tag{4.81}$$

Thus, the $(1 - \alpha) \times 100\%$ confidence interval for $\hat{a} + \hat{b}x^*$ is given by

$$\hat{a} + \hat{b}x^* \pm t_{1-\alpha/2,n-2}\, s \sqrt{\frac{1}{n} + \frac{(x^* - \bar{x})^2}{S_{xx}}}. \tag{4.82}$$

This interval can be computed for any $x = x^*$ and is shown by the red lines in Figure 4.3. It means that, with 95% chance, the model prediction lies between the red lines. Note that the red lines diverge as x^* deviates from $\bar{x}$. This implies that the confidence interval becomes larger as we go out of the range of the data.

The error of the prediction at x^* is

$$e^* = Y - \hat{Y} = Y - (\hat{a} + \hat{b}x^*). \tag{4.83}$$

This is normally distributed if all the assumptions for the linear model are satisfied. The error's mean is zero and variance is

$$\begin{aligned}
\mathrm{Var}[e^*] &= \mathrm{Var}[Y - \hat{a} + \hat{b}x^*] \\
&= \mathrm{Var}[Y] + \mathrm{Var}[\hat{a} + \hat{b}x^*] \\
&= \sigma^2 + \sigma^2 \left[\frac{1}{n} + \frac{(x^* - \bar{x})^2}{S_{xx}} \right].
\end{aligned} \tag{4.84}$$

This result implies that the confidence interval for the response variable Y at an assigned x^* is

$$\hat{a} + \hat{b}x^* \pm t_{1-\alpha/2,n-2}\, s \sqrt{1 + \frac{1}{n} + \frac{(x^* - \bar{x})^2}{S_{xx}}}. \tag{4.85}$$

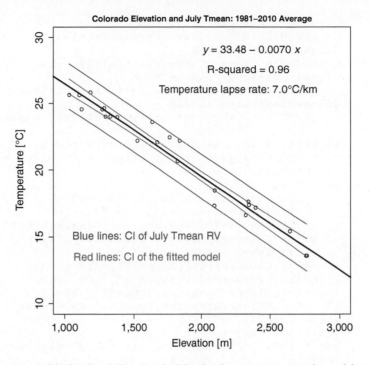

Figure 4.3 The confidence interval of the fitted model based on the Colorado July mean temperature data and their corresponding station elevations (red lines), and confidence interval of the response variable Tmean (blue lines).

The confidence interval is shown by the blue lines in Figure 4.3. It means that, with 95% chance, the Y values are between the blue lines. As expected, almost all the Colorado TLR observed data lie between the blue lines. There is only 5% chance for a data point to lie outside the blue lines. According to Eq. (4.85), the blue lines also diverge, but very slowly because

$$\frac{(x^* - \bar{x})^2}{S_{xx}} < 0.1310$$

is much smaller than $1 + \frac{1}{n} = 1.0417$.

Figure 4.3 can be produced by the following computer code.

```
#R plot Fig. 4.3: Confidence intervals of a regression model
setwd("/Users/sshen/climstats")
#Confidence interval of the linear model
x1 = seq(max(x), min(x),len=100)
n = 24
xbar = mean(x)
reg = lm(y ~ x)
SSE = sum((reg$residuals)^2)
s_squared = SSE/(length(y)-2)
s = sqrt(s_squared)
modTLR = 33.476216 + -0.006982*x1
xbar = mean(x)
Sxx = sum((x-xbar)^2)
```

```
CIupperModel= modTLR +
    qt(.975, df=n-2)*s*sqrt((1/n)+(x1-xbar)^2/Sxx)
CIlowerModel= modTLR -
    qt(.975, df=n-2)*s*sqrt((1/n)+(x1-xbar)^2/Sxx)
CIupperResponse= modTLR +
    qt(.975, df=n-2)*s*sqrt(1+(1/n)+(x1-xbar)^2/Sxx)
CIlowerResponse= modTLR -
    qt(.975, df=n-2)*s*sqrt(1+(1/n)+(x1-xbar)^2/Sxx)

setEPS() #Plot the figure and save the file
postscript("fig0403.eps", height = 8, width = 8)
par(mar=c(4.5,4.5,2.0,0.5))
plot(x,y,
     ylim=c(10,30), xlim=c(1000,3000),
     xlab="Elevation [m]",
     ylab=bquote("Temperature ["~degree~"C]"),
    main="Colorado Elevation and July Tmean: 1981-2010 Average",
     cex.lab=1.5, cex.axis=1.5)
lines(x1,CIupperModel,type="l",col='red')
lines(x1,CIlowerModel,type="l",col='red')
lines(x1,CIupperResponse,type="l",col='blue')
lines(x1,CIlowerResponse,type="l",col='blue')
abline(reg,lwd=3)
text(2280, 26,
bquote("Temperature lapse rate: 7.0"~degree~"C/km"),
cex=1.5)
text(2350, 27.5,"R-squared = 0.96", cex=1.5)
text(2350, 29, "y= 33.48 - 0.0070 x", cex=1.5)
text(1600, 15,"Blue lines: CI of July Tmean RV",
     col="blue", cex=1.5)
text(1600, 13.5,"Red lines: CI of the fitted model",
     col="red", cex=1.5)
dev.off()
```

```
#Python plot Fig. 4.3: Confidence intervals of regression
reg1 = np.array(np.polyfit(x, y, 1))#regression
abline = reg1[1] + x*reg1[0]
# get confidence intervals
yl,yu,xd,rl,ru = linregress_CIs(x,y)
# plot the figure
fig, ax = plt.subplots(figsize=(13,12))
ax.plot(x,y, 'ko');
ax.plot(x, abline, 'k-');
ax.plot(xd, yu, 'r-')
ax.plot(xd, yl, 'r-')
ax.plot(xd, ru, 'b-')
ax.plot(xd, rl, 'b-')
ax.set_title("Colorado Elevation vs. July Tmean \n\
1981 - 2010 Average",size=25, pad = 20,
             fontweight = 'bold')
ax.set_xlabel("Elevation $[m]$", size = 25, labelpad = 20)
ax.set_ylabel(r"Temperature [$\degree$C]",
              size = 25, labelpad = 20);
```

```
ax.tick_params(length=6, width=2, labelsize=20);
ax.set_xticks(np.round(np.linspace(1000, 3000, 5), 2))
ax.set_yticks(np.round(np.linspace(10, 30, 5), 2))
ax.text(1600, 25, r"Temp␣lapse␣rate:␣7.0$\degree$C/km",
        color= 'k', size = 20)
ax.text(1750, 27,
        r"$y␣=␣%1.2f␣%1.4f␣x$"%(reg1[1], reg1[0]),
        size = 20)
ax.text(1800, 26, r"$R-squared␣=␣%.2f$" % Rsqu,
        color= 'k', size = 20)
ax.text(1050, 15, "Blue␣lines:␣CI␣of␣July␣Tmean␣RV",
        color= 'b', size = 20)
ax.text(1050, 14, "Red␣lines:␣CI␣of␣the␣fitted␣model",
        color= 'r', size = 20)
plt.show()
```

4.1.3.6 When the Assumptions for the Linear Regression Model Are Violated

The computing and plotting of a linear regression is quite straightforward, and has been used widely in data analysis. However, users often are unaware of the assumptions behind the computing results. You may wish to know what the assumptions are and how they are violated.

(a) Review of the assumptions of linear regression

The linear regression model and its inference have four major assumptions:

(i) Approximate linearity: The data support an approximate linear relationship between x and Y;

(ii) Normality: Both the model error term ε and Y are normally distributed;

(iii) Constant variance: ε and Y have variance σ^2, which does not change with respect to x;

(iv) Independence of the error term: $\text{Cor}[\varepsilon_i, \varepsilon_j] = \delta_{ij}$.

Although the derivation of the estimation formulas for $\hat{a}$ and $\hat{b}$ does not need any of these assumptions, the interpretation and inferences on $\hat{a}$, $\hat{b}$, $\hat{a} + \hat{b}x$, and other quantities are often based on these assumptions. If one or more assumptions are violated, the inference may not work. There are a variety of checking methods and resolution methods in literature, such as using a nonlinear relationship (e.g., a polynomial fitting model). R has a package lmtest to test linear regression models. A residual scatter plot (x_i, e_i) may be a good first step to intuitively identify possible violations of the assumptions (see Fig. 4.4). Ideally, the plot shows uniform noise for a good model. Figure 4.4 shows the residuals of the Colorado temperature regression against elevation, and seems showing this kind of noise without an obvious pattern. We may safely conclude that the assumptions of linearity and constant variance are satisfied.

To be rigorous on statistical inferences about the results of a linear regression, a comprehensive residual analysis may be made to assess whether the studied dataset satisfies the

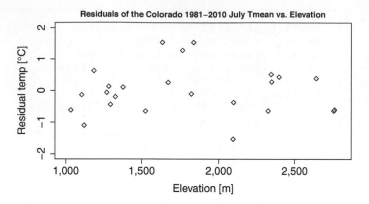

Figure 4.4 Scatter plot of the residuals of the linear regression for the Colorado July mean temperature data against the elevations. The data are from the 24 USHCN stations in Colorado, and the July Tmean is the 1981–2010 average temperature.

linear regression assumptions. If some assumptions are violated for a given dataset, we can seek other methods to remediate the problems. One way is to make a data transformation or to use a new regression function so that the assumptions are satisfied for the transformed data or the new functional model. Another way is to use a nonparametric method, which does not assume any distribution. The comprehensive residual analysis methods and the corresponding R and Python codes can be found online: this book is limited to the discussion of a few examples.

Violation of independence assumption (iv) often occurs when y is a time series, in which case there could be correlations from one time to later times. This serial correlation effect leads to fewer independent time intervals, i.e., the dof can be reduced, and the statistical inference needs special attention.

```
#R plot Fig. 4.4: Regression residuals
reg = lm(y~x)
setEPS() #Plot the figure and save the file
postscript("fig0404.eps", height = 4.5, width = 8)
par(mar=c(4.5,4.5,2.0,0.5))
plot(x, reg$residuals, pch=5,
     ylim=c(-2,2), xlim=c(1000,2800),
     xlab="Elevation [m]",
     ylab=bquote("Residual Temp ["~degree~"C]"),
     main="Residuals of the Colorado July Tmean vs. Elevation",
     cex.lab=1.5, cex.axis=1.5, cex.main = 1.2)
dev.off()
```

```
#Python plot Fig. 4.4: Regression residuals
reg1 = np.array(np.polyfit(x, y, 1))#regression
abline = reg1[1] + x*reg1[0]
# calculate residuals
```

```
r = y - abline
fig, ax = plt.subplots(figsize=(12,8))
ax.plot(x, r, 'kd')
ax.set_title("Residuals␣of␣the␣Colorado␣1981␣-␣2010␣July␣\n\
Tmean␣vs.␣Elevation",fontweight = 'bold', size=25, pad = 20)
ax.set_xlabel("Elevation␣$[m]$", size = 25, labelpad = 20)
ax.set_ylabel(r"Residual␣Temp␣[$\degree$C]",
              size = 25, labelpad = 20);
ax.tick_params(length=6, width=2, labelsize=20);
ax.set_xticks(np.linspace(1000, 3000, 5))
ax.set_yticks(np.linspace(-2,2,5))
plt.show()
```

(b) Test for normality

The normality and independence assumptions may not show in the residual scatter plot. You may use a Q-Q plot to test normality, and the Durbin–Watson test to check independence. In the Colorado TLR example, the Q-Q plot is shown in Figure 4.5. The residual data quantile points overlap fairly well with the theoretical line for the normal distribution. We may conclude that the data satisfy the normality assumption.

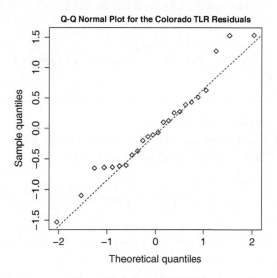

Figure 4.5 Q-Q plot of the linear regression residuals for Colorado July mean temperature data at the 24 USHCN stations. The diagonal straight line is the theoretical line for the standard normal distribution.

However, the Q-Q plot is only a subjective visual test. A quantitative test may be used, such as the Kolmogorov–Smirnov (KS) test or Shapiro–Wilk (SW) test. The KS-test statistic and its p-value for the Colorado TLR data are $D = 0.20833$ and p-value $= 0.686$. The null hypothesis is not rejected, i.e., the temperature residuals from the linear regression are not significantly different from the normal distribution. The normality assumption is valid.

Figure 4.5 can generated by the following computer code.

```
#R plot Fig. 4.5: Q-Q plot of quantiles
reg = lm(y~x)
setEPS() #Plot the figure and save the file
postscript("fig0405.eps", height = 6, width = 6)
par(mar=c(4.5,4.5,2.0,0.5))
qqnorm(reg$residuals, pch=5,
       main="QQ-Normal␣Plot␣for␣the␣Colorado␣TLR␣Residuals",
       cex.lab = 1.4, cex.axis = 1.4)
qqline(reg$residuals, lty=2)
dev.off()
```

```
#Python plot Fig. 4.5: Q-Q plot of quantiles
fig, ax = plt.subplots(figsize=(12,8))#fig setup
reg1 = np.array(np.polyfit(x, y, 1))#regression
abline = reg1[1] + x*reg1[0]
r = y - abline# calculate residuals
#Q-Q plot as a probability plot: quantiles vs quantiles
pp1 = scistats.probplot(r, dist="norm", plot=ax,)
ax.set_title("QQ-Normal␣Plot␣for␣Colorado␣TLR␣Residuals",
             fontweight = 'bold',size = 25, pad = 20);
ax.set_ylabel("Sample␣Quantiles", size = 25, labelpad = 20);
ax.set_xlabel("Theoretical␣Quantiles",
              size = 25, labelpad = 20);
ax.tick_params(length=6, width=2, labelsize=20);
ax.set_xticks(np.round(np.linspace(-2, 2, 5), 2))
plt.show()
```

(c) Serial correlation

The R package `lmtest` has a function to test for independence of the error term, using the Durbin–Watson (DW) test statistic defined as

$$DW = \frac{\sum_{i=2}^{n}(e_i - e_{i-1})^2}{\sum_{i=1}^{n} e_i^2}. \tag{4.86}$$

The DW test is meant for a time series. We thus sort the Tmean data according to the ascending order of elevation, and then compute DW.

When $\mathrm{cor}(e_i, e_j) = \sigma^2 \delta_{ij}$, we have $DW \approx 2$. For the Colorado July TLR data, $DW = 2.3072$ and $p - value = 0.7062$, which can be computed by the following code.

```
#R code for the DW-test for independence
#install.packages("lmtest")
library(lmtest)
ElevTemp=cbind(x,y, 1:24)
#Sort the data for ascending elevation
ElevTemp=ElevTemp[order(x),]
reg1=lm(ElevTemp[,2] ~ ElevTemp[,1])
dwtest(reg1)
#DW = 2.3072, p-value = 0.7062
```

```
#Python code for the DW test for independence
from statsmodels.stats.stattools import durbin_watson
orderET = np.arange(1,25)
ElevTemp = np.stack((x,y, orderET), axis = -1)
sorted_ind = np.argsort(ElevTemp[:,0],kind='mergesort')
dat1 = ElevTemp[sorted_ind]
# use the first-order polynomial fit for linear regression
reg1 = np.array(np.polyfit(dat1[:,0], dat1[:,1], 1))
abline = reg1[1] + dat1[:,0]*reg1[0]
r = dat1[:,1] - abline# calculate residuals
#perform Durbin-Watson test
durbin_watson(r)
#2.307190038542777
```

The large p-value 0.7062 implies that the null hypothesis of no serial correlation (i.e., independence) is not rejected, because DW 2.3072 is not too far away from 2, which indicates independence of the error terms. Some literatures use $1.5 < \text{DW} < 2.5$ to conclude independence without checking the p-value.

When the independence assumption is violated, the data have serial correlation, which reduces the sample size and hence enlarges the confidence interval for the regression results and makes the results not as reliable.

(d) Nonparametric trend inference and heteroscedasticity

Even when some of the four assumptions for the linear regression model are invalid, one can still estimate $\hat{a}$ and $\hat{b}$ and the linear model $\hat{a}+\hat{b}x$ in the same way and can still interpret the study subject using the domain knowledge. In this case, the statistical inference based on the normality assumption may not make any sense, and the confidence interval calculations based on the t-distribution are generally invalid. Now, you may use a nonparametric test to make an inference on the residuals. For example, a nonparametric test for the trend can be applied, such as Mann–Kendall (MK) trend test, and Theil–Sen's trend estimation and test. The trend, if presenting, can be linear or nonlinear. These tests are meant for time series. Thus, before applying these tests, the data should be sorted according to the ascending order of the explainable variable. For the Colorado TLR example, the MK test for the linear regression residual can be computed by the following code.

```
#R code for the Mann-Kendall test
#install.packages("trend")
library(trend)
ElevTemp=cbind(x, y, 1:24)
#Sort the data for ascending elevation
ElevTemp=ElevTemp[order(x),]
reg1=lm(ElevTemp[,2] ~ ElevTemp[,1])
ElevTemp[,3]=reg1$residuals
mk.test(ElevTemp[,3])
#data:  ElevTemp[, 3]
#z = 0.47128, n = 24, p-value = 0.6374
mk.test(ElevTemp[,2])
#z = -5.9779, n = 24, p-value = 2.261e-09
```

```
#Python code for the Mann-Kendall test
import pymannkendall as mk
orderET = np.arange(1,25)
ElevTemp = np.stack((x,y, orderET), axis = -1)
sorted_ind = np.argsort(ElevTemp[:,0],kind='mergesort')
dat1 = ElevTemp[sorted_ind]
# use the first order polynomial fit for linear regression
reg1 = np.array(np.polyfit(dat1[:,0], dat1[:,1], 1))
abline = reg1[1] + dat1[:,0]*reg1[0]
r = dat1[:,1] - abline# calculate residuals
#perform Durbin-Watson test
dat1[:,2] = r
print(mk.original_test(dat1[:,2]))#test for residual trend
#p=0.6374381847429542, z=0.47128365713220055
print(mk.original_test(dat1[:,1]))#test for temp trend
#p=2.260863496417187e-09, z=-5.97786112467686
```

The large p-value 0.6374 of the MK test for residuals implies that the residuals have no trend, linear or nonlinear. The small p-value $2.261\text{e-}09$ of the MK test for the sorted temperature implies that the sorted July Tmean according to elevation has a significant trend.

The linear trend is frequently used to explain the global average temperature. The temperature data with respect to time is clearly nonlinear. When the scatter plot shows a clear pattern, then one or more of the four assumptions are usually violated. When that happens, there are different ways to remediate the situation. For example, when linearity is violated, the problem may be intrinsically nonlinear and a polynomial model may be a better fit than a linear model. When the constant variance is violated, known as heteroscedastic data in statistics literature, then the data standardization (i.e., the anomaly data being divided by the data standard deviation) or logarithmic transform for the positive-valued data, may be used to transform the data and to make the data homoskedastic, i.e., constant variance.

4.2 Multiple Linear Regression

This section uses the method of matrices, which will be described in Chapter 5. If you are not familiar with matrices, you may come back to this section after reading the first two sections of Chapter 5.

4.2.1 Calculating the Colorado TLR When Taking Location Coordinates into Account

In the previous simple linear regression, the July temperature was assumed to linearly depend only on the vertical coordinate of the station: elevation. However, the temperature may also depend on the horizontal coordinates of the station: latitude and longitude. This subsection deals with the problem of more than one explanatory variable. A multivariate linear regression model (or multiple linear regression model) can be expressed as follows:

$$Y = b_0 + b_1 x_1 + b_2 x_2 + \cdots + b_m x_m + \varepsilon, \tag{4.87}$$

where $x_1, x_2, \ldots, x_m$ are m explanatory variables, which are nonrandom and deterministic variables; ε is the model error with zero mean and variance σ^2 being a constant; Y is the response variable, which is random and has its variance equal to σ^2, and expected value equal to

$$E[Y] = b_0 + b_1 x_1 + b_2 x_2 + \cdots + b_m x_m; \tag{4.88}$$

and $b_0, b_1, b_2, \ldots, b_m$ are parameters to be estimated from data $(x_{ij}, y_j), i = 1, 2, \ldots, m$, and $j = 1, 2, \ldots, n$.

In the Colorado July TLR example, we have x_1 as latitude, x_2 as longitude, x_3 as elevation, Y as the July air temperature, $m = 3$, and $n = 24$. The latitude and longitude data are as follows:

```
lat=c(
39.9919, 38.4600, 39.2203, 38.8236, 39.2425, 37.6742,
39.6261, 38.4775, 40.6147, 40.2600, 39.1653, 38.5258,
37.7717, 38.0494, 38.0936, 38.0636, 37.1742, 38.4858,
8.0392, 38.0858, 40.4883, 37.9492, 37.1786, 40.0583
)
lon=c(
-105.2667, -105.2256, -105.2783, -102.3486, -107.9631, -106.3247,
-106.0353, -102.7808, -105.1314, -103.8156, -108.7331, -106.9675,
-107.1097, -102.1236, -102.6306, -103.2153, -105.9392, -107.8792,
-103.6933, -106.1444, -106.8233, -107.8733, -104.4869, -102.2189
)
```

The elevation and temperature data are the same as those at the beginning of this chapter.

The computer code for the multiple linear regression of three variables is as follows.

```
#R code for the TLR multivariate linear regression
elev = x; temp = y #The x and y data were entered earlier
dat = cbind(lat, lon, elev, temp)
datdf = data.frame(dat)
datdf[1:2,] #Show the data of the first two stations
#      lat        lon     elev      temp
# 39.9919 -105.2667 1671.5 22.064
# 38.4600 -105.2256 1635.6 23.591

#Multivariate linear regression
reg=lm(temp ~ lat + lon + elev, data = datdf)
summary(reg)   #Display the regression results
#               Estimate   Std. Error   t value  Pr(>|t|)
#(Intercept) 36.4399561  9.4355746    3.862 0.000971 ***
#  lat        -0.4925051  0.1320096   -3.731 0.001319 **
#  lon        -0.1630799  0.0889159   -1.834 0.081564 .
#  elev       -0.0075693  0.0003298  -22.953 7.67e-16 ***
#Residual standard error: 0.6176 on 20 degrees of freedom
#Multiple R-squared:  0.979
```

```
#Python code for the TLR multivariate linear regression
from sklearn import linear_model
lat=np.array([
39.9919, 38.4600, 39.2203, 38.8236, 39.2425, 37.6742,
39.6261, 38.4775, 40.6147, 40.2600, 39.1653, 38.5258,
37.7717, 38.0494, 38.0936, 38.0636, 37.1742, 38.4858,
8.0392, 38.0858, 40.4883, 37.9492, 37.1786, 40.0583
])
lon=np.array([
-105.2667,-105.2256,-105.2783,-102.3486,-107.9631,-106.3247,
-106.0353,-102.7808,-105.1314,-103.8156,-108.7331,-106.9675,
-107.1097,-102.1236,-102.6306,-103.2153,-105.9392,-107.8792,
-103.6933,-106.1444,-106.8233,-107.8733,-104.4869,-102.2189
])
elev = x; temp = y #The x and y data were entered earlier
dat = np.stack((lat, lon, elev), axis = -1)
print(dat[0:2,:]) #Show the data of the first two stations
#[[  39.9919 -105.2667 1671.5    ]
# [  38.46   -105.2256 1635.6    ]]
#Multivariate linear regression
xdat = dat
ydat = temp
regr = linear_model.LinearRegression()
multi_reg = regr.fit(xdat, ydat)
print('Intercept:␣\n', regr.intercept_)
print('Coefficients:␣\n', regr.coef_)
```

According to this three-variable linear regression, the TLR is the regression coefficient for elevation, i.e., -0.0075694. This means $7.6°$C/km, larger than the TLR estimated earlier using only one variable: elevation.

The quantile of $t_{0.975,20} = 2.085963$. The dof is 20 now because four parameters have been estimated. The confidence interval of TLR can be computed as follows:

$$(-0.0075694 - 2.085963 \times 0.0003298, -0.0075694 + 2.085963 \times 0.0003298)$$

$$= (-0.008257351, -0.006881449). \tag{4.89}$$

The confidence interval for TLR at the 95% confidence level is $(6.9, 8.3)°$C/km.

The R^2 value is very large: 0.979, which means that the linear model data can explain 98% of the variance of observed temperature data.

4.2.2 Formulas for Estimating Parameters in the Multiple Linear Regression

The n groups of data for the explainable $x_1, x_2, \ldots, x_m$ and response variable Y are as follows:

$$x_{ij}, i = 1, 2, \ldots, m, \text{ and } j = 1, 2, \ldots, n, \tag{4.90}$$

and

$$y_1, y_2, \ldots, y_n. \tag{4.91}$$

The data and their corresponding regression coefficients are written in matrix form as follows:

$$\mathbf{y} = \begin{bmatrix} y_1 \\ y_2 \\ \vdots \\ y_n \end{bmatrix} \tag{4.92}$$

$$\mathbf{X} = \begin{bmatrix} 1 & x_{11} & x_{21} & \cdots & x_{m1} \\ 1 & x_{12} & x_{22} & \cdots & x_{m2} \\ \vdots & \vdots & \vdots & \vdots & \vdots \\ 1 & x_{1n} & x_{2n} & \cdots & x_{mn} \end{bmatrix} \tag{4.93}$$

$$\hat{\mathbf{b}} = \begin{bmatrix} \hat{b}_0 \\ \hat{b}_1 \\ \hat{b}_2 \\ \vdots \\ \hat{b}_m \end{bmatrix}. \tag{4.94}$$

The data, linear model prediction, and the corresponding residuals (i.e., the prediction errors of the linear model) can be written as follows:

$$\mathbf{y}_{n\times 1} = \mathbf{X}_{n\times(m+1)}\hat{\mathbf{b}}_{(m+1)\times 1} + \mathbf{e}_{n\times 1} \tag{4.95}$$

where the residual vector is

$$\mathbf{e}_{n\times 1} = \begin{bmatrix} e_1 \\ e_2 \\ \vdots \\ e_n \end{bmatrix}. \tag{4.96}$$

Multiplying Eq. (4.95) by $\mathbf{X}^t$ from the left and enforcing

$$(\mathbf{X}^t)_{(m+1)\times n}\mathbf{e}_{n\times 1} = \mathbf{0}, \tag{4.97}$$

you can obtain the estimate of the regression coefficients

$$\hat{\mathbf{b}}_{(m+1)\times 1} = \left[(\mathbf{X}^t)_{(m+1)\times n}\mathbf{X}_{n\times(m+1)}\right]^{-1}(\mathbf{X}^t)_{(m+1)\times n}\mathbf{y}_{n\times 1}. \tag{4.98}$$

Condition (4.97) means each column vector of the $\mathbf{X}_{n\times(m+1)}$ matrix is perpendicular to the residual vector $\mathbf{e}_{n\times 1}$. For the first column, the condition corresponds to the assumption of zero mean of the model error:

$$\sum_{i=1}^{n} e_i = 0. \tag{4.99}$$

For the remaining columns, the condition means that the residual vector is perpendicular to the data vector for each explanatory variable. This implies that the residual vectors' Euclidean distances to the data vectors are minimized, i.e., the minimizing sum of squared errors (SSE). Therefore, Eq. (4.98) is the least square estimate of the regression coefficients.

As illustrated in the previous subsection, to implement the R estimate of the regression coefficients, we put data $\mathbf{X}$ and $\mathbf{y}$ in a single matrix in the form of `datdf=data.frame(cbind(X,y))` with proper column names, such as
`colnames(datdf) <- c('x1', 'x2', 'x3', 'y')`
for the case of three explanatory variables. Then, use the R command to make the linear model estimate.

```
reg=lm(y ~ x1 + x2 + x3, data = datdf)
```

Finally, `summary(reg)` outputs all the important regression results.

The R command `reg$` allows to display the specific result, such as regression coefficients.

```
round(reg$coefficients, digits=5)
# (Intercept)          lat          lon         elev
#     36.43996    -0.49251     -0.16308     -0.00757
```

Thus, the multiple linear model for the Colorado July temperature is

$$\text{Temp}[^\circ\text{C}] = 36.43527 - 0.49255 \times \text{Lat} - 0.16314 \times \text{Lon} - 0.00757 \times \text{Elev [m]}. \tag{4.100}$$

The confidence interval for the linear model

$$\hat{Y} = b_0 + b_1 x_1 + b_2 x_2 + \cdots + b_m x_m \tag{4.101}$$

at a given point $\mathbf{x}^* = (1, x_1^*, x_2^*, \cdots, x_m^*)$ is given by the following formula:

$$\hat{\mathbf{b}}^t \mathbf{x}^* \pm t_{1-\alpha/2, n-m}\, s\sqrt{(\mathbf{x}^*)^t (\mathbf{X}^t \mathbf{X})^{-1} \mathbf{x}^*}. \tag{4.102}$$

The confidence interval for the response variable Y at a given point

$$\mathbf{x}^* = (1, x_1^*, x_2^*, \cdots, x_m^*)$$

is

$$\hat{\mathbf{b}}^t \mathbf{x}^* \pm t_{1-\alpha/2, n-m}\, s\sqrt{1 + (\mathbf{x}^*)^t (\mathbf{X}^t \mathbf{X})^{-1} \mathbf{x}^*}, \tag{4.103}$$

where s is as defined earlier for the simple linear regression:

$$s = \sqrt{\frac{SSE}{n-2}}. \tag{4.104}$$

If $m = 1$, (4.102) and (4.103) are the same as those for the confidence intervals given by Eqs. (4.82) and (4.85) for the simple linear regression.

4.3 Nonlinear Fittings Using the Multiple Linear Regression

4.3.1 Diagnostics of Linear Regression: An Example of Global Temperature

When the data have strong nonlinearity, the scatter plot of residuals will show an obvious pattern, as shown in Figure 4.6 for the linear regression of the global average annual mean

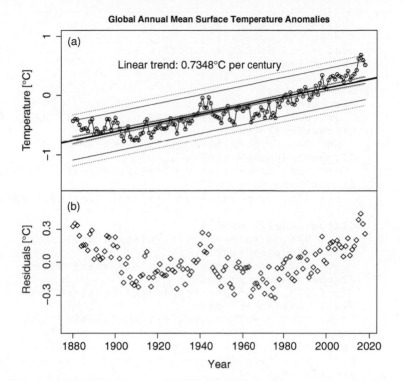

Figure 4.6 (a) Linear regression of the global average annual mean land and ocean surface air temperature anomalies with respect to the 1971–2000 climatology based on the NOAAGlobalTemp dataset (Zhang et al. 2019); (b) Scatter plot of the linear regression residuals.

surface air temperature anomalies from 1880 to 2018 with respect to the 1971–2000 climatology. The linearity assumption of the simple linear regression is clearly violated. The independence assumption is also violated.

The following computer code can plot Figure 4.6 and provide diagnostics of the linear regression.

```
#R plot Fig. 4.6: Regression diagnostics
setwd("/Users/sshen/climstats")
dtmean<-read.table(
  "data/aravg.ann.land_ocean.90S.90N.v5.0.0.201909.txt",
  header=F)
dim(dtmean)
#[1] 140    6
x = dtmean[1:139,1]
y = dtmean[1:139,2]
reg =  lm(y ~ x) #linear regression
reg
#(Intercept)        yrtime
#-14.574841     0.007348

#Confidence interval of the linear model
xbar = mean(x)
SSE = sum((reg$residuals)^2)
```

```
s_squared = SSE/(length(y)-2)
s = sqrt(s_squared)
modT = -14.574841 + 0.007348 *x
xbar = mean(x)
Sxx = sum((x-xbar)^2)
n = length(y)
CIupperModel= modT +
  qt(.975, df=n-2)*s*sqrt((1/n)+(x-xbar)^2/Sxx)
CIlowerModel= modT -
  qt(.975, df=n-2)*s*sqrt((1/n)+(x-xbar)^2/Sxx)
CIupperResponse= modT +
  qt(.975, df=n-2)*s*sqrt(1+(1/n)+(x-xbar)^2/Sxx)
CIlowerResponse= modT -
  qt(.975, df=n-2)*s*sqrt(1+(1/n)+(x-xbar)^2/Sxx)

CIupperModelr= modT +
  qt(.975, df=5)*s*sqrt((1/n)+(x-xbar)^2/Sxx)
CIlowerModelr= modT -
  qt(.975, df=5)*s*sqrt((1/n)+(x-xbar)^2/Sxx)
CIupperResponser= modT +
  qt(.975, df=5)*s*sqrt(1+(1/n)+(x-xbar)^2/Sxx)
CIlowerResponser= modT -
  qt(.975, df=5)*s*sqrt(1+(1/n)+(x-xbar)^2/Sxx)

setEPS() #Plot the figure and save the file
postscript("fig0406.eps", height = 8, width = 8)
par(mfrow=c(2,1))
par(mar=c(0,4.5,2.5,0.7))
plot(x, y,  ylim = c(-1.5, 1),
     type="o", xaxt="n", yaxt="n",
     cex.lab=1.4, cex.axis=1.4,
     xlab="Year", ylab=bquote("Temperature [~degree~"C]"),
     main="Global Annual Mean Surface Temperature Anomalies",
     cex.lab=1.4, cex.axis=1.4
)
axis(side = 2, at = c(-1.0, 0, 1.0), cex.axis = 1.4)
abline(reg, col="black", lwd=3)
lines(x,CIupperModel,type="l",col='red')
lines(x,CIlowerModel,type="l",col='red')
lines(x,CIupperResponse,type="l",col='blue')
lines(x,CIlowerResponse,type="l",col='blue')

lines(x,CIupperModelr,type="l", lty = 3, col='red')
lines(x,CIlowerModelr,type="l", lty = 3, col='red')
lines(x,CIupperResponser,type="l",lty = 3, col='blue')
lines(x,CIlowerResponser,type="l",lty = 3, col='blue')

text(1940, 0.5,
     bquote("Linear trend: 0.7348"~degree~"C per century"),
     col="black",cex=1.4)
text(1880, 0.9, "(a)", cex=1.4)
par(mar=c(4.5,4.5,0,0.7))
plot(x, reg$residuals, ylim = c(-0.6,0.6),
     pch=5, cex.lab=1.4, cex.axis=1.4,
     yaxt = 'n', xlab="Year",
     ylab=bquote("Residuals [~degree~"C]"))
```

```
axis(side = 2, at = c(-0.3, 0, 0.3), cex.axis = 1.4)
text(1880, 0.5, "(b)", cex=1.4)
dev.off()
```

```
#Python plot Fig. 4.6: Regression diagnostics
# Change current working directory
os.chdir("/Users/sshen/climstats/")
# Read the data
dtmean = np.array(read_table(
    "data/aravg.ann.land_ocean.90S.90N.v5.0.0.201909.txt",
                header = None, delimiter = "\s+"))
xDT = dtmean[0:139, 0]
yDT = dtmean[0:139, 1]
# Make a linear fit, i.e., linear regression
regLin = np.array(np.polyfit(xDT, yDT, 1))
ablinereg = regLin[1] + xDT*regLin[0]
yld,yud,xdd,rld,rud = linregress_CIs(xDT,yDT)

# Plot the figure
fig, ax = plt.subplots(2, 1, figsize=(12,12))
ax[0].plot(xDT, yDT, 'ko-')
ax[0].plot(xDT, ablinereg, 'k-')
ax[0].plot(xdd, yld, 'r-')
ax[0].plot(xdd, yud, 'r-')
ax[0].plot(xdd, rld, 'b-')
ax[0].plot(xdd, rud, 'b-')
ax[1].plot(xDT, yDT - ablinereg, 'kd')
ax[0].set_title(
    "Global␣Annual␣Mean␣Land␣and␣Ocean␣Surface␣\n\
Temperature␣Anomalies", fontweight = 'bold',
                size = 25, pad = 20)
ax[0].set_ylabel("Temperature␣[$\degree$C]",
                size = 25, labelpad = 20)
ax[0].tick_params(length=6, width=2, labelsize=20);
ax[0].set_yticks(np.round(np.linspace(-1, 1, 5), 2))
ax[0].text(1880, 0.4,
        "Linear␣temp␣trend␣0.7348␣deg␣C␣per␣century",
            color = 'k', size = 20)
ax[0].text(1877, 0.6, "(a)", size = 20)
ax[0].axes.get_xaxis().set_visible(False)
ax[1].tick_params(length=6, width=2, labelsize=20);
ax[1].set_yticks(np.round(np.linspace(-0.5, 0.5, 5), 2))
ax[1].set_ylabel("Residuals␣[$\degree$C]",
                size = 25, labelpad = 20)
ax[1].text(1877, 0.4, "(b)", size = 20)
ax[1].set_xlabel("Year", size = 25, labelpad = 20)
fig.tight_layout(pad=-1.5)
plt.show()
```

Based on the visual check of Figure 4.6(b), the constant variance assumption seems satisfied. The KS test shows that the normality assumption is also satisfied.

```
#Kolmogorov-Smirnov (KS) test for normality
library(fitdistrplus)
```

```
resi_mean = mean(reg$residuals)
resi_sd = sd(reg$residuals)
test_norm = rnorm(length(reg$residuals),
                  mean = 0, sd = 1)
testvar = (reg$residuals - resi_mean)/resi_sd
ks.test(testvar, test_norm)
#D = 0.057554, p-value = 0.9754
#The normality assumption is accepted

#Diagnostics on independence and normality
# Durbin-Watson (DW) test for independence
dwtest(reg)
#DW = 0.45235, p-value < 2.2e-16
#The independence assumption is rejected

#degrees of freedom and critical t values
rho1 = acf(y)[[1]][2] #Auto-correlation function
rho1 #[1] 0.9270817
edof = (length(y) - 2)*(1 - rho1)/(1 + rho1)
edof #[1] 5.183904 effective degrees of freedom
qt(.975, df=137) #[1] 1.977431 critical t value
qt(.975, df=5) #[1] 2.570582 critical t value
```

```
#Kolmogorov-Smirnov (KS) test for normality
import statistics
from scipy.stats import kstest
resi = yDT - ablinereg
testvar = (resi - np.mean(resi))/statistics.stdev(resi)
kstest(testvar, 'norm')#
#KstestResult(statistic=0.0751, pvalue=0.3971)
#The normality assumption is accepted

#perform Durbin-Watson test for independence
from statsmodels.stats.stattools import durbin_watson
durbin_watson(resi)
#0.45234915992769864 not in (1.5, 2.5)
#The independence assumption is rejected

#calculate autocorrelations of temp data for edof
import statsmodels.api as sm
autocorr = sm.tsa.acf(yDT)
rho1 = autocorr[1]
edof = (yDT.size - 2)*(1 - rho1)/(1 + rho1)
edof #[1] 5.183904 effective degrees of freedom

#Compare the critical t values with different edof
import scipy.stats
scipy.stats.t.ppf(q=.975, df=137)
#critical t value 1.9774312122928936
scipy.stats.t.ppf(q=.975, df=5)
##critical t value 2.5705818366147395
```

The DW test shows that the independence assumption is violated. This implies the existence of serial correlation, which in turn implies a smaller dof, and hence larger $t_{1-\alpha,\text{dof}}$. Thus, the confidence intervals for model $\hat{Y}$ are wider than the red lines in Figure 4.6(a), and those for the Y values are wider than the blue lines in the same figure. Therefore, the regression results are less reliable. In this case, you need to compute the effective dof (edof), which is less than $n-2$. For the NOAAGlobalTemp time series, the one-year time lag serial correlation is $\rho_1 = 0.9271$, and the edof under the assumption of an autoregression one process AR(1) is approximately equal to

$$\text{edof} = \frac{1-\rho_1}{1+\rho_1} \times \text{dof}. \tag{4.105}$$

Consequently, the edof is reduced to 5, compared to the nonserial correlation case dof 137. R can compute qt(.975, df=137) and yield 1.977431, and compute qt(.975, df=5) and yield 2.570582. Thus, the actual confidence interval at the 95% confidence level for the linear regression is about 25% wider. Namely, the dotted color lines are wider than the solid color lines.

The W pattern of the residuals shown in Figure 4.6(b) implies that the linearity assumption is violated. To remediate the nonlinearity, we fit the data to a nonlinear model and hope the residuals do not show clear patterns. We will use the third-order polynomial to illustrate the procedure.

4.3.2 Fit a Third-Order Polynomial

Figure 4.7(a) shows the third-order polynomial fitting to the NOAAGlobalTemp dataset:

$$y = b_0 + b_1 x + b_2 x^2 + b_3 x^3 + \varepsilon. \tag{4.106}$$

This fitting can be estimated by the method of multiple linear regression with three variables:

$$x_1 = x, x_2 = x^2, x_3 = x^3. \tag{4.107}$$

Then, a computer code for the multiple linear regression can be used to estimate the coefficients of the polynomial and plot Figure 4.7.

```
#R plot Fig. 4.7: Polynomial fitting
x1=x
x2=x1^2
x3=x1^3
dat3=data.frame(cbind(x1,x2,x3,y))
reg3 = lm(y ~ x1 + x2 + x3, data=dat3)
# simply use
# reg3 = lm(y ~ x + I(x^2) + I(x^3))
setEPS() #Plot the figure and save the file
postscript("fig0407.eps", height = 8, width = 8)
par(mfrow=c(2,1))
par(mar=c(0,4.5,2.5,0.7))
plot(x, y,  type="o", xaxt="n",
    cex.lab=1.4, cex.axis=1.4, xlab="Year",
    ylab=bquote("Temperature [" ~ degree ~ "C]"),
    main="Global Annual Mean Surface Temperature Anomalies",
```

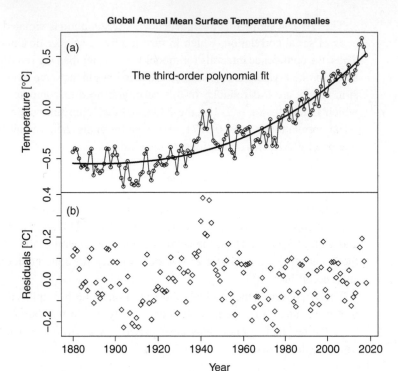

Figure 4.7 (a) Fit a third-order polynomial to the global average annual mean land and ocean surface air temperature anomalies with respect to the 1971–2000 climatology based on the NOAAGlobalTemp dataset (Zhang et al. 2019); (b) Scatter plot of the corresponding residuals.

```
     cex.lab=1.4, cex.axis=1.4)
lines(x, predict(reg3), col="black", lwd=3)
reg3
#(Intercept)            x1            x2            x3
#-1.426e+03    2.333e+00    -1.271e-03    2.308e-07
text(1940, 0.3,
     "The␣third-order␣polynomial␣fit",
     col="black",cex=1.4)
text(1880, 0.58, "(a)", cex=1.4)
par(mar=c(4.5,4.5,0,0.7))
plot(x1, reg3$residuals,
     pch=5, cex.lab=1.4, cex.axis=1.4,
     xlab="Year", ylab=bquote("Residuals␣["~degree~"C]"))
text(1880, 0.32, "(b)", cex=1.4)
dev.off()
```

```
Python plot Fig. 4.7: Polynomial fitting
os.chdir("/Users/sshen/climstats/")
# Read the data
dtmean = np.array(read_table(
```

```
          "data/aravg.ann.land_ocean.90S.90N.v5.0.0.201909.txt",
                  header = None, delimiter = "\s+"))
xDT = dtmean[0:139, 0]
yDT = dtmean[0:139, 1]
# Create trend line
reg3 = np.array(np.polyfit(xDT, yDT, 3))
ablineDT3 = reg3[3] + xDT*reg3[2] + \
            (xDT**2)*reg3[1] + (xDT**3)*reg3[0]
fig, ax = plt.subplots(2, 1, figsize=(12,12))
ax[0].plot(xDT, yDT, 'ko-')
ax[0].plot(xDT, ablineDT3, 'k-')

ax[1].plot(xDT, yDT - ablineDT3, 'kd')
ax[0].set_title("Global Annual Mean Land and Ocean \n\
            Surface Temperature Anomalies",
            fontweight = 'bold', size = 25, pad = 20)
ax[0].tick_params(length=6, width=2, labelsize=20);
ax[0].set_yticks(np.round(np.linspace(-1, 1, 5), 2))
ax[0].set_ylabel("Temperature [$\degree$C]",
                 size = 25, labelpad = 20)
ax[0].text(1900, 0.3, "The third-order polynomial fit",
            color = 'k', size = 20)
ax[0].text(1877, 0.6, "(a)", size = 20)
ax[0].axes.get_xaxis().set_visible(False)
ax[1].tick_params(length=6, width=2, labelsize=20);
ax[1].set_yticks(np.round(np.linspace(-0.4, 0.4, 5), 2))
ax[1].set_ylabel("Residuals [$\degree$C]",
                 size = 25, labelpad = 20)
ax[1].set_xlabel("Year", size = 25, labelpad = 20)
ax[1].text(1877, 0.35, "(b)", size = 20)
fig.tight_layout(pad=-1.5)
plt.show()
```

R has another simpler command for the third-order polynomial regression:

```
lm(y ~ poly(x, 3, raw=TRUE))
```

where, `raw=TRUE` means using the raw polynomial form written like formula (4.106). Another option is `raw=FALSE`, which means that the data are fitted to an orthogonal polynomial.

Comparison of Figures 4.6(b) and 4.7(b) shows that the W-shape pattern of the residuals in the third-order polynomial fitting is weaker than that for the linear regression. Nonetheless, the W-shape pattern is still clear. Figure 4.7(b) visually suggests that the constant variance assumption is satisfied. The KS test shows that the normality assumption is also satisfied. However, the DW test shows the existence of serial correlation. The computing of these tests is left as exercise problems.

You can try to fit a ninth-order orthogonal polynomial. This can eliminate the W-shape nonlinear pattern of the residuals. You can also try to fit other polynomials or functions.

4.4 Chapter Summary

This chapter has introduced three regression methods commonly used in climate science: (a) simple linear regression of a single variate, (b) multiple linear regression, and (c) nonlinear fitting. For (a) and (b) we used the data of surface air temperature and elevation of the 24 USHCN stations in the state of Colorado, USA. For (c) we used the NOAAGlobalTemp's global average annual mean temperature data from 1880 to 2018.

A linear regression model has four fundamental assumptions:

(i) Approximate linearity between x and Y,
(ii) Normal distribution of ε and Y,
(iii) Constant variance of ε and Y, and
(iv) Independence of error terms: $\mathrm{Cor}[\varepsilon_i, \varepsilon_j] = \sigma^2 \delta_{ij}$.

Assumptions (i) and (iii) can be verified by visually examining the scatter plot of residuals. Assumption (ii) may be verified by the KS test. Assumption (iv) can be verified by the DW test. When one or more assumptions are violated, you may improve your model, for example, by using a nonlinear model or transforming the data.

References and Further Reading

[1] M. Cordova, R. Celleri, C.J. Shellito et al., 2016: Near-surface air temperature lapse rate over complex terrain in the Southern Ecuadorian Andes: implications for temperature mapping. *Arctic, Antarctic, and Alpine Research*, 48, 673–684.

> Figure 3 of this paper shows the linear regression with $R^2 = 0.98$ for the temperature data of nine stations whose elevations are in the range from 2,610 to 4,200 meters. The regression implies a lapse rate of 6.88°C/km for the annual mean temperature.

[2] P. L. Fall, 1997: Timberline fluctuations and late Quaternary paleoclimates in the Southern Rocky Mountains, Colorado. *Geological Society of America Bulletin*, 109, 1306–1320.

> Figure 2a of this paper shows a lapse rate of 6.9°C/km for the mean July temperature based on the data of 104 stations and a linear regression with $R^2 = 0.86$. Figure 2b shows the annual mean temperature TLR equal to 6.0°C/km with $R^2 = 0.80$.

[3] F. A. Graybill and H. K. Iyer, 1994: *Regression Analysis: Concepts and Applications*. Duxbury Press.

> This book clearly outlines the assumptions of regression. It focuses on concepts and applications about solving practical statistical problems.

[4] M. J. Menne, C. N. Williams, and R. S. Vose, 2009: The United States Historical Climatology Network monthly temperature data Version 2. *Bulletin of the American Meteorological Society*, 90, 993–1007.

> This is one of the series of papers on the USHCN datasets prepared at the NOAA National Centers for Environmental Information and widely used since the early 1990s. The network includes 1,218 stations and has both monthly and daily data of temperature and precipitation publicly available.

```
www.ncdc.noaa.gov/ushcn
```

[5] H.-M., Zhang, B. Huang, J. Lawrimore, M. Menne, and T. M. Smith, 2019: NOAA Global Surface Temperature Dataset (NOAAGlobalTemp), Version 5.0 (Time Series). NOAA National Centers for Environmental Information. doi:10.7289/V5FN144H. Last accessed March 2021.

The NOAA Merged Land Ocean Global Surface Temperature Analysis (NOAAGlobalTemp) dataset is produced and maintained by the NOAA National Centers for Environmental Information. This is monthly anomaly data with respect to the 1971–2000 climatology. It covers time from January 1880 to present, and has a spatial resolution $5° \times 5°$. The data can be visualized at www.4dvd.org.

Exercises

4.1 (a) Compute the temperature lapse rate for August using the TOB mean temperature data from the 24 USHCN stations in Colorado during the period of 1981–2010.

(b) Plot the figure similar to Figure 4.1 for the data and results of (a).

4.2 Repeat the previous problem but for January. Compare your January results with those of July in Figure 4.1 and its related text.

4.3 (a) Compute the temperature lapse rate for the annual mean temperature based on the 24 USHCN stations in Colorado during the period of 1981–2010.

(b) Plot the figure similar to Figure 4.1 for the data and results of (a).

4.4 (a) Compute the annual mean temperature lapse rate for a high mountainous region and for a period of your choice.

(b) Plot the figure similar to Figure 4.1 for the data and results of (a).

4.5 Show that the orthogonality condition Eq. (4.27)

$$\mathbf{e} \cdot \mathbf{x}_a = 0 \qquad (4.108)$$

is equivalent to the condition of minimizing SSE given condition (4.19)

$$\sum_{i=1}^{n} e_i = 0. \qquad (4.109)$$

4.6 Show that when $m = 1$, the confidence interval formula for the multiple linear regression (4.102) is reduced to the confidence interval formula for the simple linear regression (4.82).

4.7 Examine the global average December temperature anomalies from 1880 to 2018 in the dataset of the NOAAGlobalTemp.

(a) Make a linear regression of the temperature anomalies against time.

(b) Compute the confidence intervals of the fitted model at the 95% confidence level.

(c) Compute the confidence intervals of the anomaly data at the 95% confidence level.

(d) On the same figure similar to Figure 4.6(a), plot the scatter plot of the anomaly data against time, and plot the confidence intervals computed in (b) and (c).

4.8 Make a diagnostic analysis for the regression results of the previous problem.

(a) Produce a scatter plot of the residuals against time.

(b) Visually check whether the assumptions of linearity and constant variance are satisfied.

(c) Use the KS test to check the normality assumption on residuals.

(d) Use the DW test to check the independence assumption on residuals.

(e) When serial correlation is considered, find the effective degrees of freedom (edof).

(f) Compute the confidence intervals in Steps (b) and (c) in Problem 4.7 using edof.

(g) Produce a scatter plot of the anomaly data against time, and plot the confidence intervals on the same figure using the results of Step (f), similar to Figure 4.6(a).

4.9 (a) Fit the NOAAGlobalTemp's global average annual mean data from 1880 to 2018 to a ninth-order orthogonal polynomial.

(b) Plot the data and the fitted polynomial function on the same figure.

(c) Produce a scatter plot of the residuals of the fitting against time as a different figure from (b).

4.10 Make a diagnostic analysis for the above regression and examine the regression assumptions. In particular, use the KS test to verify the normality assumption, and the DW test to verify the independence assumption.

4.11 (a) Use multiple linear regression to compute the 12th-order polynomial fitting of the NOAAGlobalTemp's global average annual mean data from 1880 to 2018:

$$T = b_0 + b_1 t + b_2 t^2 + \cdots + b_{12} t^{12} + \varepsilon. \tag{4.110}$$

(b) Plot the data and the fitted polynomial function on the same figure.

(c) Produce a scatter plot of the residuals of the fitting against time as a different figure from (b).

4.12 Make a diagnostic analysis for the previous regression and examine the regression assumptions. In particular, use the KS test to verify the normality assumption, and the DW test to verify the independence assumption.

4.13 (a) Fit the global average **January** monthly mean temperature anomaly data from 1880 to 2018 in the NOAAGlobalTemp dataset to a third-order orthogonal polynomial. The global average monthly mean NOAAGlobalTemp time series data are included in the book's master dataset named `data.zip`. You can also download the updated data from the Internet.

(b) Plot the data and the fitted polynomial function on the same figure.

(c) Produce a scatter plot of the residuals of the fitting against time in another figure.

4.14 Make a diagnostic analysis for the previous **January** data fit following the procedures in this chapter. In particular, use the KS test to verify the normality assumption, and the DW test to verify the independence assumption.

4.15 (a) Fit the global average **July** monthly mean temperature anomaly data from 1880 to 2018 in the NOAAGlobalTemp dataset to a third-order orthogonal polynomial.

(b)   Plot the data and the fitted polynomial function on the same figure.

(c)   Produce a scatter plot of the residuals of the fitting against time in another figure.

4.16 Make a diagnostic analysis for the previous **July** data fit following the procedures in this chapter. In particular, use the KS test to verify the normality assumption, and the DW test to verify the independence assumption.

4.17 Use the gridded monthly NOAAGlobalTemp dataset and make a third-order polynomial fit to the **January** monthly mean temperature anomaly data from 1880 to 2018 in a grid box that covers Tokyo, Japan. The gridded monthly NOAAGlobalTemp dataset is included in the book's master data file `data.zip`. You can also download the updated data from the Internet.

4.18 Use the gridded monthly NOAAGlobalTemp dataset and make a third-order polynomial fit to the **January** monthly mean temperature anomaly data from 1880 to 2018 in a grid box that covers Bonn, Germany.

4.19 Use the gridded monthly NOAAGlobalTemp dataset and make a fifth-order polynomial fit to the monthly mean temperature anomaly data for a grid box, a month, and a period of time of your choice. For example, you may choose your hometown grid box, the month you were born, and the period of 1900–1999.

4.20 Make a diagnostic analysis for the previous data fit following the procedures in this chapter.

4.21 Use the January time series data of an USHCN station of your choice and fit a third-order polynomial. Make a diagnostic analysis for the fit. The updated monthly USHCN station data may be downloaded from the Internet.

5   Matrices for Climate Data

Matrices appear everywhere in climate science. For examples, climate data may be written as a matrix – a 2-dimensional rectangular array of numbers or symbols – and most data analyses and multivariate statistical studies require the use of matrices. The study of matrices is often included in a course known as linear algebra. This chapter is limited to (i) describing the basic matrix methods needed for this book, such as the inverse of a matrix and the eigenvector decomposition of a matrix, and (ii) presenting matrix application examples of real climate data, such as the sea level pressure data of Darwin and Tahiti. From climate data matrices, we wish to extract helpful information, such as the spatial patterns of climate dynamics (e.g., El Niño Southern Oscillation), and temporal occurrence of the patterns. These are related to eigenvectors and eigenvalues of matrices. This chapter features the space-time data arrangement, which uses rows of a matrix for spatial locations, and columns for temporal steps. The singular value decomposition (SVD) helps reveal the spatial and temporal features of climate dynamics as singular vectors and the strength of their variability as singular values.

To better focus on matrix theory, some application examples of linear algebra, such as the balance of chemical reaction equations, are not included in the main text, but are arranged as exercise problems. We have also designed exercise problems for the matrix analysis of real climate data from both observations and models.

5.1 Matrix Definitions

Matrices have appeared earlier in this book, e.g., the matrix formulation of a multi-linear regression in Chapter 4. A matrix is a rectangular array of numbers (or even expressions), often denoted by an uppercase letter in either boldface or plain, $\mathbf{A}$ or A:

$$\mathbf{A} = \begin{bmatrix} a_{11} & a_{12} & \cdots & a_{1p} \\ a_{21} & a_{22} & \cdots & a_{2p} \\ \cdots & \cdots & \cdots & \cdots \\ \cdots & \cdots & \cdots & \cdots \\ a_{n1} & a_{n2} & \cdots & a_{np} \end{bmatrix} \tag{5.1}$$

A Sample Space-Time Dataset

Lat	Lon	1934-3	1934-4	1934-5	1934-6	1934-7	1934-8	1934-9	1934-10	1934-11	1934-12	1935-1	1935-2	1935-3	1935-4	1935-5
32.5	242.5	1.86	1.14	1.03	−0.65	0.12	−0.27	−0.30	−0.30	0.13	0.34	−0.48	−0.18	−1.43	−0.50	−0.84
32.5	247.5	3.14	2.42	2.29	−2.08	0.93	−0.09	−0.49	0.46	0.21	0.79	0.07	−0.18	−2.04	−0.44	−2.32
32.5	252.5	1.42	1.94	2.47	−0.24	1.18	1.04	0.11	1.30	0.46	0.73	0.93	−1.07	−0.09	0.18	−2.32
32.5	257.5	−0.97	0.96	0.95	1.89	1.44	1.49	0.50	2.68	1.55	0.87	2.65	−0.36	1.95	0.64	−2.18
32.5	262.5	−1.89	1.09	0.51	2.64	2.12	2.36	0.07	2.67	1.90	0.47	2.40	0.14	2.50	0.04	−1.66
32.5	267.5	−1.36	0.82	-0.06	1.36	0.89	1.46	−0.55	2.28	1.19	−0.17	2.14	0.54	3.27	0.21	−0.35
32.5	272.5	−0.98	0.61	0.26	0.82	0.36	0.38	−0.15	1.37	0.93	−0.66	1.31	0.23	2.45	0.76	0.85
32.5	277.5	−1.26	0.51	-0.24	0.75	0.65	0.39	0.71	0.91	0.36	−0.85	0.99	−0.25	2.12	0.68	0.90
32.5	282.5	0.54	0.88	0.09	0.25	0.32	0.13	0.20	1.08	0.51	0.55	0.81	0.87	1.31	0.94	0.77
32.5	287.5	0.72	0.99	0.29	0.39	0.36	0.12	0.50	1.04	0.32	0.38	0.23	0.46	0.62	0.58	0.39
32.5	292.5	0.79	0.93	0.27	0.48	0.23	0.18	0.90	1.01	0.48	0.22	−0.23	0.11	0.21	0.30	−0.04
32.5	297.5	0.68	0.59	0.26	0.33	0.17	0.17	0.82	0.69	0.50	0.26	−0.42	−0.15	−0.11	0.07	−0.37
32.5	302.5	0.63	0.42	0.33	0.35	0.46	0.21	0.65	0.48	0.34	0.27	−0.64	−0.27	−0.40	−0.13	−0.59
32.5	307.5	0.69	0.43	0.48	0.54	0.69	0.20	0.46	0.33	0.25	0.26	−0.63	−0.15	−0.37	−0.12	−0.54
32.5	312.5	0.80	0.51	0.44	0.44	0.68	0.26	0.45	0.41	0.26	0.28	−0.28	0.08	−0.16	0.05	−0.21
32.5	317.5	0.83	0.47	0.16	0.26	0.61	0.36	0.47	0.49	0.14	0.10	−0.01	0.21	0.04	0.23	0.24
32.5	322.5	0.62	0.16	−0.19	0.10	0.44	0.39	0.41	0.43	−0.04	−0.09	−0.10	0.10	0.09	0.33	0.45
32.5	327.5	0.24	−0.29	−0.54	0.05	0.27	0.29	0.21	0.23	−0.35	−0.19	−0.21	−0.03	0.09	0.40	0.52

Figure 5.1 A subset of the monthly surface air temperature anomalies from the NOAAGlobalTemp Version 4.0 dataset.

This is an $n \times p$ matrix, in which a_{ij} are called elements or entries of the matrix $\mathbf{A}$, i is the row index from 1 to n, and j is the column index from 1 to p. The dimensions of a matrix are denoted by subscript, e.g., $\mathbf{A}_{n \times p}$ indicating that the matrix $\mathbf{A}$ has n rows and p columns. The matrix may also be indicated by square brackets around a typical element $\mathbf{A} = [a_{ij}]$ or sometimes $\{A\}_{ij}$, maybe even A_{ij}. If $n = p$, then the array is a square matrix.

Figure 5.1 is an example of a space-time climate data matrix. It is a subset of the $5° \times 5°$ gridded monthly surface air temperature anomalies from the NOAA Merged Land Ocean Global Surface Temperature Analysis (NOAAGlobalTemp) (Version 4.0). The rows are indexed according to the spatial locations prescribed by the latitude and longitude of the centroid of a $5° \times 5°$ grid box (see the entries of the first two columns in boldface). The columns are indexed according to time (see the first-row entries in boldface). The other entries are the temperature anomalies with respect to the 1971–2000 monthly climatology. The anomalies are arranged according to the locations by rows and the time by columns. The units for the anomaly data are °C.

For a given month, the spatial temperature data on the Earth's surface is itself a 2-dimensional array. To make a space-time matrix, we assign each grid box a unique index s from 1 to n if the spatial region has n grid boxes. The index assignment is subjective, depending on the application needs. The commonly used way is to fix a latitude and increase the index number as the longitude increases, as indicated by the first two columns of the data matrix shown in Figure 5.1. When the longitude is finished at this latitude band, go to the next latitude band until the completion of the latitude range. This can go from south to north, or from north to south. Of course, one can fix the longitude first, and increase the index according to the ascending or descending order of latitudes.

Following this spatial index as the row number, the climate data for a given month is a column vector. If the dataset has data for p months, then the space-time data matrix has p columns. If the dataset has n grid boxes, then the data forms an $n \times p$ space-time data matrix. You can conveniently use row or column operations of a computer language to calculate statistics of the dataset, such as spatial average, temporal mean, temporal variance, etc.

For more explicit indication of space and time, you may use s for the row index and t for the column index in a space-time data matrix. Thus, $[A_{st}]$ indicates a space-time data matrix (s for space and t for time).

This space-time indexing can be extended to the data in 3D space and 1D time, as long as we can assign a unique ID s from 1 to n for a 3D grid box and a unique ID t for time. The presently popular netCDF (Network Common Data Form) data format in climate science, denoted by .nc, uses this index procedure for a 4D dataset. For example, to express the output of a 3D climate model, you can start your index longitude first for a given latitude and altitude. When longitude exhausts, count the next latitude. When the latitude exhausts, count the next altitude until the last layer of the atmosphere or ocean. Eventually, a space-time data matrix is formed $[A_{st}]$.

To visualize the row data of a space-time data matrix $[A_{st}]$, just plot a line graph of the row data against time. To visualize the column data of a space-time data matrix $[A_{st}]$, you need to convert the column vector into a 2D pixel format for a 2D domain (e.g., the contiguous United States (CONUS) region), or a 3D data array format for a 3D domain (e.g., the CONUS atmosphere domain from the 1,000 mb surface level to the 10 mb height level). This means that the climate data are represented in another matrix format, such as the surface air temperature anomaly data on a 5-degree latitude–longitude grid for the entire world for December 2015 visualized by Figure 1.8. The data behind the figure is a 36×72 data matrix on the grid whose rows are for latitude and columns for longitude. This data matrix is in space-space pixel format like the data for a photo. Each time corresponds to a new space-space pixel data matrix. Thus, the latitude–longitude-time forms a 3D data array. With elevation, then latitude–longitude-altitude-time forms a 4D data array, which is often written in the netCDF file in climate science. You can use the 4DVD data visualization tool www.4dvd.org, described in Chapter 1, to visualize the 4D Reanalysis data array as an example to understand the space-time data plot, and netCDF data structure.

The coordinates (32.5, 262.5) in the sixth row of Figure 5.1 indicate a $5° \times 5°$ grid box centered at (32.5° N, 97.5° W). This box covers part of Texas, USA. The large temperature anomalies for the summer of 1934 (2.64°C for June, 2.12°C for July, and 2.36°C for August) were in the 1930s Dust Bowl period. The hot summer of 1934 was a wave of severe drought. The disastrous dust storms in the 1930s over the American and Canadian prairies destroyed many farms and greatly damaged the ecology.

5.2 Fundamental Properties and Basic Operations of Matrices

This section provides a concise list summarizing fundamental properties and commonly used operations of matrices. We limit our material to the basics that are sufficient for this book.

(i) Zero matrix: A zero matrix has every entry equal to zero: $\mathbf{0} = [0]$, or $0 = [0]$, or explicitly

$$\mathbf{0} = \begin{bmatrix} 0 & 0 & \cdots & 0 \\ 0 & 0 & \cdots & 0 \\ \cdots & \cdots & \cdots & \cdots \\ \\ \cdots & \cdots & \cdots & \cdots \\ 0 & 0 & \cdots & 0 \end{bmatrix} \tag{5.2}$$

(ii) Identity matrix: An identity matrix is a square matrix whose diagonal entries are all equal to one and whose off-diagonal entries are all equal to zero, and is denoted by I or $\mathbf{I}$. See an expression of an identity matrix below:

$$\mathbf{I} = \begin{bmatrix} 1 & 0 & \cdots & 0 \\ 0 & 1 & \cdots & 0 \\ \cdots & \cdots & \cdots & \cdots \\ \\ \cdots & \cdots & \cdots & \cdots \\ 0 & 0 & \cdots & 1 \end{bmatrix} \tag{5.3}$$

People also use the following notation

$$\mathbf{I} = [\delta_{ij}], \tag{5.4}$$

where

$$\delta_{ij} = \begin{cases} 1 & \text{when } i = j \\ 0 & \text{otherwise} \end{cases} \tag{5.5}$$

is called the Kronecker delta.

An identity matrix may be regarded as a special case of a *diagonal matrix*, which refers to any square matrix whose off-diagonal elements are zero. Hence, a diagonal matrix has the following general expression:

$$\mathbf{D} = [d_i \delta_{ij}], \tag{5.6}$$

where $d_1, d_2, \ldots, d_n$ are the n-diagonal elements. If $d_i = 1, i = 1, 2, \ldots, n$, then the diagonal matrix becomes an identity matrix.

(iii) A transpose matrix: The *transpose* of a matrix A is obtained by interchanging the rows and columns. The new matrix is denoted by $\mathbf{A}^t$. Computing the matrix transpose is very easy: simply rotate each horizontal row clockwise to a vertical column, one row at a time. If $\mathbf{A}$ has dimension $n \times p$, its transpose has dimension $p \times n$. The elements of the transposed matrix are related to the originals by

$$(A^t)_{ij} = A_{ji} \tag{5.7}$$

For example, if

$$\mathbf{A} = \begin{bmatrix} 1 & 2 & 3 \\ 4 & 5 & 6 \end{bmatrix} \tag{5.8}$$

then

$$\mathbf{A}^t = \begin{bmatrix} 1 & 4 \\ 2 & 5 \\ 3 & 6 \end{bmatrix} \tag{5.9}$$

If $\mathbf{A}^t = \mathbf{A}$ or $a_{ij} = a_{ji}$, then the matrix $\mathbf{A}$ is said to be *symmetric*. Of course, a symmetric matrix must be a square matrix.

(iv) Equal matrices: Two matrices $\mathbf{A}$ and $\mathbf{B}$ are equal if every pair of corresponding entries is equal, i.e., the equation $\mathbf{A} = \mathbf{B}$ means $a_{ij} = b_{ij}$ for all i and j.

(v) Matrix addition: The sum of two matrices $\mathbf{A}$ and $\mathbf{B}$ is defined by the sum of corresponding entries, i.e., $\mathbf{A} + \mathbf{B} = [a_{ij} + b_{ij}]$.

(vi) Matrix subtraction: The difference of two matrices $\mathbf{A}$ and $\mathbf{B}$ is defined by the difference of corresponding entries, i.e., $\mathbf{A} - \mathbf{B} = [a_{ij} - b_{ij}]$. The equation of $\mathbf{A} = \mathbf{B}$ is equivalent to $\mathbf{A} - \mathbf{B} = 0$.

(vii) Row vector: A row vector is of dimension p is a $1 \times p$ matrix:

$$\mathbf{u} = \begin{bmatrix} u_1 & u_2 & \cdots & u_p \end{bmatrix}. \tag{5.10}$$

Matrix $\mathbf{A}_{n \times p}$ may be regarded as a stack of n row vectors $\mathbf{a}_{i:}, i = 1, 2, \ldots, n$, each of which is a p-dimensional row vector. Hence,

$$\mathbf{A} = \begin{bmatrix} \mathbf{a}_{1:} \\ \mathbf{a}_{2:} \\ \cdots \\ \mathbf{a}_{n:} \end{bmatrix}. \tag{5.11}$$

Here, : in the second position of the double-index subscript means the inclusion of all the columns.

(viii) Column vector: A column vector of dimension n is an $n \times 1$ matrix

$$\mathbf{v} = \begin{bmatrix} v_1 \\ v_2 \\ \cdots \\ v_n \end{bmatrix}. \tag{5.12}$$

The transpose of a column vector becomes a row vector, and vice versa.

Matrix $\mathbf{A}_{n \times p}$ may be regarded as an array of p column vectors $\mathbf{a}_{:j}, j = 1, 2, \ldots, p$, each of which is an n-dimensional column vector. Hence,

$$\mathbf{A} = \begin{bmatrix} \mathbf{a}_{:1} & \mathbf{a}_{:2} & \cdots & \mathbf{a}_{:p} \end{bmatrix}. \tag{5.13}$$

Here, : in the first position of the double-index subscript means the inclusion of all the rows.

(ix) Dot product of two vectors: Two vectors of the same dimension can form a *dot product* that is equal to the sum of the products of the corresponding entries:

$$\mathbf{u} \cdot \mathbf{v} = u_1 v_1 + u_2 v_2 + \cdots + u_n v_n. \tag{5.14}$$

The dot product is also called an *inner product*.

For example, if

$$\mathbf{u} = \begin{bmatrix} 1 & 2 & 3 \end{bmatrix}, \quad \mathbf{v} = \begin{bmatrix} 4 & 5 & 6 \end{bmatrix}, \tag{5.15}$$

then

$$\mathbf{u} \cdot \mathbf{v} = 1 \times 4 + 2 \times 5 + 3 \times 6 = 32. \tag{5.16}$$

The *amplitude of vector* $\mathbf{u}$ of dimension n is defined as

$$|\mathbf{u}| = \sqrt{u_1^2 + u_2^2 + \cdots + u_n^2}. \tag{5.17}$$

Sometimes, the amplitude is also called length, or Euclidean length, or magnitude. Please do not mix the concept of Euclidean length of a vector with the dimensional length of a vector. The latter means the number of entries of a vector, i.e., n.

If the Euclidean length of $\mathbf{u}$ is equal to one, we say that the $\mathbf{u}$ is a *unit vector*. If every element of $\mathbf{u}$ is zero, then we say that $\mathbf{u}$ is a *zero vector*.

By the definition of dot product, we have

$$|\mathbf{u}|^2 = \mathbf{u} \cdot \mathbf{u}. \tag{5.18}$$

If $\mathbf{u} \cdot \mathbf{v} = 0$, we say that $\mathbf{u}$ and $\mathbf{v}$ are *orthogonal*. Further, if $\mathbf{u} \cdot \mathbf{v} = 0$ and $|\mathbf{u}| = |\mathbf{v}| = 1$, then we say that $\mathbf{u}$ and $\mathbf{v}$ are *orthonormal*.

(x) Matrix multiplication: The product of matrix $\mathbf{A}_{n \times p}$ and matrix $\mathbf{B}_{p \times m}$ is an $n \times m$ matrix $\mathbf{C}_{n \times m}$ whose element c_{ij} is the dot product of the ith row vector of A and jth column vector of $\mathbf{B}$:

$$c_{ij} = \mathbf{a}_{i:} \cdot \mathbf{b}_{:j}. \tag{5.19}$$

We denote

$$\mathbf{C}_{n \times m} = \mathbf{A}_{n \times p} \mathbf{B}_{p \times m}, \tag{5.20}$$

or simply

$$\mathbf{C} = \mathbf{AB}. \tag{5.21}$$

Note that the number of columns of $\mathbf{A}$ and the number of rows of $\mathbf{B}$ must be the same before the multiplication $\mathbf{AB}$ can be made, because the dot product $\mathbf{a}_{i:} \cdot \mathbf{b}_{:j}$ requires this condition. This is referred to as the dimension-matching condition for matrix multiplication. If this condition is violated, the two matrices cannot be multiplied. For example, for the following two matrices

$$\mathbf{A}_{3 \times 2} = \begin{bmatrix} 1 & 0 \\ 0 & 4 \\ 3 & 2 \end{bmatrix} \quad \mathbf{B}_{2 \times 2} = \begin{bmatrix} 0 & -1 \\ 1 & 2 \end{bmatrix}, \tag{5.22}$$

we can compute

$$\mathbf{A}_{3\times2}\mathbf{B}_{2\times2} = \begin{bmatrix} 0 & -1 \\ 4 & 8 \\ 2 & 1 \end{bmatrix}. \tag{5.23}$$

However, the expression

$$\mathbf{B}_{2\times2}\mathbf{A}_{3\times2} \tag{5.24}$$

is not defined, because the dimensions do not match. Thus, for matrix multiplication of two matrices, their order is important. The product $\mathbf{BA}$ may not be equal to $\mathbf{AB}$ even when both are defined. That is, the commutative law does not hold for matrix multiplication.

The dot product of two vectors can be written as the product of two matrices. If both $\mathbf{u}$ and $\mathbf{v}$ are n-dimensional column vectors, then

$$\mathbf{u} \cdot \mathbf{v} = \mathbf{u}^t \mathbf{v}. \tag{5.25}$$

The right-hand side is a $1 \times n$ matrix times an $n \times 1$ matrix, and the product is a 1×1 matrix, whose element is the result of the dot product. Computer programs usually calculate a dot product using this process of matrix multiplication.

A scaler can always multiply a matrix, which is defined as follows. Given a scalar c and a matrix $\mathbf{A}$, their product is

$$c\mathbf{A} = [ca_{ij}] = \mathbf{A}c. \tag{5.26}$$

The scaler multiplication can be extended to multiple vectors

$$(\mathbf{u}_1, \mathbf{u}_2, \cdots, \mathbf{u}_p)$$

or matrices to form a *linear combination*:

$$\mathbf{u} = c_1\mathbf{u}_1 + c_2\mathbf{u}_2 + \cdots + c_p\mathbf{u}_p, \tag{5.27}$$

where $c_1, c_2, \ldots, c_p$ are coefficients of the linear combination and at least one of the coefficients is nonzero. Multivariate linear regression discussed at the end of previous chapter is a linear combination. This is a very useful mathematical expression in data science.

(xi) Matrix inversion: For a given square matrix $\mathbf{A}$, if there is a matrix $\mathbf{B}$ such that

$$\mathbf{BA} = \mathbf{AB} = \mathbf{I}, \tag{5.28}$$

then $\mathbf{B}$ is called the *inverse matrix* of $\mathbf{A}$, denoted by $\mathbf{A}^{-1}$, i.e.,

$$\mathbf{A}^{-1}\mathbf{A} = \mathbf{AA}^{-1} = \mathbf{I}. \tag{5.29}$$

Not all the matrices have an inverse. If a matrix has an inverse, then the matrix is said to be *invertible*. Equivalently, $\mathbf{A}^{-1}$ exists.

As an example of the matrix inversion, given

$$\mathbf{A} = \begin{bmatrix} 1 & -1 \\ 1 & 2 \end{bmatrix}, \tag{5.30}$$

we have

$$\mathbf{A}^{-1} = \begin{bmatrix} 2/3 & 1/3 \\ -1/3 & 1/3 \end{bmatrix}. \tag{5.31}$$

Hand calculation for the inverse of a small matrix is already very difficult, and that for the inverse of a large matrix is almost impossible. Computers can do the calculations for us, as will be shown in examples later.

According to the definition of inverse, we have the following formula for the inverse of the product of two matrices:

$$(\mathbf{AB})^{-1} = \mathbf{B}^{-1}\mathbf{A}^{-1}, \tag{5.32}$$

if both $\mathbf{A}$ and $\mathbf{B}$ are invertible matrices. Please note the order switch of the matrices.

With the definition of an inverse matrix, we can define the *matrix division* by

$$\mathbf{A}/\mathbf{B} = \mathbf{AB}^{-1} \tag{5.33}$$

when $\mathbf{B}^{-1}$ exists. In matrix operations, we usually do not use the concept of matrix division, but always use the matrix inverse and matrix multiplication.

(xii) More properties of the matrix transpose:

$$(\mathbf{A}^t)^t = \mathbf{A} \tag{5.34}$$

$$(\mathbf{A}+\mathbf{B})^t = \mathbf{A}^t + \mathbf{B}^t \tag{5.35}$$

$$(\mathbf{AB})^t = \mathbf{B}^t\mathbf{A}^t \tag{5.36}$$

$$(\mathbf{A}^{-1})^t = (\mathbf{A}^t)^{-1}. \tag{5.37}$$

(xiii) Orthogonal matrices: An *orthogonal matrix*[1] is one whose row vectors are orthonormal. In this case, the inverse matrix can be easily found: it is its transpose. That is, if $\mathbf{A}$ is an orthogonal matrix, then

$$\mathbf{A}^{-1} = \mathbf{A}^t. \tag{5.38}$$

The proof of this claim is very simple. The orthonormal property of the row vectors of $\mathbf{A}$ implies that

$$\mathbf{A}\mathbf{A}^t = \mathbf{I}. \tag{5.39}$$

By the definition of matrix inverse, $\mathbf{A}^t$ is the inverse matrix of $\mathbf{A}$.

If $\mathbf{A}$ is an orthogonal matrix, its row vectors are also orthonormal. This can be proved by multiplying both sides of the above by $\mathbf{A}^t$ from the left:

$$\mathbf{A}^t\mathbf{A}\mathbf{A}^t = \mathbf{A}^t\mathbf{I}. \tag{5.40}$$

Then multiply both sides of this equation by $(\mathbf{A}^t)^{-1}$ from the right:

$$\mathbf{A}^t\mathbf{A}(\mathbf{A}^t(\mathbf{A}^t)^{-1}) = \mathbf{A}^t\mathbf{I}(\mathbf{A}^t)^{-1}, \tag{5.41}$$

which yields

$$\mathbf{A}^t\mathbf{A} = \mathbf{I}. \tag{5.42}$$

---

[1] Although *orthogonal matrix* is a standard mathematical terminology, it is acceptable if you call it *orthonormal matrix*.

This implies that the column vectors of **A** are orthonormal.

As an example, the following matrix

$$\mathbf{T} = \begin{bmatrix} \cos\theta & -\sin\theta \\ \sin\theta & \cos\theta \end{bmatrix}, \tag{5.43}$$

is an orthogonal matrix for any given real number θ. You can easily verify this using the trigonometrical identity $\sin^2\theta + \cos^2\theta = 1$. We thus have

$$\mathbf{T}^{-1} = \begin{bmatrix} \cos\theta & \sin\theta \\ -\sin\theta & \cos\theta \end{bmatrix}, \tag{5.44}$$

You can easily verify that $\mathbf{T}^{-1}\mathbf{T} = \mathbf{I}$ by hand calculation of the product of the two matrices in this equation.

5.3 Some Basic Concepts and Theories of Linear Algebra

According to Encyclopedia.com, "Linear algebra originated as the study of linear equations." Linear algebra deals with vectors, matrices, and vector spaces. Before the 1950s, it was part of Abstract Algebra (Tucker 1993). In 1965, the Committee on the Undergraduate Program in Mathematics, Mathematical Association of America, outlined the following topics for a stand-alone linear algebra course: linear systems, matrices, vectors, linear transformations, unitary geometry with characteristic values. The vectors and matrices have been dealt with in the previous two sections of this chapter. This section deals with linear systems of equations and linear transformations, and next with characteristic values.

5.3.1 Linear Equations

A meteorologist needs to make a decision on what instruments to order under the following constraint. She is given a budget of $1,000 to purchase 30 instruments for her observational sites. Her supplier has two products for the instrument: the first is $30 per set, and the second $40 per set. She would like to buy as many of the second type of instrument as possible under the budget constraint. Then, the question is how many instruments of the second kind she can buy? This problem leads to the following linear system of two equations:

$$30x_1 + 40x_2 = 1000, \tag{5.45}$$

$$x_1 + x_2 = 30. \tag{5.46}$$

The solution to these linear equations is $x_1 = 20$ and $x_2 = 10$.

This system of linear equations can be expressed in a matrix and two vectors as follows:

$$\mathbf{Ax} = \mathbf{b}, \tag{5.47}$$

where

$$\mathbf{A} = \begin{bmatrix} 30 & 40 \\ 1 & 1 \end{bmatrix} \tag{5.48}$$

$$\mathbf{x} = \begin{bmatrix} x_1 \\ x_2 \end{bmatrix} \tag{5.49}$$

$$\mathbf{b} = \begin{bmatrix} 1000 \\ 30 \end{bmatrix} \tag{5.50}$$

$$\tag{5.51}$$

Then, the solution of this system may be tightly expressed in the following way:

$$\mathbf{x} = \mathbf{A}^{-1}\mathbf{b}. \tag{5.52}$$

This expression is convenient for mathematical proofs, but is rarely used for solving a linear system, because finding an inverse matrix is computationally costly. A way to solve a linear system is to use Gauss elimination. The corresponding computing procedure is called the row operation on a matrix. There are numerous ways of solving a linear system. Some are particularly efficient for a certain system, such as a sparse matrix or a matrix of a diagonal band of width equal 3 or 5. Efficient algorithms for a linear system, particularly an extremely large system, are forever a research topic. In this book, we use a computer to solve a linear system without studying the algorithm details. The R and Python commands are as follows.

```
solve(A, b) #This is the R code for finding x
numpy.linalg.solve(A, b) #This is the Python code
```

5.3.2 Linear Transformations

A *linear transformation* is to convert vector $\mathbf{x}_{n \times 1}$ into $\mathbf{y}_{m \times 1}$ using the multiplication of a matrix $\mathbf{T}_{m \times n}$:

$$\mathbf{y}_{m \times 1} = \mathbf{T}_{m \times n}\mathbf{x}_{n \times 1}. \tag{5.53}$$

For example, the matrix

$$\mathbf{T} = \begin{bmatrix} -0.1 & 4 \\ 0.1 & -3 \end{bmatrix} \tag{5.54}$$

transforms the vector

$$\begin{bmatrix} 1000 \\ 30 \end{bmatrix} \tag{5.55}$$

into

$$\begin{bmatrix} 20 \\ 10 \end{bmatrix}. \tag{5.56}$$

This is the solution of the linear system in the previous subsection.

Usually, the linear transformation $\mathbf{Tx}$ changes both direction and magnitude of $\mathbf{x}$. However, if $\mathbf{T}$ is an orthogonal matrix, then $\mathbf{Tx}$ does not change the magnitude of $\mathbf{x}$, and changes only the direction. This claim can be simply proved by the following formula:

$$|\mathbf{Tx}|^2 = (\mathbf{Tx})^t \mathbf{Tx} = \mathbf{x}^t \mathbf{T}^t \mathbf{Tx} = \mathbf{x}^t (\mathbf{T}^t \mathbf{T})\mathbf{x} = \mathbf{x}^t \mathbf{Ix} = |\mathbf{x}|^2. \tag{5.57}$$

Thus, if $\mathbf{T}$ is an orthogonal matrix, then $\mathbf{Tx}$ is a rotation of the vector $\mathbf{x}$.

5.3.3 Linear Independence

The vectors $\mathbf{x}_1, \mathbf{x}_2, \ldots, \mathbf{x}_p$ are *linearly dependent* if no vector can be represented by a linear combination of other $p-1$ vectors in this group. Otherwise, the group of vectors are linearly dependent.

If it is not linearly independent, then there must be a vector which can be represented by the other vectors through a linear combination. Suppose this vector is $\mathbf{x}_1$, then

$$\mathbf{x}_1 = d_2\mathbf{x}_2 + \cdots + d_p\mathbf{x_p}, \tag{5.58}$$

where at least one of the coefficients $d_2, d_3, \cdots, d_p$ is non-zero. Thus, the linear system of equations for $c_1, c_2, c_3, \cdots, c_p$

$$c_1\mathbf{x}_1 + c_2\mathbf{x}_2 + \cdots + c_p\mathbf{x_p} = 0 \tag{5.59}$$

has a non-zero solution. This system can be written as a matrix form

$$\mathbf{Xc} = \mathbf{0}, \tag{5.60}$$

where column vectors $\mathbf{x}_1, \mathbf{x}_2, \cdots, \mathbf{x_p}$ form the matrix $\mathbf{X}$

$$\mathbf{X} = \begin{bmatrix} \mathbf{x}_1 & \mathbf{x}_2 & \cdots & \mathbf{x_p} \end{bmatrix}, \tag{5.61}$$

the unknown vector is $\mathbf{c}$

$$\mathbf{c} = \begin{bmatrix} c_1 \\ c_2 \\ \vdots \\ c_p \end{bmatrix}, \tag{5.62}$$

and $\mathbf{0}$ is the p-dimensional zero column vector. The solution of this matrix equation is

$$\mathbf{c} = \mathbf{X}^{-1}\mathbf{0} = \mathbf{0}. \tag{5.63}$$

However, $\mathbf{c}$ must not be zero. This contradiction implies that $\mathbf{X}^{-1}$ does not exist if the column vectors are linearly dependent. In other words, if $\mathbf{X}^{-1}$ exists, then its column vectors are linearly independent.

Consider vectors in a 3-dimensional space. Any two column vectors $\mathbf{x}_2$ and $\mathbf{x}_3$ define a plane. If $\mathbf{x}_1$ can be written as a linear combination of $\mathbf{x}_2$ and $\mathbf{x}_3$, then it must lie in the same plane. So, $\mathbf{x}_1$, $\mathbf{x}_2$ and $\mathbf{x}_3$ are linearly dependent. The matrix $[\mathbf{x}_1\ \mathbf{x}_2\ \mathbf{x}_3]_{3\times 3}$ is not invertible.

5.3.4 Determinants

For a square matrix $\mathbf{A}$, a convenient notation and concept is its *determinant*. It is a scaler and is denoted by $\det[\mathbf{A}]$ or $|\mathbf{A}|$. For a 2×2 matrix

$$\mathbf{A} = \begin{bmatrix} a & b \\ c & d \end{bmatrix},$$

(5.64)

its determinant is

$$\det[\mathbf{A}] = ad - cb.$$

(5.65)

For a high-dimensional matrix, the determinant computation is quite complex and is computationally expensive. We usually do not need to calculate the determinant of a large matrix, say $\mathbf{A}_{172 \times 172}$. The computer command for computing the determinant of a small square matrix is as follows:

```
det(A) #R command for determinant
np.linalg.det(a) #Python command for determinant
```

Two 2-dimensional column vectors $\mathbf{x}_1$ and $\mathbf{x}_2$ can span a parallelogram, whose area S is equal to the absolute value of the determinant of the matrix consisting of the two vectors $\mathbf{A} = [\mathbf{x}_1 \ \mathbf{x}_2]$:

$$S = \left| \det[\mathbf{x}_1 \ \mathbf{x}_2] \right|.$$

(5.66)

Three 3-dimensional column vectors $\mathbf{x}_1, \mathbf{x}_2$, and $\mathbf{x}_3$ can span a parallelepiped, whose volume V is equal to the absolute value of the determinant of the matrix consisting of the three vectors $\mathbf{A} = [\mathbf{x}_1 \ \mathbf{x}_2 \ \mathbf{x}_3]$:

$$V = \left| \det[\mathbf{x}_1 \ \mathbf{x}_2 \ \mathbf{x}_3] \right|.$$

(5.67)

A few commonly used properties of determinant are listed below

(a) The determinant of a diagonal matrix is the product of its diagonal elements.
(b) If a determinant has a zero row or column, the determinant is zero.
(c) The determinant does not change after a matrix transpose, i.e., $\det[\mathbf{A}^t] = \det[\mathbf{A}]$.
(d) The determinant of the product of two matrices: $\det[\mathbf{AB}] = \det[\mathbf{A}]\det[\mathbf{B}]$.
(e) The determinant of the product of a matrix with a scaler: $\det[c\mathbf{B}] = c^n \det[\mathbf{A}]$, if $\mathbf{A}$ is an $n \times n$ matrix.
(f) The determinant of an orthogonal matrix is equal to 1 or -1.

5.3.5 Rank of a Matrix

The *rank* of $\mathbf{A}$ is the greatest number of columns of the matrix that are linearly independent, and is denoted by $r[\mathbf{A}]$.

For a 3×3 matrix $\mathbf{A}$, we treat each column as a 3-dimensional vector. If all three lie along a line (i.e., colinear), the rank of $\mathbf{A}$ is one. If all three of the vectors lie in a plane, but are not collinear, the rank is two. If none of the three are collinear or lie in a plane, the rank is three.

If the rank of a square matrix is less than its dimension, then at least one column can be a linear combination of other columns, which implies that the determinant vanishes. If $\mathbf{A}$ has rank r, it is possible to find r linearly independent columns, and all the other columns are linear combinations of these r independent columns.

Some properties about the matrix rank are listed below:

(a) If $\det[\mathbf{A}_{n \times n}] \neq 0$, then $r[\mathbf{A}] = n$, and the matrix $\mathbf{A}_{n \times n}$ is invertible and is said to be *nonsingular*.

(b) If $\det[\mathbf{A}] = 0$, then the rank of $\mathbf{A}$ is less than n, and $\mathbf{A}$ is not invertible and is said to be *singular*.

(c) If $\mathbf{B}$ is multiplied by a nonsingular matrix $\mathbf{A}$, the product has the same rank as $\mathbf{B}$.

(d) $0 \leq r[\mathbf{A}_{n \times p}] \leq \min(n, p)$.

(e) $r[\mathbf{A}] = r[\mathbf{A}^t]$.

(f) $r[\mathbf{A}\,\mathbf{B}] \leq \min(r[\mathbf{A}], r[\mathbf{B}])$.

(g) $r[\mathbf{A}\,\mathbf{A}^t] = r[\mathbf{A}^t\mathbf{A}] = r[\mathbf{A}]$.

(h) $r[\mathbf{A} + \mathbf{B}] \leq r[\mathbf{A}] + r[\mathbf{B}]$.

Computers can easily demonstrate the matrix computations following the theories presented in this chapter so far. The computer code is below.

```
#R code: Computational examples of matrices
A = matrix(c(1,0,0,4,3, 2), nrow = 3, byrow = TRUE)
B = matrix(c(0,1,-1,2), nrow = 2) #form a matrix by columns
C = A%*%B #matrix multiplication
C
#[1,]      0    -1
#[2,]      4     8
#[3,]      2     1
t(C) # transpose matrix of C
#[1,]      0     4     2
#[2,]     -1     8     1

A = matrix(c(1, -1, 1, 2), nrow =2, byrow = TRUE)
solve(A) #compute the inverse of A
#[1,]   0.6666667 0.3333333
#[2,]  -0.3333333 0.3333333
A%*%solve(A) #verify the inverse of A
#[1,] 1.000000e+00     0
#[2,] 1.110223e-16     1

#Solve linear equations
A = matrix(c(30, 40, 1, 1), nrow =2, byrow = TRUE)
b = c(1000, 30)
solve(A,b)
#[1] 20 10
solve(A)%*%b #Another way to solve the equations
det(A) #compute the determinant
#[1] -10

library(Matrix)
rankMatrix(A) #Find the rank of a matrix
#[1] 2 #rank(A) = 2
```

```
#Orthogonal matrices
p = sqrt(2)/2
Q = matrix(c(p,-p,p,p), nrow=2)
Q #is an orthogonal matrix
#             [,1]        [,2]
#[1,]   0.7071068 0.7071068
#[2,] -0.7071068 0.7071068
Q%*%t(Q) #verify O as an orthogonal matrix
#       [,1] [,2]
#[1,]     1    0
#[2,]     0    1
det(Q) #The determinant of an orthogonal matrix is 1 or -1
#[1] 1
```

```
#Matrix multiplication
A = [[1, 0],  [0, 4],[3, 2]]
B = [[0,-1],[1,2]]
C = np.matmul(A,B) #Or C = np.dot(A,B)
print('C=', C)
#C= [[ 0 -1]
# [ 4  8]
# [ 2  1]]
print('Transpose matrix of C =', C.transpose())
#Transpose matrix of C = [[ 0  4  2]
# [-1  8  1]]

#matrix inversion
A = [[1,-1],[1,2]]
np.linalg.inv(A)# compute the inverse of A
#array([[ 0.66666667,  0.33333333],
#       [-0.33333333,  0.33333333]])

#Solve a system of linear equations
A = [[30, 40],[1, 1]]
b = [[1000],[30]]
x = np.linalg.solve(A,b)
print('x=',x)
#x= [[20.]
# [10.]]

#Compute determinant of the previous matrix A
print('Determinant det(A)=', np.linalg.det(A))
#Determinant det(A)= -10.000000000000002

#An orthogonal matrix
p = np.sqrt(2)/2
Q = [[p,p],[-p,p]]
print('Orthogonal matrix Q=', np.round(Q,2))
T = np.transpose(Q)
print('Q times transpose of Q= ', np.matmul(Q,T))
print('Determinant of Q=', np.linalg.det(Q))
#Orthogonal matrix Q= [[ 0.71  0.71]
# [-0.71  0.71]]
```

```
#Q times transpose of Q =  [[1. 0.]
# [0. 1.]]
#Determinant of Q = 1.0
```

5.4 Eigenvectors and Eigenvalues

5.4.1 Definition of Eigenvectors and Eigenvalues

The linear transform $\mathbf{y} = \mathbf{A}\mathbf{u}$ usually results in $\mathbf{y}$ not parallel to $\mathbf{u}$. For example,

$$\begin{bmatrix} 1 & 2 \\ 2 & 1 \end{bmatrix} \times \begin{bmatrix} 1 \\ 0 \end{bmatrix} = \begin{bmatrix} 1 \\ 2 \end{bmatrix} \tag{5.68}$$

The vectors $\mathbf{u} = (1,0)$ and $\mathbf{A}\mathbf{u} = (1,2)$ are not parallel in the 2-dimensional space. See the blue vectors in Figure 5.2.

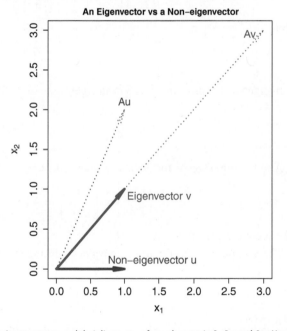

Figure 5.2 An eigenvector **v**, a non-eigenvector **u**, and their linear transforms by matrix **A**: **Av** and **Au**. Here, **Av** and **v** are parallel, and **Au** and **u** are not parallel.

However, there exist some special vectors $\mathbf{v}$ such that $\mathbf{A}\mathbf{v}$ is parallel to $\mathbf{v}$. For example, $\mathbf{v} = (1,1)$ is such a vector, since $\mathbf{A}\mathbf{v} = (3,3)$ is in the same direction as $\mathbf{v} = (1,1)$. See the red vectors in Figure 5.2. If two vectors are parallel, then one vector is a scalar multiplication of the other, e.g., $(3,3) = 3(1,1)$. We denote this scalar by λ. Thus,

$$\mathbf{A}\mathbf{v} = \lambda\mathbf{v}. \tag{5.69}$$

These vectors v are special to A, maintain their own orientation when multiplied by A, and are called *eigenvectors*. Here, "eigen" is from German, meaning "self," "own," "particular," or "special."[2] The corresponding scalars λ are called *eigenvalues*. The formula (5.69) is a mathematical definition of the eigenvalue problem for matrix $\mathbf{A}$.

If $\mathbf{v}$ is an eigenvector, then its multiplication to a scalar c is also an eigenvector, since

$$\mathbf{A}(c\mathbf{v}) = c\mathbf{A}\mathbf{v} = c\lambda\mathbf{v} = \lambda(c\mathbf{v}).$$

Namely, all the vectors in the same direction $\mathbf{v}$ are also eigenvectors. The eigenvectors of length 1 are called unit eigenvectors, or unitary eigenvectors, and are unique up to a positive or negative sign. Most computer programs output unit eigenvectors. Thus, eigenvector describes a direction or an orientation. Each square matrix has its own special orientations.

The aforementioned vector

$$\mathbf{v} = \left[\begin{array}{c} 1 \\ 1 \end{array} \right] \tag{5.70}$$

is an eigenvector that maintains its own direction after being multiplied by $\mathbf{A}$:

$$\left[\begin{array}{cc} 1 & 2 \\ 2 & 1 \end{array} \right] \times \left[\begin{array}{c} 1 \\ 1 \end{array} \right] = \left[\begin{array}{c} 3 \\ 3 \end{array} \right] = 3 \left[\begin{array}{c} 1 \\ 1 \end{array} \right]. \tag{5.71}$$

Thus,

$$\mathbf{v} = \left[\begin{array}{c} 1 \\ 1 \end{array} \right]$$

is an eigenvector of $\mathbf{A}$ and $\lambda = 3$ is an eigenvalue of $\mathbf{A}$. The corresponding unit eigenvector is

$$\mathbf{e} = \mathbf{v}/|\mathbf{v}| = \left[\begin{array}{c} 1/\sqrt{2} \\ 1/\sqrt{2} \end{array} \right].$$

Another eigenvector for the above matrix $\mathbf{A}$ is $\mathbf{v}_2 = (1, -1)$:

$$\left[\begin{array}{cc} 1 & 2 \\ 2 & 1 \end{array} \right] \times \left[\begin{array}{c} 1 \\ -1 \end{array} \right] = -1 \times \left[\begin{array}{c} 1 \\ -1 \end{array} \right]. \tag{5.72}$$

The second eigenvalue is $\lambda_2 = -1$.

The computer code for the eigenvalues and eigenvectors of the above matrix $\mathbf{A}$ is as follows.

```
#R code for eigenvectors and eigenvalues
A = matrix(c(1, 2, 2, 1), nrow=2)
eigen(A)
#$values
#[1]   3  -1
#$vectors
#[,1]         [,2]
#[1,] 0.7071068 -0.7071068
#[2,] 0.7071068  0.7071068
```

---

[2] The "eigen" part in the word "eigenvector" is from German or Dutch and means "self" or "own," as in "one's own." Thus, an eigenvector v is A's "own" vector. In English books, the word "eigenvector" is the standard translation of the German word "eigenvektor." The word "eigenvalue" is translated from the German word "eigenwert," as "wert" means "value." Instead of eigenvector, some English publications use "characteristic vector," which indicates "characteristics of a matrix," or "its own property of a matrix." German mathematician David Hilbert (1862–1943) was the first to use "eigenvektor," and "eigenwert" in his 1904 article about a general theory of linear integral equations.

```
#Python code for eigenvectors and eigenvalues
A = [[1,2], [2,1]]
np.linalg.eig(A)
#(array([ 3., -1.]), array([[ 0.70710678, -0.70710678],
#          [ 0.70710678,  0.70710678]]))
```

If $\mathbf{A}$ is an $N \times N$ matrix, then it has N eigenvalues and eigenvectors $(\lambda_n, v_n), n = 1, 2, \ldots, N$ for the following reason. The eigenvector $\mathbf{v}$ satisfies

$$(\mathbf{A} - \lambda \mathbf{I})\mathbf{v} = 0. \tag{5.73}$$

The nonzero solution $\mathbf{v}$ of this equation requires that

$$\det(\mathbf{A} - \lambda \mathbf{I}) = 0. \tag{5.74}$$

Expanding the determinant out leads to an Nth degree polynomial in λ, which has exactly N roots. Some roots may be repeated and hence counted multiple times toward N. Some roots may be complex numbers, which are not discussed in this book. Each root is an eigenvalue and corresponds to an eigenvector.

This concise determinant expression is tidy and useful for mathematical proofs, but is not used as a computer algorithm for calculating eigenvalues or eigenvectors because it is computationally costly or even impossible for a large matrix. Nonetheless, some old textbooks of linear algebra defined eigenvalues using Eq. (5.74).

Figure 5.2 may be generated by the following computer code.

```
#R plot Fig. 5.2: An eigenvector v vs a non-eigenvector u
setwd('/Users/sshen/climstats')
setEPS() #Plot the figure and save the file
postscript("fig0502.eps", width = 6)
par(mar=c(4.5,4.5,2.0,0.5))
plot(9,9,
     main = 'An eigenvector vs a non-eigenvector',
     cex.axis = 1.4, cex.lab = 1.4,
     xlim = c(0,3), ylim=c(0,3),
     xlab = bquote(x[1]), ylab = bquote(x[2]))
arrows(0,0, 1,0, length = 0.25,
       angle = 8, lwd = 5, col = 'blue')
arrows(0,0, 1,2, length = 0.3,
       angle = 8, lwd = 2, col = 'blue',  lty = 3)
arrows(0,0, 1,1, length = 0.25,
       angle = 8, lwd = 5, col='red')
arrows(0,0, 3,3, length = 0.3,
       angle = 8, lwd = 2, col='red', lty = 3)
text(1.4,0.1, 'Non-eigenvector u', cex =1.4, col = 'blue')
text(1.0,2.1, 'Au', cex =1.4, col = 'blue')
text(1.5,0.9, 'Eigenvector v', cex =1.4, col = 'red')
text(2.8, 2.95, 'Av', cex =1.4, col = 'red')
dev.off()
```

```
#Python plot Fig. 5.2: An eigenvector vs a non-eigenvector
import matplotlib.patches as patches
fig = plt.figure(figsize = (12,12))

plt.axes().set_xlim(-0.1,3.1)
plt.axes().set_ylim(-0.1,3.1)
plt.axes().set_aspect(1)
style = "Simple, tail_width=0.5,head_width=8,head_length=18"
kw1 = dict(arrowstyle=style, color="blue")
kw2 = dict(arrowstyle=style, color="red")

a1 = patches.FancyArrowPatch((0,0), (1,0), **kw1,
                             linewidth =5)
a2 = patches.FancyArrowPatch((0, 0), (1,2),**kw1)
a3 = patches.FancyArrowPatch((0, 0), (1,1), **kw2,
                             linewidth = 5)
a4 = patches.FancyArrowPatch((0, 0), (3,3),**kw2)
for a in [a1, a2, a3, a4]:
    plt.gca().add_patch(a)
plt.title('An eigenvector v vs a non-eigenvector u')
plt.xlabel(r'$x_1$', fontsize = 25)
plt.ylabel(r'$x_2$', fontsize = 25)
plt.text(0.6, 0.1, 'Non-eigenvector u',
         color = 'blue', fontsize =25)
plt.text(0.8, 2.0, 'Au',
         color = 'blue', fontsize =25)
plt.text(1.03,0.85, 'Eigenvector v',
         color = 'red', fontsize =25)
plt.text(2.7, 2.9, 'Av',
         color = 'red', fontsize =25)
plt.show()
```

5.4.2 Properties of Eigenvectors and Eigenvalues for a Symmetric Matrix

A covariance matrix or a correlation matrix is a symmetric matrix and is often used in climate science. For a symmetric matrix $\mathbf{A}$, its eigenvalues and eigenvectors have the following properties:

(a) Eigenvalues of a symmetric matrix are real numbers.

(b) The n different unit eigenvectors of a symmetric matrix $\mathbf{A}_{n \times n}$ are independent and form an orthonormal set, i.e., $\mathbf{e}^{(\ell')} \cdot \mathbf{e}^{(\ell)} = \delta_{\ell \ell'}$, where $\mathbf{e}^{(\ell')}$ and $\mathbf{e}^{(\ell)}$ are any two different unit eigenvectors. We can use these unit vectors to express any n-dimensional vector using a linear combination:

$$\mathbf{x} = c_1 \mathbf{e}^{(1)} + c_2 \mathbf{e}^{(2)} + \cdots + c_n \mathbf{e}^{(n)}, \tag{5.75}$$

where $c_1, c_2, \ldots, c_n$ are scaler coefficients of the linear combination.

(c) If all the eigenvalues of $\mathbf{A}_{n \times n}$ are positive, then the *quadratic form*

$$Q(\mathbf{x}) = \mathbf{x}^t \mathbf{A} \mathbf{x} = \sum_{i,j=1}^{n} a_{ij} x_i x_j \qquad (5.76)$$

is also a positive scaler for any nonzero vector $\mathbf{x}$. A quadratic form may be used to express kinetic energy or total variance in climate science. The kinetic energy in climate science is always positive. For all unit vectors $\mathbf{x}$, the maximum quadratic form is equal to the largest eigenvalue of $\mathbf{A}$. The maximum is achieved when $\mathbf{x}$ is the corresponding eigenvector of $\mathbf{A}$.

(d) The rank of matrix $\mathbf{A}$ is equal to the number of nonzero eigenvalues, where the multiplicity of repeated eigenvalues is counted.

(e) Eigenvalues of a diagonal matrix are equal to the diagonal elements.

(f) For a symmetric matrix $\mathbf{A}_{n \times n}$, its n unit column eigenvectors form an orthogonal matrix $\mathbf{Q}_{n \times n}$ such that $\mathbf{A}_{n \times n}$ can be diagonalized by $\mathbf{Q}_{n \times n}$ in the following way:

$$\mathbf{Q}_{n \times n}^t \mathbf{A}_{n \times n} \mathbf{Q}_{n \times n} = \mathbf{D}, \qquad (5.77)$$

where $\mathbf{D}$ is a diagonal matrix whose diagonal elements are eigenvalues $\lambda_1, \lambda_2, \ldots, \lambda_n$ of $\mathbf{A}$. Further, the column vectors of $\mathbf{Q}$ are the unit eigenvectors of $\mathbf{A}$. This is a matrix diagonalization process.

For the symmetric matrix $\mathbf{A}$ in Eq. (5.68),

$$\mathbf{A} = \begin{bmatrix} 1 & 2 \\ 2 & 1 \end{bmatrix}, \qquad (5.78)$$

its eigenvalues and eigenvectors are

$$\lambda_1 = 3, \qquad \lambda_2 = -1, \qquad (5.79)$$

$$\mathbf{v}_1 = \begin{bmatrix} \sqrt{2}/2 \\ \sqrt{2}/2 \end{bmatrix}, \qquad \mathbf{v}_2 = \begin{bmatrix} -\sqrt{2}/2 \\ \sqrt{2}/2 \end{bmatrix}. \qquad (5.80)$$

Thus, the orthogonal matrix $\mathbf{Q}$ and the diagonal matrix $\mathbf{D}$ are as follows:

$$\mathbf{Q} = \begin{bmatrix} \sqrt{2}/2 & \sqrt{2}/2 \\ \sqrt{2}/2 & -\sqrt{2}/2 \end{bmatrix}, \qquad \mathbf{D} = \begin{bmatrix} 3 & 0 \\ 0 & -1 \end{bmatrix}. \qquad (5.81)$$

With $\mathbf{Q}$, $\mathbf{A}$, and $\mathbf{D}$, you can easily verify Eq. (5.77) by hand calculation or by computer coding.

Equation (5.77) can be written in the following way:

$$\mathbf{A}_{n \times n} = \mathbf{Q}_{n \times n} \mathbf{D}_{n \times n} \mathbf{Q}_{n \times n}^t \qquad (5.82)$$

or

$$\mathbf{A}_{n \times n} = \sum_{k=1}^{n} \lambda_k \left(\mathbf{q}^{(k)} \right)_{n \times 1} \left((\mathbf{q}^{(k)})^t \right)_{1 \times n}. \qquad (5.83)$$

This is a process of matrix decomposition by orthogonal matrices or by eigenvectors. Further, if all the eigenvalues are nonnegative, then the matrix is said to be *positive semi-definite*, and the formula can be written as

$$\mathbf{A}_{n \times n} = \sum_{k=1}^{n} \left(\mathbf{v}^{(k)} \right)_{n \times 1} \left((\mathbf{v}^{(k)})^t \right)_{1 \times n}, \tag{5.84}$$

where $\mathbf{v}^{(k)} = \sqrt{\lambda_k} \mathbf{q}^{(k)} (k = 1, 2, \ldots, n)$. The sample covariance matrix in climate science satisfies the positive eigenvalue assumption, and will be discussed in more details in the next chapter. Equation (5.84) means that a positive semi-definite symmetric matrix $\mathbf{A}_{n \times n}$ can be decomposed into a sum of n outer products of eigenvectors.

Matrix $\mathbf{A}$,

$$\mathbf{A} = \begin{bmatrix} 1 & 2 \\ 2 & 1 \end{bmatrix}, \tag{5.85}$$

is not positive semi-definite since its second eigenvalue is -1, but matrix $\mathbf{C}$,

$$\mathbf{C} = \begin{bmatrix} 2 & 1 \\ 1 & 2 \end{bmatrix}, \tag{5.86}$$

is positive semi-definite since its eigenvalues are 3 and 1, which are positive. The matrix $\mathbf{C}$ is actually positive definite. The eigenvectors are the same those of $\mathbf{A}$. Thus,

$$\mathbf{C} = \mathbf{Q} \mathbf{D}_c \mathbf{Q}^t, \tag{5.87}$$

where $\mathbf{Q}$ is given by Eq. (5.81), and $\mathbf{D}_c$ is

$$\mathbf{D}_c = \begin{bmatrix} 3 & 0 \\ 0 & 1 \end{bmatrix}. \tag{5.88}$$

This can be verified by the following computer code.

```
#Verify diagonalization and decomposition: R code
C = matrix(c(2,1,1,2), nrow = 2)
eigen(C)
#$values
#[1] 3 1
#$vectors
#    [,1]         [,2]
#[1,] 0.7071068 -0.7071068
#[2,] 0.7071068  0.7071068
Q = eigen(C)$vectors
D = t(Q)%*%C%*%Q #Matrix diagonalization
D
#[1,]     3     0
#[2,]     0     1
Q%*%D%*%t(Q) #Matrix decomposition
#[1,]     2     1
#[2,]     1     2
D[1,1]*Q[,1]%*%t(Q[,1]) + D[2,2]*Q[,2]%*%t(Q[,2])
#[1,]     2     1
#[2,]     1     2
```

```
#Verify diagonalization and decomposition: Python code
C = [[2,1],[1,2]]
valC, Q = np.linalg.eig(C)
print('eigenvalues␣of␣C␣=', valC)
#eigenvalues of C = [3. 1.]
print('eigenvectors␣of␣C␣=', Q)
#eigenvectors of C = [[ 0.70710678 -0.70710678]
# [ 0.70710678  0.70710678]]
D = Q.transpose(1,0).dot(C).dot(Q)
print('D␣=␣', D)
#D =   [[3. 0.]
# [0. 1.]]
# Matrix C is decomposed into three matrices: C = Q D Q'
Q.dot(D).dot(Q.transpose(1,0))
#array([[2., 1.],
#        [1., 2.]])
D[0][0]*np.outer(Q[:][0],Q.transpose()[:][0]) + \
D[1][1]*np.outer(Q[:][1],Q.transpose()[:][1])
#array([[1., 2.],
#        [2., 1.]]) #matrix C is recovered from vectors
```

5.5 Singular Value Decomposition

The previous section shows an eigenvector-eigenvalue decomposition of a symmetric square matrix. A similar decomposition can be made for a rectangular matrix. The decomposition using unit eigenvectors and eigenvalues for a general rectangular matrix is called the *singular value decomposition* (SVD). Singular value is another name for eigenvalue. Although the basic mathematical theory of SVD was developed almost 200 years ago (Stewart 1993), the modern algorithm of efficient SVD computing was only developed by Gene H. Golub (1932–2007) and his colleagues in the 1970s. Now, SVD has become an important data analysis tool for every field: climate science is not an exception. A space-time climate data matrix is often a rectangular matrix, since the number of sites is not likely to be equal to the number of temporal observations at those sites. We may denote a space-time climate data matrix by $\mathbf{X}_{n \times m}$, where n is the number of sites, and m is the total number of temporal observations. For $\mathbf{X}_{n \times m}$, we can also interpret n as the number of grid boxes of a climate model output, and m as the number of time steps in the output.

5.5.1 SVD Formula and a Simple SVD Example

If $m \leq n$, then matrix $\mathbf{A}_{n \times m}$ has the following SVD decomposition:

$$\mathbf{A}_{n \times m} = \mathbf{U}_{n \times m} \mathbf{D}_{m \times m} (\mathbf{V}^t)_{m \times m}. \tag{5.89}$$

Here, **U** may be interpreted as a spatial matrix, consisting of m spatial orthonormal column vectors that are unit eigenvectors of $\mathbf{AA}^t$; **V** may be interpreted as a temporal matrix, consisting of m temporal orthonormal column vectors that are unit eigenvectors of $\mathbf{A}^t\mathbf{A}$; and **D** is a diagonal matrix whose elements are the square root of the eigenvalues of $\mathbf{AA}^t$ and may be interpreted as standard deviations. Figure 5.3 may help you understand this formula. The diagonal elements of **D** are called singular values, and the column vectors of **U** and **V** are called singular vectors. Here, the word "singular" may be understood as special (or opposite to "general"), or distinguished or out of ordinary.

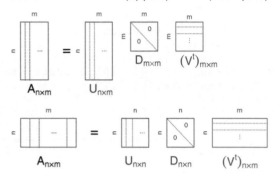

SVD: A = UDV^t when n > m (top panel) or n < m (bottom panel)

Figure 5.3 Schematic diagrams of SVD: $A = UDV^t$. Case 1: $n > m$ (top panel). Case 2: $n < m$ (bottom panel).

If $m \geq n$, then matrix $\mathbf{A}_{n \times m}$ has the following SVD decomposition:

$$\mathbf{A}_{n \times m} = \mathbf{U}_{n \times n}\mathbf{D}_{n \times n}(\mathbf{V}^t)_{n \times m}. \tag{5.90}$$

The following computer code shows a simple SVD example of a 2×3 matrix.

```
#SVD example for a 2-by-3 matrix: R code
A=matrix(c(-1,1,0,2,-2,3),nrow=2)
A #Show the 2-by-3 matrix
#     [,1] [,2] [,3]
#[1,]   -1    0   -2
#[2,]    1    2    3
svdA=svd(A) #Compute the SVD of A and put the results in svdA
svdA #Show SVD results: d, U, and V
round(svdA$d, digits=2) #Show only the singular values
#[1] 4.22 1.09
round(svdA$u, digits=2) #Show only matrix U
#     [,1] [,2]
#[1,] -0.48 0.88
#[2,]  0.88 0.48
round(svdA$v, digits=2)#Show only matrix V
#     [,1]  [,2]
#[1,] 0.32 -0.37
#[2,] 0.42  0.88
#[3,] 0.85 -0.29
sqrt(eigen(A%*%t(A))$values)
#[1] 4.221571 1.085514
```

```
#SVD example for a 2-by-3 matrix: Python code
A = [[-1,0, -2],[1,2,3]]
UsvdA, DsvdA, VsvdA = np.linalg.svd(A)
print('Singular␣values=␣', np.round(DsvdA,2))
#Singular values=  [4.22 1.09]
print('Spatial␣singular␣vectors=␣', np.round(UsvdA,2))
#Spatial singular vectors=  [[-0.48  0.88]
# [ 0.88  0.48]]
print('Temporal␣singular␣vectors=␣', np.round(VsvdA,2))
#Temporal singular vectors=  [[ 0.32  0.42  0.85]
# [-0.37  0.88 -0.29]
# [-0.87 -0.22  0.44]]

B = np.array(A)
C = np.matmul(B, B.T) #B times B transpose
valC, vecC = np.linalg.eig(C)
np.sqrt(valC)
#array([1.0855144 , 4.22157062])
```

The computer code shows the following SVD results expressed in following mathematical formulas:

(a) The vector form:

$$A = \begin{bmatrix} -1 & 0 & -2 \\ 1 & 2 & 3 \end{bmatrix}$$

$$= 4.22 \begin{bmatrix} -0.48 \\ 0.88 \end{bmatrix} \times \begin{bmatrix} 0.32 & 0.42 & 0.85 \end{bmatrix} +$$

$$1.09 \begin{bmatrix} 0.88 \\ -0.48 \end{bmatrix} \times \begin{bmatrix} -0.37 & 0.88 & -0.29 \end{bmatrix} \tag{5.91}$$

and

(b) The matrix form:

$$A = \begin{bmatrix} -1 & 0 & -2 \\ 1 & 2 & 3 \end{bmatrix} = \begin{bmatrix} -0.48 & 0.88 \\ 0.88 & 0.48 \end{bmatrix} \begin{bmatrix} 4.22 & 0 \\ 0 & 1.09 \end{bmatrix} \begin{bmatrix} 0.32 & 0.42 & 0.85 \\ -0.37 & 0.88 & -0.29 \end{bmatrix} \tag{5.92}$$

If we use only the first singular vectors to approximate A from the two triplets of singular vectors and singular values, the result is as follows.

```
#Data reconstruction by singular vectors: R code
round(svdA$d[1]*svdA$u[,1]%*%t(svdA$v[,1]),
      digits=1)
#      [,1] [,2] [,3]
#[1,] -0.7 -0.8 -1.7
#[2,]  1.2  1.5  3.2
```

It is quite close to A.

```
#Data reconstruction by singular vectors: Python code
np.round(DsvdA[0]*np.outer(UsvdA[:][0], VsvdA[:][0]),1)
#array([[-0.7, -0.8, -1.7],
#        [ 1.2,  1.5,  3.2]])
```

If we use both singular vectors to reconstruct *A*, then the reconstruction is exact without errors, as expected.

```
round(svdA$d[1]*svdA$u[,1]%*%t(svdA$v[,1]) +
  svdA$d[2]*svdA$u[,2]%*%t(svdA$v[,2]),
  digits =2)
#      [,1] [,2] [,3]
#[1,]    -1    0   -2
#[2,]     1    2    3
```

```
A1 = DsvdA[0]*np.outer(UsvdA[:][0], VsvdA[:][0])
A2 = DsvdA[1]*np.outer(UsvdA[:][1], VsvdA[:][1])
np.round(A1 + A2, 2)
#array([[-1.,  0., -2.],
#        [ 1.,  2.,  3.]])
```

Figure 5.3 for the schematic diagram of SVD may be plotted by the following computer code.

```
#R plot Fig. 5.3:  Schematic diagram of SVD
setwd('/Users/sshen/climstats')
setEPS() #Plot the figure and save the file
postscript("fig0503.eps", width = 11)
par(mar=c(0,0,0,0))
plot(200, axes = FALSE,
     xlab = "", ylab = "",
     xlim = c(-3,28), ylim = c(-3,16))
text(13,15.5, cex=2.2,
     bquote("SVD:" ~ A==UDV^t~ "when n > m or n < m"))
#Space-time data matrix A when n>m
segments(x0 = c(0,0,3,3),
         y0 = c(6,12,12,6) +1,
         x1 = c(0,3,3,0),
         y1 = c(12,12,6,6) +1,
         col = c('blue','red','blue','red'),lwd =3)
segments(x0 = c(0.5,1.0),
         y0 = c(6,6)+1,
         x1 = c(0.5,1.0),
         y1 = c(12,12)+1,
         lwd =1.3, lty = 3)
text(-.8, 9+1, 'n', srt=90, col ='blue', cex = 1.4)
text(1.5, 12.8+1, 'm', col = 'red',  cex = 1.4)
text(2.0, 9+1, '...',  cex = 1.4)
text(2, 5+1, bquote(A[n%*%m]),  cex = 2.5)
text(5, 9+1, '=',  cex = 3)
```

```
#Spatial matrix U
segments(x0 = c(7,7,10,10),
        y0 = c(6,12,12,6)+1,
        x1 = c(7,10,10,7),
        y1 = c(12,12,6,6)+1,
        col = c('blue','blue','blue','blue'), lwd =3)
segments(x0 = c(7.5,8),
        y0 = c(6,6)+1,
        x1 = c(7.5,8),
        y1 = c(12,12)+1,
        lwd =1.3, lty = 3, col = 'blue')
text(6.2, 9+1, 'n', srt=90, col ='blue', cex = 1.4)
text(8.5, 12.8+1, 'm', col = 'red',  cex = 1.4)
text(9, 9+1, '...',  cex = 1.4, col='blue')
text(8.7, 5.0+1, bquote(U[n%*%m]),  cex = 2.5, col= 'blue')
#Singular value diagonal matrix D
segments(x0 = c(12,12,15,15),
        y0 = c(9,12,12,9)+1,
        x1 = c(12,15,15,12),
        y1 = c(12,12,9,9)+1,
        col = c('brown','brown','brown','brown'), lwd =3)
segments(x0 = 12, y0 = 12+1, x1 = 15, y1 = 9+1, lty=3,
        col = c('brown'), lwd =1.3)#diagonal line
text(11.2, 10.5+1, 'm', srt=90, col ='red', cex = 1.4)
text(13.5, 12.8+1, 'm', col = 'red',  cex = 1.4)
text(14.1, 11.3+1, '0', col = 'brown',  cex = 1.4)
text(12.9, 10.0+1, '0', col = 'brown',  cex = 1.4)
text(13.9, 8.0+1, bquote(D[m%*%m]),  cex = 2.5, col='brown')
#Temporal matrix V
segments(x0 = c(17,17,20,20),
        y0 = c(9,12,12,9)+1,
        x1 = c(17,20,20,17),
        y1 = c(12,12,9,9)+1,
        col = c('red','red','red','red'), lwd =3)
segments(x0 = c(17,17),
        y0 = c(11.5,10.8)+1,
        x1 = c(20,20),
        y1 = c(11.5,10.8)+1,
        col = c('red','red'), lty=3, lwd =1.3)
text(16.2, 10.5+1, 'm', srt=90, col ='red', cex = 1.4)
text(18.5, 12.5+1, 'm', col = 'red',  cex = 1.4)
text(19.5, 8+1, bquote((V^t)[m%*%m]),  cex = 2.5, col='red')
text(18.5, 10+1, '...',  col='red', srt=90, cex =1.4)
#Space-time data matrix B when n < m
segments(x0 = c(0,0,6,6),
        y0 = c(0,3,3,0),
        x1 = c(0,6,6,0),
        y1 = c(3,3,0,0),
        col = c('blue','red','blue','red'), lwd =3)

segments(x0 = c(1,2,5),
        y0 = c(0,0,0),
        x1 = c(1,2,5),
        y1 = c(3,3,3),
        lwd =1.3, lty = 3)
text(-0.8, 1.5, 'n', srt=90, col ='blue', cex = 1.4)
text(3, 3.8, 'm', col = 'red',  cex = 1.4)
```

```
text(3.5, 1.5, '...',  cex = 1.4)
text(3, -1.5, bquote(A[n%*%m]),  cex = 2.5)
text(8, 1.5, '=',  cex = 3)
#Spatial matrix U
segments(x0 = c(11,11,14,14),
         y0 = c(0,3,3,0),
         x1 = c(11,14,14,11),
         y1 = c(3,3,0,0),
         col = c('blue','blue','blue','blue'), lwd =3)
segments(x0 = c(11.5,12.2),
         y0 = c(0,0),
         x1 = c(11.5,12.2),
         y1 = c(3,3),
         lwd =1.3, lty = 3, col = 'blue')
text(10.2, 1.5, 'n', srt=90, col ='blue', cex = 1.4)
text(12.5, 3.8, 'n', col = 'blue',  cex = 1.4)
text(13.2, 1.5, '...',  cex = 1.4, col='blue')
text(12.5, -1.5, bquote(U[n%*%n]),  cex = 2.5, col= 'blue')
#Singular value diagonal matrix D
segments(x0 = c(16,16,19,19),
         y0 = c(0,3,3,0),
         x1 = c(16,19,19,16),
         y1 = c(3,3,0,0),
         col = c('brown','brown','brown','brown'), lwd =3)
segments(x0 = 16, y0 = 3, x1 = 19, y1 = 0, lty=3,
         col = c('brown'), lwd =1.3)#diagonal line
text(15.2, 1.5, 'n', srt=90, col ='blue', cex = 1.4)
text(17.5, 3.8, 'n', col = 'blue',  cex = 1.4)
text(18.1, 2.3, '0', col = 'brown',  cex = 1.4)
text(16.9, 1.0, '0', col = 'brown',  cex = 1.4)
text(17.5, -1.5, bquote(D[n%*%n]),  cex = 2.5, col='brown')
#Temporal matrix V
segments(x0 = c(21,21,27,27),
         y0 = c(0,3,3,0),
         x1 = c(21,27,27,21),
         y1 = c(3,3,0,0),
         col = c('red','red','red','red'),
         lwd =3)
segments(x0 = c(21,21),
         y0 = c(2.5,1.8),
         x1 = c(27,27),
         y1 = c(2.5,1.8),
         col = c('red','red'), lty=3, lwd =1.3)
text(20.2, 1.5, 'n', srt=90, col ='blue', cex = 1.4)
text(24, 3.8, 'm', col = 'red',  cex = 1.4)
text(24, -1.5, bquote((V^t)[n%*%m]),  cex = 2.5, col='red')
text(24, 1, '...',  col='red', srt=90, cex =1.4)
dev.off()
```

```
#Python plot Fig. 5.3: Schematic diagram of SVD
import matplotlib.patches as patches
import numpy as np
import pylab as pl
```

```python
from matplotlib import collections  as mc
lines = [[(0, 7), (0, 13)], [(0, 13), (3, 13)],
         [(3, 13), (3, 7)], [(3, 7), (0,7)]]
c = np.array(['b', 'r', 'b', 'r'])
lc = mc.LineCollection(lines, colors=c, linewidths=3)
fig, ax = pl.subplots()
ax.set_xlim([-3, 28])
ax.set_ylim([-3, 16])
ax.add_collection(lc)
ax.margins(0.1)
plt.plot([0.5,0.5], [7, 13],
         linestyle='dotted', color = 'k')
plt.plot([1, 1], [7, 13],
         linestyle='dotted', color = 'k')
plt.text(13, 15.5,
         r'SVD:  $A==UDV^t$ when n > m or n < m',
         fontsize = 30)
plt.text(-1.2, 10, 'n', color = 'blue',
         fontsize =25, rotation=90)
plt.text(1.1, 13.3, 'm', color = 'red',
         fontsize =25, rotation=0)
plt.text(0.0, 5.5, r'$A_{n\times m}$',
         fontsize =35, rotation=0)
plt.text(1.5, 10, '...',
         fontsize =25, rotation=0)
plt.axis('off')
plt.show()
```

This Python code generates the top-left rectangular box in Figure 5.3. The remaining code for other boxes is highly repetitive and can be found from the book website.

5.6 SVD for the Standardized Sea Level Pressure Data of Tahiti and Darwin

The Southern Oscillation Index (SOI) is an indicator for El Niño or La Niña. It is computed as the difference of sea level pressure (SLP) of Tahiti ($17.75°$ S, $149.42°$ W) minus that of Darwin ($12.46°$ S, $130.84°$ E). An SVD analysis of the SLP data can substantiate this calculation formula.

The following shows the data matrix of the standardized SLP anomalies of Tahiti and Darwin from 2009 to 2015, and its SVD.

```r
#R SVD analysis for the weighted SOI from SLP data
setwd("/Users/sshen/climmath")
Pda<-read.table("data/PSTANDdarwin.txt", header=F)
dim(Pda)
#[1] 65 13 #Monthly Darwin data from 1951-2015
pdaDec<-Pda[,13] #Darwin Dec standardized SLP anomalies data
Pta<-read.table("data/PSTANDtahiti.txt", header=F)
ptaDec=Pta[,13] #Tahiti Dec standardized SLP anomalies
```

```
ptada1 = cbind(pdaDec, ptaDec) #space-time data matrix

#Space-time data format
ptada = t(ptada1[59:65,]) #2009-2015 data
colnames(ptada)<-2009:2015
rownames(ptada)<-c("Darwin", "Tahiti")
ptada #6 year of data for two stations
#         2009 2010 2011 2012 2013 2014 2015
#Darwin   0.5 -2.3 -2.2  0.3  0.3  0.1 -0.4
#Tahiti  -0.7  2.5  1.9 -0.7  0.4 -0.8 -1.3
svdptd = svd(ptada) #SVD for the 2-by-6 matrix
U=round(svdptd$u, digits=2)
U
#[1,] -0.66 0.75
#[2,]  0.75 0.66
D=round(diag(svdptd$d), digits=2)
D
#[1,]   4.7 0.00
#[2,]   0.0 1.42
V =round(svdptd$v, digits=2)
t(V)
#[1,] -0.18  0.72  0.61 -0.15 0.02 -0.14 -0.15
#[2,] -0.06 -0.06 -0.28 -0.17 0.34 -0.32 -0.82
```

```
#Python SVD analysis of the Darwin and Tahiti SLP data
import os
os.chdir("/Users/sshen/climstats")
PDA = np.array(read_table("data/PSTANDdarwin.txt", \
                          header = None, delimiter = "\s+"))
PTA = np.array(read_table("data/PSTANDtahiti.txt", \
                          header = None, delimiter = "\s+"))
pdata = np.stack([PDA[58:65,12], PTA[58:65,12]], axis=0)
print('The Darwin and Tahiti SLP data 2009-2015 =', pdata)
#The Darwin and Tahiti Standardized SLP anomalies =
#[[ 0.5 -2.3 -2.2  0.3  0.3  0.1 -0.4]
# [-0.7  2.5  1.9 -0.7  0.4 -0.8 -1.3]]
u, d, v = np.linalg.svd(pdata)
print('Spatial singular vectors EOFs U =', np.round(u,2))
#Spatial singular vectors EOFs U = [[-0.66  0.75]
# [ 0.75  0.66]]
print('Diagonal matrix D =', np.round(np.diag(d),2))
#Diagonal matrix D = [[4.7  0.  ]
# [0.   1.42]]
print('Temporal singular vectors PCs V=', np.round(v,2))
#Temporal singular vectors PCs V=
#[[-0.18  0.72  0.61 -0.15  0.02 -0.14 -0.15]
# [-0.06 -0.06 -0.28 -0.17  0.34 -0.32 -0.82] ...]
```

One can verify that

$$\mathbf{UDV}^t \tag{5.93}$$

approximately recovers the original data matrix.

The first column vector $(-0.66, 0.75)$ of the spatial pattern matrix $\mathbf{U}$ may be interpreted to be associated with the SOI, which puts a negative weight -0.66 on Darwin, and a positive weight 0.75 on Tahiti. The weighted sum is approximately equal to the difference of Tahiti's SLP minus that of Darwin, which is the definition of SOI. The index measures large-scale ENSO dynamics of the tropical Pacific (Trenberth 2020). The magnitude of the vector $(-0.66, 0.75)$ is approximately 1, because U is a unitary matrix. The corresponding first temporal singular vector has a distinctly large value 0.72 in December 2010, which was a strong La Niña month. In this month, the Darwin had a strong negative SLP anomaly, while Tahiti had a strong positive SLP anomaly. This situation enhanced the easterly trade winds in the tropical Pacific and caused abnormally high precipitation in Australia in the 2010–2011 La Niña period.

The second column vector $(0.75, 0.66)$ of the spatial pattern matrix $\mathbf{U}$ also has climate implications. The weighted sum with two positive equal weights measures the small scales tropical Pacific dynamics (Trenberth 2020).

5.7 Chapter Summary

A matrix can be regarded as a 2-dimensional $n \times m$ rectangular array of numbers or symbols, or as m column vectors, or as n row vectors. This may be interpreted as climate data at n locations with m time steps. Many mathematical properties of a matrix of climate data have climate interpretations. For example, SVD decomposes a space-time climate data matrix into an orthogonal spatial pattern matrix, an orthogonal temporal pattern matrix, and a diagonal "energy-level" matrix that measures the standard deviation of the temporal pattern:

$$\mathbf{A} = \mathbf{UDV}^t. \tag{5.94}$$

The column vectors of the spatial matrix $\mathbf{U}$ are spatial singular vectors, also called EOFs, while those of the temporal matrix $\mathbf{V}$ are temporal singular vectors, also called PCs. The first few EOFs often have climate dynamic interpretations, such as El Niño Southern Oscillation (ENSO). If EOF1 corresponds to El Niño and shows some typical ENSO properties, such as the opposite signs of SLP anomalies of Darwin and Tahiti, then PC1 shows a temporal pattern, e.g., the extreme values of PC1 indicating both the occurrence time and the strength of El Niño. The diagonal elements of matrix $\mathbf{D}$ are singular values, also known as eigenvalues.

An eigenvector $\mathbf{v}$ of a square matrix $\mathbf{C}$ is a special vector such that $\mathbf{C}$'s action on $\mathbf{v}$ does not change its orientation, i.e., $\mathbf{Cv}$ is parallel to $\mathbf{v}$. This statement implies the existence of a scaler λ, called eigenvalue (also known as singular value or characteristic value), such that

$$\mathbf{Cv} = \lambda \mathbf{v}. \tag{5.95}$$

We have also discussed the matrix method of solving a system of linear equations

$$\mathbf{Ax} = \mathbf{b}, \tag{5.96}$$

linear independence of vectors, linear transform, and other basic matrix methods. These methods are useful for the chapters on covariance, EOFs, spectral analysis, regression analysis, and machine learning. You may focus on the computing methods and computer code of the relevant methods. The mathematical proofs of this chapter, although helpful for exploring new mathematical methods, are not necessarily needed to read the other chapters of this book. If you are interested in an in-depth mathematical exploration of matrix theory, you may wish read the books by Horn and Johnson (1985) and Strang (2016).

References and Further Reading

[1] G. H. Golub and C. Reinsch, 1970: Singular value decomposition and least squares solutions. *Numerische Mathematik*, 14, 403–420.

> This seminal paper established an important method, known as the Golub–Reinsch algorithm, to compute the eigenvalues of a covariance matrix from a space-time matrix A without actually first computing the covariance matrix AA^t. This algorithm makes the SVD computation very efficient, which helps scientists consider SVD as a genuine linear algebra method, not a traditionally regarded statistical method based on a covariance matrix.

[2] R. A. Horn and C. R. Johnson, 1985: *Matrix Analysis*. Cambridge University Press.

> This is a comprehensive book on matrix theory and is a good reference for a researcher in climate statistics. It assumes knowledge of a first course of linear algebra.

[3] G. Strang, 2016: *Introduction to Linear Algebra*. 5th ed., Wellesley-Cambridge Press.

> Gilbert Strang (1934–) is an American mathematician and educator. His textbooks and pedagogy have been internationally influential. This text is one of the very few basic linear algebra books that includes excellent materials on SVD, probability, and statistics.

[4] G. W. Stewart, 1993: On the early history of the singular value decomposition. *SIAM Review*, 35, 551–566.

> This paper describes the contributions from five mathematicians in the period of 1814–1955 to the development of the basic SVD theory.

[5] K. Trenberth, and National Center for Atmospheric Research Staff (eds.), 2020: The Climate Data Guide: Southern Oscillation Indices: Signal, Noise and Tahiti/Darwin SLP (SOI).

https://climatedataguide.ucar.edu/climate-data/southern-
oscillation-indices-signal-noise-and-tahitidarwin-slp-soi

> This site describes the optimal indices for large- and small-scale dynamics.

[6] A. Tucker, 1993: The growing importance of linear algebra in undergraduate mathematics. *College Mathematics Journal*, 24, 3–9.

> This paper describes the historical development of linear algebra, such as the term "matrix" being coined by J. J. Sylvester in 1848, and pointed out that "tools of linear algebra find use in almost all academic fields and throughout modern society." The use of linear algebra in the big data era is now even more popular.

Exercises

5.1 Write a computer code to

(a) Read the NOAAGlobalTemp data file, and

(b) Generate a 4×8 space-time data matrix for the December mean surface air temperature anomaly data of four grid boxes and eight years. *Hint: You may find the NOAA Global Surface Temperature (NOAAGlobalTemp) dataset online. You can use either netCDF format or CSV format.*

5.2 Write a computer code to find the inverse of the following matrix.

```
#        [,1]  [,2]  [,3]
#[1,]    1.7  -0.7   1.3
#[2,]   -1.6  -1.4   0.4
#[3,]   -1.5  -0.3   0.6
```

5.3 Write a computer code to solve the following linear system of equations:

$$\mathbf{Ax} = \mathbf{b}, \tag{5.97}$$

where

$$\mathbf{A} = \begin{bmatrix} 1 & 2 & 3 \\ 4 & 5 & 6 \\ 7 & 8 & 0 \end{bmatrix} \quad \mathbf{x} = \begin{bmatrix} x_1 \\ x_2 \\ x_3 \end{bmatrix} \quad \mathbf{b} = \begin{bmatrix} 1 \\ -1 \\ 0 \end{bmatrix}. \tag{5.98}$$

5.4 The following equation

$$\begin{bmatrix} 1 & 2 & 3 \\ 4 & 5 & 6 \\ 7 & 8 & 9 \end{bmatrix} \begin{bmatrix} x_1 \\ x_2 \\ x_3 \end{bmatrix} = \begin{bmatrix} 0 \\ 0 \\ 0 \end{bmatrix} \tag{5.99}$$

has infinitely many solutions, and cannot be directly solved by a simple computer command, such as `solve(A, b)`.

(a) Show that the three row vectors of the coefficient matrix are not linearly independent.

(b) Because of the dependence, the linear system has only two independent equations. Thus, reduce the linear system to two equations by treating x_3 as an arbitrary value while treating x_1 and x_2 as variables.

(c) Solve the two equations for x_1 and x_2 and express them in terms of x_3. The infinite possibilities of x_3 imply infinitely many solutions of the original system.

5.5 Ethane is a gas similar to the greenhouse gas methane and can burn with oxygen to form carbon dioxide and water:

$$C_2H_6 + O_2 \longrightarrow CO_2 + H_2O. \tag{5.100}$$

Given two ethane molecules, how many molecules of oxygen, carbon dioxide and water are required for this chemical reaction equation to be balanced? *Hint: Assume x, y, and z molecules of oxygen, carbon dioxide, and water, respectively, and use the balance of the number of atoms of carbon, hydrogen, and oxygen to form linear equations. Solve the system of linear equations.*

5.6 Carry out the same procedure as the previous problem but for the burning of methane CH_4.

5.7 (a) Use matrix multiplication to show that the vector

$$\mathbf{u} = \begin{bmatrix} 1 \\ 1 \end{bmatrix} \tag{5.101}$$

is not an eigenvector of the following matrix:

$$\mathbf{A} = \begin{bmatrix} 0 & 4 \\ -2 & -7 \end{bmatrix} \tag{5.102}$$

(b) Find all the unit eigenvectors of matrix $\mathbf{A}$ in (a).

5.8 Use hand calculation to compute the matrix multiplication of $\mathbf{UDV}^t$ where the data of relevant matrices are given by the following R output.

```
A=matrix(c(1,-1,1,1),nrow=2)
A
#[1,]    1    1
#[2,]   -1    1
svd(A)
#$d
#[1] 1.414214 1.414214
#$u
#[1,] -0.7071068 0.7071068
#[2,]  0.7071068 0.7071068
#$v
#[1,]   -1    0
#[2,]    0    1
```

5.9 Use a computer to find the matrices $\mathbf{U}, \mathbf{D}$ and $\mathbf{V}$ of the SVD for the following data matrix:

$$\mathbf{A} = \begin{bmatrix} 1 & 2 & 3 & 4 & 5 \\ 6 & 7 & 8 & 9 & 10 \\ 11 & 12 & 13 & 14 & 15 \end{bmatrix}. \tag{5.103}$$

5.10   Use the first singular vectors from the result of Exercise 5.9 to approximately reconstruct the data matrix $\mathbf{A}$ using

$$B = d_1 \mathbf{u}_1 \mathbf{v}_1^t. \tag{5.104}$$

Describe the goodness of the approximation using text, limited to 20 to 100 words.

5.11   For the following data matrix

$$\mathbf{A} = \begin{bmatrix} 1.2 & -0.5 & 0.9 & -0.6 \\ 1.0 & -0.7 & -0.4 & 0.9 \\ -0.2 & 1.1 & 1.6 & -0.4 \end{bmatrix}, \tag{5.105}$$

(a)   Use a computer to find the eigenvectors and eigenvalues of matrix $\mathbf{AA}^t$.

(b)   Use a computer to find the eigenvectors and eigenvalues of matrix $\mathbf{A}^t\mathbf{A}$.

(c)   Use a computer to calculate SVD of $\mathbf{A}$.

(d)   Compare the singular vectors and singular values in Step (c) with the eigenvalues and eigenvectors computed in Steps (a) and (b). Use text to discuss your comparison.

(e)   Use a computer to verify that the column vectors of $\mathbf{U}$ and $\mathbf{V}$ in the SVD of Step (c) are orthonormal to each other.

5.12   Conduct an SVD analysis of the December standardized anomalies data of sea level pressure (SLP) at Darwin and Tahiti from 1961 to 2010. You can find the data from the Internet or from the website of this book.

(a)   Write a computer code to organize the data into a 2×50 space-time data matrix.

(b)   Make the SVD calculation for this space-time matrix.

(c)   Plot the first singular vector in $\mathbf{V}$ against time from 1961 to 2010 as a time series curve, which is called the first principal component, denoted by PC1.

(d)   Interpret the first singular vector in $\mathbf{U}$, which is called the first empirical orthogonal function (EOF1), as weights of Darwin and Tahiti stations.

(e)   Check the historical El Niño events between 1961 and 2010 from the Internet, and interpret the extreme values of PC1.

5.13   Plot PC2 against time from the SVD analysis in the previous problem. Discuss the singular values λ_1 and λ_2. Interpret PC2 in reference to Trenberth (2020).

5.14   Conduct an SVD analysis similar to the previous two problems for the January standardized SLP anomalies data at Darwin and Tahiti from 1961 to 2010.

5.15   Conduct an SVD analysis similar to the previous problem for the monthly standardized SLP anomalies data at Darwin and Tahiti from 1961 to 2010. This problem includes anomalies for every month. The space-time data is a 2×600 matrix.

5.16   For the observed data of the monthly surface air temperature at five stations of your interest from January 1961 to December 2010, form a space-time data matrix, compute the SVD of this matrix, and interpret your results from the perspective of climate science. Plot PC1, PC2, and PC3 against time. You may find your data from the internet, e.g., the NOAA Climate Data Online website www.ncdc.noaa.gov/cdo-web

5.17   Do the same analysis as the previous problem, but for the monthly precipitation data at the same stations in the same time period.

5.18   For a Reanalysis dataset, conduct an SVD analysis similar to the previous problem for the monthly surface temperature data over 10 grid boxes of your choice in the time period of 1961–2010. The space-time data is a 10×600 matrix. You can use your preferred Reanalysis dataset and download the data from the Internet, e.g., NCEP/NCAR Reanalysis, and ECMWF ERA.

5.19 Let $\mathbf{C} = \mathbf{AA}^t$ and $\mathbf{A}$ is any real-valued rectangular matrix. Show that

(a) the eigenvalues of $\mathbf{C}$ are nonnegative;

(b) if $\mathbf{v} = \mathbf{A}^t\mathbf{u}$, then $\mathbf{v}$ is a eigenvector of $\mathbf{C}' = \mathbf{A}^t\mathbf{A}$.

5.20 Given that

$$\mathbf{A} = \begin{bmatrix} 2 & -1 \\ -1 & 3 \end{bmatrix}, \quad \mathbf{x} = \begin{bmatrix} x_1 \\ x_2 \end{bmatrix}, \tag{5.106}$$

write down the second-order polynomial corresponding to the following matrix expression:

$$P(x_1, x_2) = \mathbf{x}^t\mathbf{AA}^t\mathbf{x}. \tag{5.107}$$

This is known as the quadratic form of the matrix $\mathbf{AA}^t$.

5.21 If $\mathbf{x}$ is a unit vector, use calculus to find the maximum value of $P(x_1, x_2)$ in the previous problem. How is your solution related to the eigenvalue of $\mathbf{AA}^t$?

## 6

Covariance Matrices, EOFs, and PCs

Covariance of climate data at two spatial locations is a scaler. The covariances of climate data at many locations form a matrix. The eigenvectors of a covariance matrix are defined as empirical orthogonal functions (EOFs), which vary in space. The orthonormal projections of the climate data on EOFs yield principal components (PCs), which vary in time. These EOFs and PCs are commonly used in climate data analysis. Physically they may be interpreted as the spatial and temporal patterns or dynamics of a climate process. The eigenvalues of the covariance matrix represent the variances of the climate field for different EOF patterns. The EOFs defined by a covariance matrix are mathematically equivalent to the SVD definition of EOFs, and the SVD definition is computationally more convenient when the space-time data matrix is not too big. The covariance definition of EOFs provides ways of interpreting climate dynamics, such as how variance is distributed across the different EOF components.

This chapter describes covariance and EOFs for both climate data and stochastic climate fields. The EOF patterns can be thought of as statistical modes. The chapter not only includes the rigorous theory of EOFs and their analytic representations, but also discusses commonly encountered problems in EOF calculations and applications, such as area-factor, time-factor, and sampling errors of eigenvalues and EOFs. We pay particular attention to independent samples, North's rule of thumb, and mode mixing.

6.1 From a Space-Time Data Matrix to a Covariance Matrix

A space-time data matrix can be used to define a covariance matrix. Let $\mathbf{X}_{np}$ be a space-time data matrix with n spatial stations or grid boxes, and p time steps:

$$\mathbf{X} = \begin{bmatrix} x_{11} & x_{12} & \cdots & x_{1p} \\ x_{21} & x_{22} & \cdots & x_{2p} \\ \vdots & \vdots & \vdots & \vdots \\ x_{n1} & x_{n2} & \cdots & x_{np} \end{bmatrix}. \tag{6.1}$$

Each row vector $\mathbf{x}_{i:}$ is the time series data for a station i or a region or a grid box (either 2D or 3D), and each column vector $x_{:t}$ is the spatial map data for a given time step t.

Here, ":" means all the elements of the ith row or tth column. Each pair of stations i, j has a covariance

$$C_{ij} = \text{Cov}(\mathbf{x}_i, \mathbf{x}_j) \tag{6.2}$$

These C_{ij}, or denoted by $C(i,j)$, form the covariance matrix, which is apparently symmetric.

For simplicity, suppose the sample mean of the station data is zero (averaging through time for each station):

$$\bar{x}_i = \frac{1}{p}\sum_{t=1}^{p} x_{it} = 0. \tag{6.3}$$

In other words, we are dealing with *anomaly* data, which means the station means have been removed. This last step simplifies the notation. We have the sample covariance

$$S_{ij} = \frac{1}{p}\sum_{t=1}^{p} x_{it}x_{jt} = \sum_{t=1}^{t} \frac{x_{it}}{\sqrt{p}}\frac{x_{jt}}{\sqrt{p}}. \tag{6.4}$$

Here $1/\sqrt{p}$ is called the *time-factor* of climate data in the time interval of length p.

Similarly, the space-factor of climate data can be introduced since each station or grid box may represent different sizes of areas in the case of a surface domain, volumes in the case of a 3D domain, and lengths in case of a line segment domain. If the ith grid box has an area A_i, then the area-factor for this box is $\sqrt{A_i}$. The climate data x_{it} with the area-factor is then

$$\sqrt{A_i}\,x_{it}. \tag{6.5}$$

The climate data matrix with both space- and time-factors can be defined as

$$\mathbf{X}_f = \left[\sqrt{\frac{A_i}{p}}x_{it}\right], \tag{6.6}$$

where

$$\sqrt{\frac{A_i}{p}} \tag{6.7}$$

are the *space-time factors*. As an example, for a $5° \times 5°$ grid and 50 years of data, for the analysis of annual data we have the space-time factor as

$$\sqrt{\frac{A_i}{p}} = C\sqrt{\cos(\phi_i)} \tag{6.8}$$

where ϕ_i is the latitude (in radians, not in degrees) of the centroid of the grid box i, and C is a constant, approximated by

$$C \approx \frac{5 \times 6400}{180} \times \frac{1}{\sqrt{50}} \ [\text{km yr}^{-1/2}]. \tag{6.9}$$

Here, the Earth is approximately regarded as a sphere with its radius equal to 6,400 km. The practical climate data analyses often drop the dimensional constant and use only the dimensionless scaling factors $w_i = \sqrt{\cos(\phi_i)}$, since a constant multiplication to a covariance matrix does not change the eigenvectors and other useful properties of the matrix. Therefore, climate data analyses often use the dimensionless space-time scaling factors instead of the dimensional space-time factors:

$$\mathbf{X}_f = [w_i x_{it}]. \tag{6.10}$$

Finally the sample covariance matrix can be written as

$$\mathbf{S} = \mathbf{X}_f \mathbf{X}_f^t. \tag{6.11}$$

Example 6.1 Compute and visualize the sample covariance matrix with space-time factor for the December SST temperature anomalies using the NOAAGlobalTemp data (anomalies with respect to 1971–2000 monthly climatology) over a zonal band at 2.5° S from 0° E to 360° (Figure 6.1(a)) and a meridional band at 2.5° E from 90° S to 90° N (Figure 6.1(b)). The monthly NOAAGlobalTemp data are the anomalies on a grid 5° × 5° from −87.5° to 87.5° in latitude and from 2.5° to 375° in longitude.

The zonal band under our consideration consists of 72 grid boxes of 5-deg. Thus, the sample covariance matrix is a 72 × 72 matrix as shown in Figure 6.1(a).

The meridional band has only 36 grid boxes. Thus, the sample covariance matrix is a 36 × 36 matrix as shown in Figure 6.1(b). The white region of high latitude indicates missing data.

Figures 6.1(a) and (b) show not only the symmetry around the upward diagonal line, but also the spatial variation of elements of covariance matrices. The figure allows us to identify the locations of large covariance. For example, the red region of Figure 6.1(a) shows the large covariance over the eastern Tropical Pacific with longitude approximately between 180° and 280°. Figure 6.1(c) shows the December 1997 warm El Niño SAT anomalies over the eastern Tropical Pacific. The covariance matrices of Figures 6.1(a) and (b) were computed from 50 years of December SAT anomaly data from 1961 to 2010, with SAT of December 1997 being one of the 50 time series data samples.

Figure 6.1 can be plotted by the following computer code.

```
#R plot Fig. 6.1a, b, c: Covariance map
#Covariance from NOAAGlaobalTemp data: December SAT anomalies
setwd('/Users/sshen/climstats')
library(ncdf4)
nc=ncdf4::nc_open("data/air.mon.anom.nc") #Read data
nc #Check metadata of the dataset
nc$dim$lon$vals # 2.5 - 357.5
nc$dim$lat$vals # -87.5 - 87.5
nc$dim$time$vals
lon <- ncvar_get(nc, "lon")
lat <- ncvar_get(nc, "lat")
time<- ncvar_get(nc, "time")
library(chron)
month.day.year(29219,c(month = 1, day = 1, year = 1800))
```

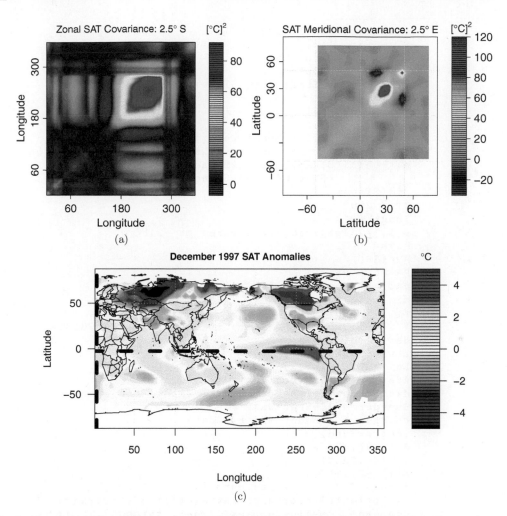

Figure 6.1   The December SAT covariance matrix computed from the NOAAGlobalTemp data from 1961 to 2010: (a) The covariance of the zonal band at 2.5° S, and (b) the covariance of the meridional band at 2.5° E. Panel (c) shows the positions of the zonal band and meridional band indicated by the thick dashed black lines. Data source: Smith et al. 2008; Vose et al. 2012. https://psl.noaa.gov/data/gridded/data.noaaglobaltemp.html

```
#1880-01-01
sat<- ncvar_get(nc, "air")
dim(sat)
#[1] 72    36 1674
#1674 months = 1880-01 to 2019-06, 139 years 6 mons

Dec = seq(12, 1674, by=12)
Decsat=sat[,, Dec]
N = 72*36
P = length(Dec)
STsat = matrix(0, nrow=N, ncol=P)
for (k in 1:P){STsat[,k]=as.vector(Decsat[,,k])}
```

```
colnames(STsat)<-1880:2018
STsat[1:4,1:4]
LAT=rep(lat, each=72)
LON=rep(lon,36)
STanom=cbind(LAT, LON, STsat)
dim(STanom)
#[1] 2592  141

#Plot Fig. 6.1a: Zonal covariance matrix
#Select only the data for the equatorial band -2.5S
n1<-which(STanom[,1]>-4&STanom[,1]<0
          &STanom[,2]>0&STanom[,2]<360)
dim(STanom)
#[1] 2592  141
length(n1)
#[1] 72 longitude grid boxes at -2.5S
P1=84
P2=133
P= P2 - P1 + 1
dat1=STanom[n1,P1:P2] #1961-2010
dim(dat1)
#[1] 72 50
#72 grid boxes, 50 years of Dec from 1961-2010.
dat1[1:3,48:50]
Lat1=STanom[n1,1]
Lon1=STanom[n1,2]
AreaFac = sqrt(cos(Lat1*pi/180))
dat2 = dat1 - rowMeans(dat1) #Minus the mean
Banddat = AreaFac*dat2
covBand = Banddat%*%t(Banddat)
max(covBand)
#[1] 90.67199
min(covBand)
#[1] -6.191734
int=seq(-7, 92,length.out=81)
rgb.palette=colorRampPalette(c('blue', 'darkgreen',
 'green', 'yellow','pink','red','maroon'),
                                  interpolate='spline')
setEPS() #Plot the figure and save the file
postscript("fig0601a.eps", width = 7, height = 5.5)
par(mar=c(4.2,5.0,1.8,0.0))
par(cex.axis=1.8,cex.lab=1.8, cex.main=1.7)
ticks = c(60,  180, 300)
filled.contour(Lon1, Lon1, covBand,
               color.palette=rgb.palette, levels=int,
               plot.title=title(main=
 expression("Zonal SAT Covariance: 2.5"*degree*S),
    xlab="Longitude", ylab="Longitude"),
    plot.axes = { axis(1, at= ticks); axis(2, at= ticks, las = 0)},
            key.title=title(main=expression("["*degree*"C]"^2))
      )
dev.off()

#Plot Fig. 6.1b: Meridional covariance matrix
#Select only the data for the meridional band 2.5E
n1<-which(STanom[,1]>-90&STanom[,1]<90
```

```
                    &STanom[,2]>0&STanom[,2]<4)
P1=84 #time index for Dec 1961
P2=133 #Dec 2010
P= P2 - P1 + 1 #=50 Decembers
dat1=STanom[n1,P1:P2] #1961-2010
Lat1=STanom[n1,1]
Lon1=STanom[n1,2]
AreaFac = sqrt(cos(Lat1*pi/180))
dat2 = dat1 - rowMeans(dat1)
Banddat = AreaFac*dat2
covBand = Banddat%*%t(Banddat)
max(covBand, na.rm=TRUE)
#[1] 115.6266
min(covBand, na.rm=TRUE)
#[1] -33.72083
int=seq(-35,120,length.out=81)
rgb.palette=colorRampPalette(c('blue', 'darkgreen','green',
                               'yellow','pink','red','maroon'),
                             interpolate='spline')
ticks = seq(-90, 90, by=30)
setEPS() #Plot the figure and save the file
postscript("fig0601b.eps", width = 7, height = 5.5)
par(mar=c(4.2,5.0,1.8,0.0))
par(cex.axis=1.8, cex.lab=1.8, cex.main=1.7)
filled.contour(Lat1, Lat1, covBand,
               color.palette=rgb.palette, levels=int,
               plot.title=title(main=
      expression("SAT Meridional Covariance: 2.5"*degree*E),
                   xlab="Latitude", ylab="Latitude"),
plot.axes={axis(1, at=ticks); axis(2, at=ticks, las =0);
           grid()},
      key.title=title(main=expression("["*degree*"C]"^2)))
dev.off()


#plot Fig. 6.1(c)
month.day.year(time[1416],c(month = 1, day = 1, year = 1800))
#Dec 1997
mapmat= sat[,,1416]
mapmat=pmax(pmin(mapmat,5),-5)
int=seq(-5, 5,length.out=51)
rgb.palette=colorRampPalette(c('black','blue',
  'darkgreen','green', 'white','yellow','pink',
  'red','maroon'), interpolate='spline')
setEPS() #Plot the figure and save the file
postscript("fig0601c.eps", width = 7, height = 3.5)
par(mar=c(4.2,5.0,1.8,0.0))
par(cex.axis=0.9,cex.lab=0.9, cex.main=0.8)
library(maps)
filled.contour(lon, lat, mapmat,
               color.palette=rgb.palette, levels=int,
          plot.title=title(main="December 1997 SAT Anomalies",
                           xlab="Longitude",ylab="Latitude"),
                plot.axes={axis(1); axis(2);
                  map('world2', add=TRUE);grid()},
                key.title=title(main=expression(degree*"C")))
segments(x0=-20,y0=-2.5, x1=255, y1=-2.5, lwd = 5, lty = 2)
```

```
segments(x0=-15,y0=-90, x1=-15, y1=90, lwd = 5, lty = 2)
dev.off()
```

```python
#Python plot Fig. 6.1a: Covariance plots
# Go to current working directory
os.chdir("/Users/sshen/climstats/")
import netCDF4 as nc
#import .nc data
nc = nc.Dataset("data/air.mon.anom.nc")
# extrat data for variables
lon = nc.variables['lon'][:]
lat = nc.variables['lat'][:]
time = nc.variables['time'][:]
sat = nc.variables['air'][:]
#sat = surface air temp anomalies: mon, 5deg grid
# 1674 months from Jan 1880-June 2019: 139yrs 6mons
print(np.shape(sat)) #(1674, 36, 72)
#print(nc)
Dec = list(np.arange(11,1674,12))#Dec index
Decsat = sat[Dec,:,:] #extract Dec data
# space-time data matrix STsat
N = 72 * 36
P = len(Dec)
STsat = np.zeros((N,P))
for k in range(P):
    STsat[:,k] = Decsat[k,:,:].flatten()
print(STsat.shape)#(2592, 139) 139 Decembers
#print(STsat[0:4, 0:4])

#Space-time dataframe with lat, lon, time
# repeat each value in lat 72 times
LAT = np.repeat(lat,72).reshape(2592,1)
# repeat lon 36 times
LON = np.tile(lon,36).reshape(2592,1)
STanom0 = np.hstack((LAT, LON, STsat))
yrlist = list(np.arange(1880, 2019))
cnames = ["LAT", "LON"] + yrlist
STanom = pd.DataFrame(STanom0, columns=cnames)

#Select the zonal band
n1 = np.where((STanom["LAT"] > -4) & (STanom["LAT"] < 0)\
        & (STanom["LON"] > 0) & (STanom["LON"] < 360))
P1 = 81 #Dec 1961 in STsat space-time data matrix
P2 = 130 #Dec 2010
P = P2 - P1 + 1 #50 years
print(STsat.shape)
da0 = STsat[n1,P1:(P2 + 1)]
da1 = da0[0,:,:]
print(da0.shape)
print(da1[0:3,47:50])

# define AreaFrac as a float for future manipulation
AreaFrac = np.sqrt(np.cos(LAT[n1]*np.pi/180))
```

```
#AreaFrac = float(np.sqrt(np.cos(Lat1[0]*np.pi/180)))

# subtract row means from dat1 to get math anomalies
dat2 = da1 - da1.mean(axis=1, keepdims=True)
# multiply each value by AreaFrac
Banddat = AreaFrac * dat2
#print(Banddat.shape)
covBand = np.dot(Banddat, Banddat.transpose())
# check max and min values
print(np.max(covBand))
print(np.min(covBand))
#Python plot the figure: Fig. 6.1a
int = np.linspace(-7,92,81) # define levels for plot
fig, ax = plt.subplots()
# define color bar
myColMap = LinearSegmentedColormap.from_list(name='my_list',
            colors=['navy','darkgreen','lime','yellow',
                    'pink','red','maroon'], N=100)
# plot covariance plot
Lon1 = np.arange(2.5, 360, 5)
colormap = plt.contourf(Lon1,Lon1,covBand,
                        levels = int, cmap = myColMap)
plt.title("Covariance of SAT Anomalies on \
Zonal Band 2.5$\degree$S",
            size = 21, pad = 15)
plt.xlabel("Longitude", labelpad = 15)
plt.ylabel("Longitude", labelpad = 15)
# plot color bar
cbar = plt.colorbar(colormap, shrink=0.9, pad = 0.05)
plt.text(372, 348, '[$\degree$C]$^2$', size = 20)
cbar.set_ticks([0,20,40,60,80])
plt.savefig("fig0601a.eps")# save figure
```

6.2 Definition of EOFs and PCs

6.2.1 Defining EOFs and PCs from the Sample Covariance Matrix

In addition to the SVD definition of EOFs and PCs, we can also define EOFs as the eigenvectors $\mathbf{u}_k$ of the sample spatial covariance matrix $\mathbf{S}$:

$$\mathbf{S}\mathbf{u}_k = \lambda_k\mathbf{u}_k, \quad k = 1,2,\ldots,s. \tag{6.12}$$

Here, λ_k are eigenvalues and s is the rank of $\mathbf{S}$.

We have seen in the previous chapter that the eigenvectors of a symmetric matrix are orthogonal to one another if eigenvalues are distinct. Sometimes an eigenvalue is a repeated root from the eigen-equation (6.12). This same eigenvalue corresponds to several eigenvectors, referred to as a degenerate multiplet of eigenvectors, which form a subspace of dimension more than one. Any linear combination of the eigenvectors in this subspace is also an eigenvector, i.e., any vector in this subspace is an eigenvector. Thus, the eigenvector's orientation in this subspace is arbitrary. Usually one chooses a set of unit orthogonal eigenvectors that span the subspace.

The EOFs are conventionally normalized into unit vectors, i.e., orthonormal vectors with

$$\mathbf{u}_k \cdot \mathbf{u}_l = \delta_{kl}, \tag{6.13}$$

where

$$\delta_{kl} = \begin{cases} 1, & \text{if } k = l, \\ 0, & \text{if } k \neq l \end{cases} \tag{6.14}$$

is the Kronecker delta.

The eigenvalues λ_k are nonnegative, which can be shown as follows. Substituting the data matrix $\mathbf{X}_f$ with space-time factor into the EOF definition equation (6.12) yields

$$(\mathbf{X}\mathbf{X}^t)\mathbf{u}_k = \lambda_k \mathbf{u}_k. \tag{6.15}$$

Here, for simplicity, we have dropped the subscript f for the space-time factor and used $\mathbf{X}$ to represent the data matrix $\mathbf{X}_f$ with space-time factor. Multiplying this equation by the row vector $\mathbf{u}_k^t$ leads to

$$\lambda_k = |\mathbf{X}^t \mathbf{u}_k|^2 \geq 0, \tag{6.16}$$

since

$$\mathbf{u}_k^t \mathbf{u}_k = 1. \tag{6.17}$$

The expression $\mathbf{X}^t \mathbf{u}_k$ can be regarded as the projection of the space-time data onto EOF $\mathbf{u}_k$. Since $\mathbf{X}$ is anomaly data,

$$|\mathbf{X}^t \mathbf{u}_k|^2$$

may be regarded as the variance associated with eigenvector $\mathbf{u}_k$. Thus, Equation (6.16) implies that the eigenvalue λ_k is equal to the variance explained by the corresponding kth EOF mode.

When $\lambda_k > 0$, the temporal vector

$$\mathbf{v}_k = \frac{1}{\sqrt{\lambda_k}} \mathbf{X}^t \mathbf{u}_k, \quad k = 1, 2, \dots, s. \tag{6.18}$$

is the kth *principal component* (PC) of the space-time data. The PCs defined this way are also orthonormal:

$$\begin{aligned}
\mathbf{v}_k^t \mathbf{v}_l &= \left(\frac{1}{\sqrt{\lambda_k}} \mathbf{X}^t \mathbf{u}_k \right)^t \frac{1}{\sqrt{\lambda_l}} \mathbf{X}^t \mathbf{u}_l, \\
&= \frac{1}{\sqrt{\lambda_k}\sqrt{\lambda_l}} \mathbf{u}_k^t \left(\mathbf{X}\mathbf{X}^t \right) \mathbf{u}_l \\
&= \frac{1}{\sqrt{\lambda_k}\sqrt{\lambda_l}} \mathbf{u}_k^t \lambda_l \mathbf{u}_l \\
&= \delta_{kl}.
\end{aligned} \tag{6.19}$$

From the perspective of the space-time decomposition of the data matrix **X**, the principal component matrix may be regarded as the temporal coefficients of the spatial EOFs. Moreover, principal components are orthogonal with respect to time. This may be interpreted to mean that (i) individual terms making up the principal component time series are statistically independent from each other, and (ii) the variance associated with the time series of EOF k is equal to the eigenvalue λ_k.

The EOFs and PCs are exactly the column vectors of the spatial eigenvector matrix U and the temporal eigenvector matrix V of the SVD in the previous chapter. The space-time data matrix X can be exactly represented by all the nonzero eigenvalues and their corresponding EOFs and PCs:

$$\mathbf{X} = \sum_k \sqrt{\lambda_k}\mathbf{u}_k\mathbf{v}_k^t. \tag{6.20}$$

This is the vector form of SVD.

The corresponding SVD for the sample covariance matrix is the square matrix diagonalization:

$$\mathbf{S} = \sum_k \lambda_k\mathbf{u}_k\mathbf{u}_k^t = \mathbf{U}\Lambda\mathbf{U}^t, \tag{6.21}$$

where Λ is the square diagonal matrix $\lambda_k\delta_{kl}$. Be careful that this SVD is not a space-time decomposition since the matrix **S** is the spatial sample covariance matrix whose rows and columns are both for space. Thus, this is a space-space decomposition. Similarly, you can construct a temporal sample covariance matrix, and make a time-time SVD for this matrix. This idea can help with the SVD computation when the space-time data matrix is extremely large, e.g., more than 200 GB, and the time dimension is relatively small, e.g., 500.

6.2.2 Percentage Variance Explained

Very often in climate science, the EOFs of the largest scales, such as the northern hemisphere–southern hemisphere differences and El Niño Southern Oscillation, also carry the largest variance. In other words, the variance tapers off in a way that is correlated with the size of the spatial scales. The eigenvalues decrease as the mode number k increases:

$$\lambda_1 \geq \lambda_2 \geq \lambda_3 \geq \cdots.$$

Correspondingly, the spatial length scale associated with the EOFs decreases, and so does the temporal length scale with the PCs. Analogous to the normal modes of a drum head, the largest scale corresponds to the fewest number of regions separated by zero iso-lines, and hence the simplest pattern. The smaller scales correspond to more complex patterns.

The percentage of the variance explained by EOF k, denoted by q_k, is often used to measure the relative importance of the EOF and PC when interpreting climate dynamic patterns corresponding to the EOF and PC:

$$q_k = \frac{\lambda_k}{\sum_{k=1}^N \lambda_k} \times 100\%, \quad k = 1,2,\ldots,N, \tag{6.22}$$

where N is the rank of the covariance matrix. The $k - q_k$ relationship forms a curve, called the *scree plot* (see Figure 6.2).

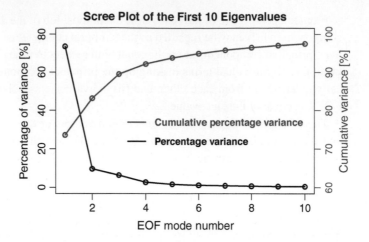

A scree plot for the eigenvalues of the covariance matrix shown in Figure 6.1(a) for the SAT anomalies over an equatorial zonal band at 2.5° S.

The total amount of variance explained by the first K EOF modes, denoted by Q_K, is

$$Q_K = \sum_{k=1}^{K} q_k = \frac{\sum_{k=1}^{K} \lambda_k}{\sum_k^N \lambda_k} \times 100\%, \quad K = 1, 2, \ldots, N. \tag{6.23}$$

Figure 6.2 shows a scree plot for the NOAAGlobalTemp December SAT anomalies in 1961–2010 over an equatorial zonal band at 2.5° S. The first EOF mode explains 74% of the total variance. This implies the importance of the large-scale tropic dynamics. The latitude of the zonal band becomes higher; the first EOF mode explains less variance. For example, for the zonal band 47.5° N, the first mode explains only 30% of the total variance, while the second one explains 24%.

Depending on the practical climate problems, you may choose to include the first K modes in your analysis so that up to 80% or 90% variance is explained.

Figure 6.2 can be plotted by the following computer code.

```
#R plot Fig. 6.2: Scree plot
K = 10
eigCov =eigen(covBand)
#covBand is for equatorial zonal band in Fig 6.1(a)
lam = eigCov$values
lamK=lam[1:K]
setEPS() #Plot the figure and save the file
postscript("fig0602.eps", width = 6, height = 4)
par(mar=c(4,4,2,4), mgp=c(2.2,0.7,0))
plot(1:K, 100*lamK/sum(lam), ylim=c(0,80), type="o",
     ylab="Percentage of Variance [%]",
     xlab="EOF Mode Number",
     cex.lab=1.2, cex.axis = 1.1, lwd=2,
     main="Scree Plot of the First 10 Eigenvalues")
legend(3,30, col=c("black"),lty=1, lwd=2.0,
       legend=c("Percentage Variance"),bty="n",
       text.font=2,cex=1.0, text.col="black")
par(new=TRUE)
```

```
plot(1:K,cumsum(100*lamK/sum(lam)),
     ylim = c(60,100), type="o",
     col="blue",lwd=2, axes=FALSE,
     xlab="",ylab="")
legend(3,80, col=c("blue"),lty=1,lwd=2.0,
        legend=c("Cumulative Percentage Variance"),bty="n",
        text.font=2,cex=1.0, text.col="blue")
axis(4, col="blue", col.axis="blue", mgp=c(3,0.7,0))
mtext("Cumulative Variance [%]",col="blue",
     cex=1.2, side=4,line=2)
dev.off()
```

```
#Python plot Fig. 6.2: Scree plot
values, vectors = np.linalg.eig(covBand)
lam = values
K = 10
lamK = lam[0:10]

fig, ax = plt.subplots()

ax.plot(np.linspace(1,K,K), 100*lamK/np.sum(lam),
        'ko-', linewidth = 3)
ax.set_title("Scree Plot of the First 10 Eigenvalues",
        pad = 15)
ax.set_xlabel("EOF Mode Number", labelpad = 15)
ax.set_ylabel("Percentage of Variance [%]", labelpad = 15)

ax1 = ax.twinx()
ax1.plot(np.linspace(1,K,K),np.cumsum(100*lamK/np.sum(lam)),
        'bo-', linewidth = 3)
ax1.set_ylabel("Cumulative Variance [%]",
            color = 'blue', labelpad = 10)
ax1.tick_params(length=6, width=2,
            labelsize=21, color = 'b', labelcolor = 'b')
ax1.set_yticks([60,70,80,90,100])
ax1.spines['right'].set_color('b')

ax.text(4,30, "????  Cumulative Percentage Variance",
        fontsize = 20, color = "blue")
ax.text(4,20, "????  Percentage Variance", fontsize = 20)
fig.tight_layout()
plt.savefig("fig0602.eps")# save figure
```

6.2.3 Temporal Covariance Matrix

In practice, the $N \times P$ anomaly data matrix $\mathbf{X}$ often has a large number of spatial grid boxes or locations N, but a relatively small time steps P. For example, the dataset of 100 years of January surface temperature anomalies from the NCEP/NCAR Reanalysis on a $2.5° \times 5°$ grid has $P = 100$ and $N = 144 \times 73 = 10,512$. Here, N is much greater than P. The spatial covariance matrix is a large $N \times N$ matrix. However, you can transpose space and time to make a smaller $P \times P$ temporal sample covariance matrix, which has the same eigenvalues as the spatial sample covariance matrix, and whose eigenvectors are PCs. Usually, we prefer to deal with a smaller matrix.

The $P \times P$ temporal sample covariance matrix $\mathbf{T}$ defined for the $N \times P$ anomaly data matrix with scaling factors $\mathbf{X}_f$ is as follows:

$$\mathbf{T} = \mathbf{X}^t \mathbf{X}. \tag{6.24}$$

Again, we have dropped the subscript f in this formula for simplicity. This temporal sample covariance matrix can be diagonalized by PCs $\mathbf{v}_k$:

$$\mathbf{T} = \sum_k \lambda_k \mathbf{v}_k \mathbf{v}_k^t = \mathbf{V} \mathbf{\Lambda} \mathbf{V}^t. \tag{6.25}$$

Mathematically, this is the same as the spatial covariance matrix diagonalized by EOFs $\mathbf{u}_k$.

Representing the time-space data matrix $\mathbf{X}^t$ by a set of independent PC vectors also leads to the SVD of $\mathbf{X}^t$:

$$\mathbf{X}^t = \sum_k \sqrt{\lambda_k} \mathbf{v}_k \mathbf{u}_k^t. \tag{6.26}$$

The equivalence of eigenvalue problems for the spatial and temporal covariance matrices can be easily shown. Multiplying the eigenvalue problem for the spatial covariance matrix (6.15) by the data matrix $\mathbf{X}^t$ leads to

$$\mathbf{X}^t (\mathbf{X} \mathbf{X}^t) \mathbf{u}_k = \lambda_k \mathbf{X}^t \mathbf{u}_k. \tag{6.27}$$

Applying the PC definition (6.18) to this equation yields

$$\mathbf{X}^t \mathbf{X} \mathbf{v}_k = \lambda_k \mathbf{v}_k. \tag{6.28}$$

This is the eigenvalue problem of the temporal covariance matrix:

$$\mathbf{T} \mathbf{v}_k = \lambda_k \mathbf{v}_k. \tag{6.29}$$

Solving this eigenvalue problem of a smaller matrix $\mathbf{T}$ yields orthonormal PCs $\mathbf{v}_k (k = 1, 2, \ldots, s)$. The corresponding EOFs $\mathbf{u}_k$ can be computed from the PC $\mathbf{v}_k$ by a simple matrix multiplication:

$$\mathbf{u}_k = \frac{1}{\sqrt{\lambda_k}} \mathbf{X} \mathbf{v}_k. \tag{6.30}$$

Whether to use the spatial or temporal sample covariance matrix depends on the need and the matrix size. One would often choose a smaller matrix. However, if your only purpose is to compute EOFs, you may efficiently use SVD to compute the EOFs directly for the space-time data matrix without computing either the spatial or temporal sample covariance matrix.

6.3 Climate Field and Its EOFs

6.3.1 SVD for a Climate Field

A climate field refers to a climate variable, such as the 2-meter surface air temperature $T(\mathbf{r},t)$, defined over a spatial region A and a time interval $[0,P]$. Here, $\mathbf{r}$ is the spatial position vector in region A, and t is the time variable in the interval $[0,P]$.

The SVD for a space-time data matrix

$$X = \sum_{m=1}^{P} \sqrt{\lambda_m} \mathbf{u}_m (\mathbf{v}_m)^t \tag{6.31}$$

has a counterpart for a space-time climate field $T(\mathbf{r},t)$:

$$T(\mathbf{r},t) = \sum_{m=1}^{\infty} \sqrt{\lambda_m} \psi_m(\mathbf{r}) T_m(t). \tag{6.32}$$

The summation goes to infinity, instead of P or N for a space-time data matrix of finite order, because a continuous field may be considered as being composed of infinitely many points. Here, $\psi_m(\mathbf{r})$ are the continuous field version of EOFs, known in the literature as the Karhunin–Loeve functions, and satisfy the orthonormal condition in the sense of integration:

$$\int_A \psi_m(\mathbf{r}) \psi_n(\mathbf{r}) dA = \delta_{mn}, \tag{6.33}$$

where A is the spatial domain of the climate field.

A summation like Eq. (6.32) is referred to as the spectral expansion, with $\psi_m(\mathbf{r})$ as basis functions, and λ_m as spectra. When $\psi_m(\mathbf{r})$ are sine or cosine functions, the expansion Eq. (6.32) is traditionally called a Fourier series. Today, even if $\psi_m(\mathbf{r})$ are not sine or cosine functions, people still call the expansion a Fourier series.

If A is divided into N subdomains or grid boxes $\Delta A_i, i = 1, 2, \ldots, N$, then our integral Eq. (6.33) can be approximated by a numerical integration:

$$\sum_{i=1}^{N} \left[\psi_m(\mathbf{r}_i) \sqrt{\Delta A_i} \right] \left[\psi_n(\mathbf{r}_i) \sqrt{\Delta A_i} \right] \approx \delta_{mn}. \tag{6.34}$$

Here ΔA_i is the area (or volume in the case of a 3D domain) of the ith subdomain or grid box, and $\sqrt{\Delta A_i}$ is the area-factor (or the volume-factor in the case of a 3D domain). The EOFs with the area-factor are denoted by

$$E_m(\mathbf{r}_i) = \left[\psi_m(\mathbf{r}_i) \sqrt{\Delta A_i} \right]. \tag{6.35}$$

The spatial projection of climate data onto EOFs also has a climate field counterpart:

$$T_m(t) = \frac{1}{\lambda_m} \int_A T(\mathbf{r},t) \psi_m(\mathbf{r}) dA. \tag{6.36}$$

A discretized version of this integral can be

$$T_m(t) \approx \frac{1}{\lambda_m} \sum_{i=1}^{N} \left(T(\mathbf{r}_i, t)\sqrt{\Delta A_i} \right) \left(\psi_m(\mathbf{r}_i)\sqrt{\Delta A_i} \right). \tag{6.37}$$

Here, $T(\mathbf{r}_i, t)\sqrt{\Delta A_i}$ are the data with area-factor.

6.3.2 Stochastic Climate Field and Covariance Function

A climate field $T(\mathbf{r}, t)$ with $\mathbf{r} \in A$ and $t \in [0, P]$ is often stochastic because a climate variable at a point or over a grid box or a region is considered to be an ensemble mean and has uncertainties. Thus, $T(\mathbf{r}, t)$ follows a probabilistic distribution. When $T(\mathbf{r}, t)$ is a stationary time series, the covariance function of $T(\mathbf{r}, t)$ is defined as

$$K(\mathbf{r}, \mathbf{r}') = \text{Cov}[T(\mathbf{r}, t), T(\mathbf{r}', t)] \tag{6.38}$$

and is independent of time. In applications, this ensemble mean is often approximated by a temporal mean:

$$K(\mathbf{r}, \mathbf{r}') \approx \frac{1}{P} \sum_{t=1}^{P} T(\mathbf{r}, t)T(\mathbf{r}', t). \tag{6.39}$$

If this approximation is valid, we say that the stochastic process is ergodic (i.e., averaging over time is equal to averaging over a large ensemble).

The corresponding space-time weighted covariance matrix K_f over a discretized domain is

$$K_f(\mathbf{r}_i, \mathbf{r}_j) = \sum_{t=1}^{P} \left[\frac{\sqrt{\Delta A_i}}{\sqrt{P}} T(\mathbf{r}_i, t) \right] \left[\frac{\sqrt{\Delta A_j}}{\sqrt{P}} T(\mathbf{r}_j, t) \right]. \tag{6.40}$$

Here, $\sqrt{\Delta A_i}/\sqrt{P}$ is an area-time factor, also called space-time factor. If it is a 3D domain, then this should correspond to a volume-time factor.

The continuous EOF fields $\psi_m(\mathbf{r})$ are defined by an integral equation:

$$\int_A K(\mathbf{r}, \mathbf{r}')\psi_m(\mathbf{r}')\mathrm{d}A' = \lambda_m \psi_m(\mathbf{r}). \tag{6.41}$$

The corresponding discretized form from sample data is

$$\sum_{j=1}^{N} \sum_{t=1}^{P} \left[\frac{\sqrt{\Delta A_i}}{\sqrt{P}} T(\mathbf{r}_i, t) \right] \left[\frac{\sqrt{\Delta A_j}}{\sqrt{P}} T(\mathbf{r}_j, t) \right] \left[\psi_m(\mathbf{r}_j)\sqrt{\Delta A_j} \right] = \frac{\lambda_m}{P} \left[\psi_m(\mathbf{r}_i)\sqrt{\Delta A_i} \right] \tag{6.42}$$

or

$$\sum_{j=1}^{N} K_f(\mathbf{r}_i, \mathbf{r}_j)E_m(\mathbf{r}_j) = \lambda'_m E_m(\mathbf{r}_i), \tag{6.43}$$

where

$$\lambda'_m = \frac{\lambda_m}{P}. \tag{6.44}$$

This eigenvalue equation (6.43) is equivalent to Eq. (6.12).

In summary, to compute the sample covariance matrix, we compute the data matrix with space-time factor $\mathbf{X}_f$, and then use $\mathbf{S} = \mathbf{X}_f\mathbf{X}_f^t$ to compute the covariance matrix. To compute EOFs and PCs, we directly apply SVD to $\mathbf{X}_f$. The EOFs $E_m(\mathbf{r}_i)$ have the area-factors and are normalized as a vector

$$\sum_{i=1}^{N} (E_m(\mathbf{r}_i))^2 = 1. \tag{6.45}$$

These EOFs $E_m(\mathbf{r}_i)$ are called the geometric EOFs, in contrast to the physical EOFs $\psi_m(\mathbf{r}_i)$ for climate dynamical interpretations. To obtain the physical EOFs $\psi_m(\mathbf{r}_i)$, you just remove the area-factor from the geometric EOFs $E_m(\mathbf{r}_i)$ by a division:

$$\psi_m(\mathbf{r}_i) = \frac{1}{\sqrt{A_i}} E_m(\mathbf{r}_i). \tag{6.46}$$

The time-factor rescales eigenvalues $\lambda_m/\sqrt{P}$, but does not change the percentage of variance explained by the mth EOF mode. In data analysis, we often only need to know the percentage of variance explained, not the eigenvalues themselves. Thus, this justifies ignoring the time-factor when computing the covariance matrix or EOFs. For the same reason, if all area-factors are approximately the same size, the area-factor can also be ignored, and one can directly compute the EOFs from the space-time data.

6.4 Generating Random Fields

Suppose we want to generate a sequence of *independent* realizations of a random field Y_{xr} with certain properties, say $\sim N(0,\Sigma)$. We use index r instead of t to emphasize that the R realizations are independent of one another, and index x for the N spatial locations.

One way to generate the field is to use the PCs and EOFs. We can use a random number generator to generate PCs, and use the following EOF-PC matrix structure to generate a random field as an array of random numbers:

$$y_{xr} = \sum_{m=1}^{M} \sqrt{\lambda_m} B_m(r) E_m(x). \tag{6.47}$$

Here, $E_m(x)$ are given orthonormal EOF vectors and satisfy the orthonormal condition:

$$\sum_{x=1}^{N} E_m(x)E_n(x) = \begin{cases} 1 & m=n \\ 0 & m \neq n \end{cases}. \tag{6.48}$$

We use M EOFs with $M \leq N$. The symbol λ_m represents the prescribed variance associated with $E_m(x)$, and $B_m(r)$ represents PCs generated by a random generator. The PC vectors also satisfy the orthonormal condition:

$$\sum_{r=1}^{M} B_m(r)B_n(r) = \begin{cases} 1 & m=n \\ 0 & m \neq n \end{cases}. \tag{6.49}$$

Of course, the actual numerical results satisfy this condition only approximately because the $B_m(r)$ are an approximation from a random generator, such as the R command `rs <- rmvnorm(Ms, zeromean, univar)`, in the example of this section.

We can interpret $B_m(r)$ as the loading of the random field along the EOF direction $E_m(x)$ for the rth realization. The orthonormal condition here is interpreted as the independence of PCs from each other, i.e., when PCs are treated as random variables, they strictly satisfy the following condition:

$$E[B_m B_n] = \delta_{mn}. \tag{6.50}$$

In this way, we can construct simulations of a random field Y_{xr} with the known EOFs and the prescribed spectra of eigenvalues $\lambda_m, m = 1, 2, \ldots, M$. An example follows.

Example 6.2   Generate a random field with random and independent PCs.

Consider a random field based on the prescribed EOFs over a spatial line segment $[0, \pi]$:

$$\psi_m(s) = \sqrt{\frac{2}{\pi}} \sin(ms), m = 1, 2, \ldots, M, \ \ 0 \le s \le \pi. \tag{6.51}$$

The prescribed eigenvalues $\lambda_m, m = 1, 2, \ldots, 5$ are

$$10.0, 9.0, 4.1, 4.0, 2.0.$$

We wish to generate a random array with these given EOFs and eigenvalues.

Given positive integers M and N, discretize the line segment $[0, \pi]$ by N intervals, each of which has a length equal to π/N. The following defines M vectors in N dimensional Euclidean space $\mathbb{R}^N$ for different m:

$$E_m(x_i) = \left(\sqrt{\frac{\pi}{N}} \right) \sqrt{\frac{2}{\pi}} \sin(mx_i), \tag{6.52}$$

where

$$x_i = i\frac{\pi}{N}, \ \ i = 1, 2, \ldots, N, \ \ m = 1, 2, \ldots, M. \tag{6.53}$$

These M vectors are approximately orthonormal in $\mathbb{R}^N$ if N is sufficiently large. The length factor is $\sqrt{\pi/N}$ and normalizes the discrete EOF $E_m(x_i)$.

The following $N \times N$ matrix C

$$C_{ij} = \sum_{m=1}^{M} \lambda_m E_m(x_i) E_m(x_j), \ \ i, j = 1, 2, \ldots, N, \tag{6.54}$$

can be regarded as a synthetic covariance matrix, which has $E_m(x_i)$ as its EOFs.

Then, we can use random PCs and the given variances to generate a random field with these specified EOFs. The following computer codes generate M random PCs with M_s samples.

```
#R code for generating a random space-time field
#Step 1: Generate EOFs
N <- 100 #Number of spatial points
eof  <-  function(n, x) (sin(n*x)/sqrt(pi/2))*sqrt((pi/N))
```

```
x    <-    seq(0,pi, len=N)
sum(eof(4,x)^2)
#[1] 0.99  #Verify the normalization condition
sum(eof(1,x)*eof(2,x))
#[1] 3.035766e-18  #Verify orthogonality

#Step 2: Generate PCs
#install.packages('mvtnorm')
library(mvtnorm) #Multivariate normal
Mode <- 5
Ms <- 1000
univar <- diag(rep(1,Mode))
zeromean <- rep(0, Mode)
rs <- rmvnorm(Ms, zeromean, univar)
pcm=matrix(0, nrow=Ms, ncol=Mode)
for(m in 1:Mode){pcm[,m] = rs[,m]*sqrt(1/Ms)}
t(pcm[,1])%*%pcm[,1]
#1.010333 #Approximately normalized
t(pcm[,1])%*%pcm[,2]
#0.008040772 #Approximately independent/orthogonal

#Step 3: Generate an independent random field
lam=c(10, 9, 4.1, 4, 2)
sqrlam = diag(sqrt(lam))
eofm = matrix(0, nrow=N, ncol=Mode)
for(m in 1:Mode){eofm[,m]=eof(m,x)}
Yxr    <-    eofm%*%sqrlam%*%t(pcm)
dim(Yxr)
#[1] 100 1000
```

```
#Python code for generating a random space-time field
# generate EOFs
N = 100 # number of spatial points

# define function for EOF
def eof(x,n):
    return((np.sin(n*x)/np.sqrt(np.pi/2))*np.sqrt(np.pi/N))

x = np.linspace(0,np.pi,N)
print(sum(eof(4,x)**2)) # verify the normalization condition
# 0.99
print(sum((eof(1,x) * eof(2,x)))) # verify orthogonality
# 2.534322578184866e-18

# generate PCs
Mode = 5
Ms = 1000
univar = np.diag(np.repeat(1,Mode))
zeromean = np.repeat(0,Mode)
rs = np.asmatrix(np.random.multivariate_normal(zeromean,
                                          univar, Ms))
pcm = np.asmatrix(np.zeros([Ms, Mode]))
```

```
for m in range(Mode):
    pcm[:,m] = rs[:,m]*np.sqrt(1/Ms)

print(np.dot(pcm[:,0].T, pcm[:,0])) #check normalized
print(np.dot(pcm[:,0].T, pcm[:,1])) #check orthogonal

# generate an independent random field
lam = [10, 9, 4.1, 4, 2]
sqrlam = np.diag(np.sqrt(lam))
eofm = np.zeros([N, Mode])

for m in range(Mode):
    eofm[:,m] = eof(m,x)
a = np.dot(np.asmatrix(eofm), np.asmatrix(sqrlam))
Yxr = np.dot(a, np.asmatrix(pcm).T)
Yxr.shape
#(100, 1000) #100 points, 1000 samples
```

Figure 6.3 shows this random field with the first 100 realizations. For a given rth realization, the field in $[0, \pi]$ shows the approximate characteristics of EOFs. For a given x value, the samples along r are not autocorrelated because the PCs are randomly generated. The following computer code uses the Durbin–Watson test to verify that Y_{xr} at $x = -2.5704$ is not auto-correlated.

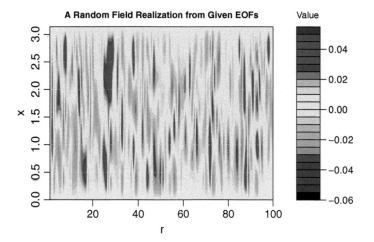

A Random Field Realization from Given EOFs

Figure 6.3   A realization of a random field over $[0, \pi]$ with the first 100 independent samples.

```
#R code for Durbin-Watson (DW) test for no serial correlation
library(lmtest)
Ms = 1000
r = 1:Ms
regYxr = lm(Yxr[10,] ~ r)
dwtest(regYxr)
```

```
#DW = 2.0344, p-value = 0.696
#Implying no significant serial correlation
```

```
#Python code for Durbin-Watson (DW) test for independence
import pandas as pd
from sklearn.linear_model import LinearRegression
from statsmodels.stats.stattools import durbin_watson
Ms = 1000
#use DataFrame to make the data correct shapes for reg
X = pd.DataFrame(np.linspace(1, Ms, num = Ms))
y = pd.DataFrame(Yxr[10,:])
y = y.transpose()
reg = LinearRegression().fit(X,y)
residuals = y - reg.predict(X) #compute residuals
#perform Durbin-Watson test
durbin_watson(residuals)
#1.96237986 in (1.5, 2.5) implies independence
```

The same DW test procedure can also be applied to PCs to conclude nonexistence of significant autocorrelation.

Because of the nonexistence of autocorrelation in PCs, Figure 6.3 does not show any wave propagation. A real climate field over a line segment and in a time period is called Hovmöller diagram (see Fig. 1.10) and usually shows patterns of wave propagation or other coherent structures, due to serial correlations. The space-time data are not independent in time. For a given spatial point, the temporal samples (i.e., a realization of a time series) are autocorrelated. The effective degree of freedom is less than that of the number of temporal samples.

Figure 6.3 can be plotted by the following computer code.

```
#R plot Fig. 6.3: The first 100 realization of a random field
r=1:100
setEPS()
postscript("fig0603.eps", width = 6, height = 4)
par(mar=c(3.5,3.5,2,0), mgp=c(2.0,0.7,0))
rgb.palette=colorRampPalette(
  c('black','blue','green',
    'yellow','pink','red','maroon'),
  interpolate='spline')
filled.contour(r,x, t(Yxr[,1:100]),
        color=rgb.palette, xlab="r", ylab="x", cex.lab=1.2,
          main="A random field realization from given EOFs",
          plot.axes={axis(1, cex.axis =1.1);
            axis(2, las = 0, cex.axis= 1.2); grid()},
          key.title={par(cex.main=0.9); title(main="Value")}
)
dev.off()
```

```
#Python plot Fig. 6.3: Plot a random field
# define color map
myColMap = LinearSegmentedColormap.from_list(name='my_list',
          colors=['black','blue','lime','yellow',
                  'pink','red','maroon'], N=100)
clev2 = np.linspace(Yxr.min(), Yxr.max(), 501)
r = np.linspace(2,100,99)
x = np.linspace(0,np.pi,N)
# find dimensions
print((Yxr[:,1:100]).shape)
print(r.shape)
print(x.shape)

# plot Fig. 6.3
contf = plt.contourf(r, x, Yxr[:,1:100], clev2,
                     cmap=myColMap);
colbar = plt.colorbar(contf, drawedges=False,
        ticks = [-0.06,-0.04,-0.02,0,0.02,0.04])
plt.title("A random field realization from given EOFs",
        pad = 20)
plt.xlabel("r", labelpad = 20)
plt.ylabel("x", labelpad = 20)
plt.text(104, 3.2, "Value", size = 17)
plt.savefig("fig0603.eps") # save figure
```

6.5 Sampling Errors for EOFs

A stationary climate anomaly field $T(\mathbf{r},t)$ over region A with N grid boxes or subregions has a covariance matrix, which is defined as an ensemble mean:

$$\Sigma_{ij} = \mathrm{E}[T(\mathbf{r}_i,t)T(\mathbf{r}_j,t)]. \tag{6.55}$$

Thus defined, Σ_{ij} may be regarded as the "true" covariance matrix. The exact "true" covariance matrix can never be known, but it can be approximated by a mathematical model. In practice, the approximation is a sample covariance matrix $\hat{\Sigma}_{ij}$ computed from the sample data as demonstrated earlier. The covariance Σ_{ij} does not depend on time t because of the stationarity assumption (see Eq. (6.38)).

Then, our question becomes: what is the error when using the sample covariance matrix to approximate the "true" covariance matrix? Specifically, what are the corresponding errors of EOFs and their corresponding eigenvalues? This section provides partial answers to these questions in both mathematical theory and numerical examples. Our common sense is that the larger the size R of the independent realizations, the better the approximation. When the sample size goes to infinity, the sample covariance by definition becomes the "true" covariance matrix. The ideas and method of this section follow that of North et al. (1982).

6.5.1 Sampling Error of Mean and Variance of a Random Variable

This subsection reviews the concepts of sample mean, sample variance, and their standard errors. Then we extend these concepts to the sample eigenvalues and sample EOFs.

Let $x_1, x_2, \ldots, x_R$ be the R independent samples of a normally distributed random variable X with mean μ and standard deviation σ. Then, the sample mean

$$\bar{x} = \frac{\sum_{i=1}^{R} x_i}{R} \tag{6.56}$$

is an unbiased estimator of the true population mean:

$$E[\bar{x}] = \mu. \tag{6.57}$$

The standard deviation of the mean estimator is

$$SE[\bar{x}] = \frac{\sigma}{\sqrt{R}}. \tag{6.58}$$

This is because

$$Var[\bar{x}] = \frac{\sigma^2}{R}. \tag{6.59}$$

The unbiased sample variance is

$$s^2 = \frac{\sum_{i=1}^{R} (x_i - \bar{x})^2}{R - 1} \tag{6.60}$$

with

$$E[s] = \sigma. \tag{6.61}$$

The standard error of s^2

$$SE[s^2] = \frac{\sigma^2}{\sqrt{(R-1)/2}}. \tag{6.62}$$

This is because

$$\frac{(R-1)s^2}{\sigma^2} \sim \chi^2_{R-1} \tag{6.63}$$

follows a chi-square distribution with $R - 1$ degrees of freedom and has a variance equal to $2(R - 1)$:

$$Var\left[\frac{(R-1)s^2}{\sigma^2}\right] = 2(R-1), \tag{6.64}$$

or

$$Var[s^2] = \frac{2\sigma^4}{R - 1}. \tag{6.65}$$

6.5.2 Errors of the Sample Eigenvalues

Let us take R independent samples from a population of M dimensional Gaussian vectors with a given covariance whose distinct eigenvalues are $\lambda_1, \lambda_2, \ldots, \lambda_M$. Here, $M \leq N$ and N is the order of the covariance matrix. The sample covariance matrix has eigenvalues $\hat{\lambda}_1, \hat{\lambda}_2, \ldots, \hat{\lambda}_M$. With the sample data, we can compute the sample covariance matrix and its eigenvalues. Then, are the sample eigenvalues $\hat{\lambda}_m$ close to the true eigenvalues λ_m, $m = 1, 2, \ldots, M$ when the sample size R is large? We would expect that the sample eigenvalues approach the true ones as $R \to \infty$. Lawley (1956) developed a theory to quantify this approximation as follows (see also von Storch and Zwiers 1998, p. 302):

$$E[\hat{\lambda}_m] = \lambda_m \left(1 - \frac{1}{R} \sum_{\substack{\ell=1 \\ \ell \neq m}}^{M} \frac{\lambda_\ell}{\lambda_\ell - \lambda_m} \right) + O\left(\frac{1}{R^2} \right), \tag{6.66}$$

$$\mathrm{Var}[\hat{\lambda}_m] = \frac{2\lambda_m^2}{R} \left(1 - \frac{1}{R} \sum_{\substack{\ell=1 \\ \ell \neq m}}^{M} \left(\frac{\lambda_\ell}{\lambda_\ell - \lambda_m} \right)^2 \right) + O\left(\frac{1}{R^3} \right). \tag{6.67}$$

These two formulas can be interpreted in a computational procedure. Samples of size R taken from an independent and identically distributed normal population of a N-dimensional random vector $\mathbf{X}$ form a data matrix Y_{xr}. This data matrix forms a sample covariance matrix $\mathbf{S}$. The eigenvalues $\hat{\lambda}_k$ of $\mathbf{S}$ may be considered as a sample statistic, namely an estimator of the true eigenvalues λ_k of the covariance matrix of a random climate variable $\mathbf{X}$. Then, asymptotically as R goes to infinity, the expected value of the $\hat{\lambda}_k$ is approximately related to the true eigenvalues λ_k in the following way:

$$E[\hat{\lambda}_k] = \lambda_k \left(1 - \frac{1}{R} \sum_{\ell=1, \ell \neq k}^{M} \frac{\lambda_\ell}{\lambda_\ell - \lambda_k} \right). \tag{6.68}$$

The approximation error is in the order of $O\left(\frac{1}{R^2} \right)$. This formula is the expected value of the $\hat{\lambda}_k$ that is not only related to λ_k, but also to all the other eigenvalues. If the λ_k are arranged in descending order, $\lambda_1 \geq \lambda_2 \geq \lambda_2 \geq \cdots \geq \lambda_M \geq 0$, then

$$E[\hat{\lambda}_1] \geq \lambda_1, \tag{6.69}$$

i.e., $E[\hat{\lambda}_1]$ is an overestimate of λ_1 since the $\frac{1}{R}$ term in Eq. (6.68) is nonnegative.

Note that this is a result for expected value. An individual realization may not always result in an overestimate although it is more likely than not.

Similarly,

$$E[\hat{\lambda}_M] \leq \lambda_M \tag{6.70}$$

is an underestimate of λ_M since the $\frac{1}{R}$ term in Eq. (6.68) is nonpositive.

Equation (6.67) implies that the standard error of the estimated eigenvalue is approximately

$$SE(\hat{\lambda}_k) = \sqrt{\frac{2}{R}} \lambda_k, \quad k = 1, 2, \cdots, M. \tag{6.71}$$

6.5.3 North's Rule-of-Thumb: Errors of the Sample Eigenvectors

This subsection estimates the errors of eigenvectors for given eigenvalues and sample size. Again, we use the asymptotic approximation method with a "small" parameter ε (Anderson 1963):

$$\varepsilon = \sqrt{\frac{2}{R}}. \tag{6.72}$$

The eigenvalue problem for the true covariance matrix for the ℓth mode is

$$\Sigma \mathbf{u}_\ell = \lambda_\ell \mathbf{u}_\ell. \tag{6.73}$$

That for the sample covariance is

$$S \hat{\mathbf{u}}_\ell = \hat{\lambda}_\ell \hat{\mathbf{u}}_\ell. \tag{6.74}$$

The asymptotic method, also called the perturbation method, is to write $\hat{\lambda}_\ell$ and $\hat{\mathbf{u}}_\ell$ as an expansion according to powers of ε:

$$\hat{\lambda}_\ell = \lambda_\ell + \varepsilon \lambda_\ell^{(1)} + \varepsilon^2 \lambda_\ell^{(2)} + \cdots, \tag{6.75}$$

$$\hat{\mathbf{u}}_\ell = \mathbf{u}_\ell + \varepsilon \mathbf{u}_\ell^{(1)} + \varepsilon^2 \mathbf{u}_\ell^{(2)} + \cdots. \tag{6.76}$$

After steps of derivation, the first order errors of eigenvalue and eigenvectors become

$$\delta \lambda = \varepsilon \lambda_\ell^{(1)} \approx \sqrt{\frac{2}{R}} \lambda_\ell. \tag{6.77}$$

The corresponding approximation for the eigenvector error is

$$\delta \mathbf{u}_\ell = \varepsilon \mathbf{u}_\ell^{(1)} \approx \frac{\delta \lambda_\ell}{\Delta \lambda_\ell} \mathbf{u}_{\ell'}, \tag{6.78}$$

where

$$\Delta \lambda_\ell = \lambda_\ell - \lambda_{\ell'} \tag{6.79}$$

is the eigenvalue gap, and $\lambda_{\ell'}$ is the eigenvalue closest to λ_ℓ. Because λ_ℓ is a non-increasing sequence, either $\ell' = \ell + 1$ or $\ell' = \ell - 1$. In the derivation of Eq. (6.78), we have ignored all the terms

$$\frac{\lambda_k}{\lambda_\ell - \lambda_k}$$

except $k = \ell'$.

The convenient estimations of Eqs. (6.77) and (6.78) are known as *North's rule of thumb* for variance and EOFs, following the original paper North et al. (1982). The rule basically says that when two eigenvalues are close to each other, the corresponding EOF can have a large error, amplified by the inverse of the small gap between two eigenvalues, but the error can be reduced by a sufficiently large number of independent samples. Further, the sampling error vector of an EOF $\mathbf{u}_\ell$ is approximately a scalar multiplication of its neighbor EOF $\mathbf{u}_{\ell'}$.

In practice, many estimates of EOFs are taken from time series wherein there is correlation of nearby temporal terms, known as serial correlation. This will tend to reduce the number of independent degrees of freedom in the time segment. The serial correlation is relevant to the time series associated with a particular EOF, typically long correlation times corresponding to large spatial EOF modes. As in regression problems, this consideration of serial correlation may help correct the estimation of an erroneously small confidence interval.

Two synthetic examples will be presented in the next two subsections to illustrate the EOF errors.

6.5.4 EOF Errors and Mode Mixing: A 1D Example

The true parameter field or data from nature are rarely known but only estimated. We thus use synthetic data as examples to illustrate the examples of EOFs when two eigenvalues are close to each other. Let the eigenvalues λ_k be

$$10.0, \ 9.0, \ 4.1, \ 4.0, \ 2.0.$$

So, λ_1 and λ_2 are close, and λ_3 and λ_4 are close. We consider two sample sizes: $R = 100$, and $R = 1,000$. The scree plot and the associated bars of standard errors are shown in Figure 6.4. The figure shows that with 100 samples, λ_1 and λ_2 are not well separated, and neither are λ_3 and λ_4. With 1,000 samples, λ_1 and λ_2 are well separated, but λ_3 and λ_4 are still not.

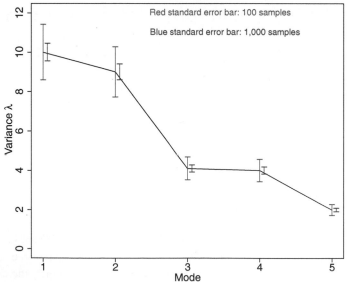

Figure 6.4   A scree plot for the specified eigenvalues and the bars of standard errors $\lambda_\ell \pm \lambda_\ell \sqrt{2/R}$.

The scree plot Figure 6.4 can be generated by the following computer code.

```
#R plot Fig. 6.4: Scree plot
Mode = 1:5
lam=c(10.0, 9.0, 4.1, 4.0, 2.0)
samp = rep("S100", 5)
sd = sqrt(2/100)*lam
sd2 = sqrt(2/1000)*lam
#par(mar=c(4.5, 4.5, 1, 0.5))
setEPS()
postscript("fig0604.eps", width = 10, height = 8)
par(mar=c(3.5, 4.5, 0.2, 0.5), mgp=c(2.5, 1.0,0))
plot(Mode, lam, type="l", ylim = c(0,12),
     xlab="Mode",
     ylab=expression("Variance␣␣␣"*lambda),
     cex.lab=1.8, cex.axis=1.8)
points(Mode,lam + sd, pch="-", col="red", cex=2)
points(Mode,lam - sd, pch="-",col="red", cex=2)
segments(Mode,lam + sd, Mode,lam - sd, col="red")
points(Mode+0.06,lam + sd2, pch="-", col="blue", cex=2)
points(Mode+0.06,lam - sd2, pch="-",col="blue", cex=2)
segments(Mode +0.06,lam + sd2, Mode + 0.06,lam - sd2,
         col="blue")
text(3.45,12, "Red␣standard␣error␣bar:␣100␣samples",
         cex = 1.5, col="red")
text(3.5,11, "Blue␣standard␣error␣bar:␣1000␣samples",
     cex = 1.5, col="blue")
dev.off()
```

```
#Python plot Fig. 6.4: Scree plot
plt.figure(figsize=(10, 8))
Mode = np.linspace(1,5,5)
lam = [10, 9, 4.1, 4, 2]
samp = np.repeat(100,5)
sd = []
for i in range(5):
    sds = (np.sqrt(2/100)*lam[i])
    sd.append(sds)
sd2 = []
for i in range(5):
    sds2 = (np.sqrt(2/1000)*lam[i])
    sd2.append(sds2)
plus = []
for i in range(5):
    values = lam[i] + sd[i]
    plus.append(values)
minus = []
for i in range(5):
    values = lam[i] - sd[i]
    minus.append(values)
plus2 = []
for i in range(5):
    values = lam[i] + sd2[i]
    plus2.append(values)
minus2 = []
```

```
for i in range(5):
    values = lam[i] - sd2[i]
    minus2.append(values)

plt.plot(Mode, lam, color = 'k')
plt.ylim(0,12)
plt.xlabel("Mode", labelpad = 20, fontsize = 25)
plt.ylabel("Variance␣$\lambda$", labelpad = 20,
          fontsize = 25)
plt.scatter(Mode, plus, color = "red")
plt.scatter(Mode, minus, color = "red")
plt.scatter(Mode+0.06, plus2, color = "blue")
plt.scatter(Mode+0.06, minus2, color = "blue")
# plot line segments
plt.arrow(1,minus[0], dx = 0, dy = plus[0]-minus[0],
          color = "red")
plt.arrow(2,minus[1], dx = 0, dy = plus[1]-minus[1],
          color = "red")
plt.arrow(3,minus[2], dx = 0, dy = plus[2]-minus[2],
          color = "red")
plt.arrow(4,minus[3], dx = 0, dy = plus[3]-minus[3],
          color = "red")
plt.arrow(5,minus[4], dx = 0, dy = plus[4]-minus[4],
          color = "red")
plt.arrow(1.06,minus2[0], dx = 0, dy = plus2[0]-minus2[0],
          color = "blue")
plt.arrow(2.06,minus2[1], dx = 0, dy = plus2[1]-minus2[1],
          color = "blue")
plt.arrow(3.06,minus2[2], dx = 0, dy = plus2[2]-minus2[2],
          color = "blue")
plt.arrow(4.06,minus2[3], dx = 0, dy = plus2[3]-minus2[3],
          color = "blue")
plt.arrow(5.06,minus2[4], dx = 0, dy = plus2[4]-minus2[4],
          color = "blue")
plt.text(2.2,11.2, "Red␣standard␣error␣bar:␣100␣samples",
          color = "red", fontsize = 19)
plt.text(2.2,10.4, "Blue␣standard␣error␣bar:␣1000␣samples",
          color = "blue", fontsize = 19)
plt.savefig("fig0604.eps") # save figure
```

The exact orthonormal EOF modes are defined in the spatial domain $[0, \pi]$:

$$E_k(i) = \sqrt{\frac{2}{N}} \sin(kx_i), \tag{6.80}$$

where $x_i = (i/N)\pi$, $i = 1, 2, \ldots, N$ are the N sampling locations in the spatial domain $[0, \pi]$.

We use random generators to generate a random sequence of length R as PCs. The PCs, eigenvalues and EOFs form an $N \times R$ random matrix, which is considered R independent samples of a "true" field in the spatial domain $[0, \pi]$. We then use SVD for the data matrix to compute sample EOFs from these generated sample data and compare the sample EOFs with the exact EOFs to show the EOF errors. The main feature of the EOF errors is the

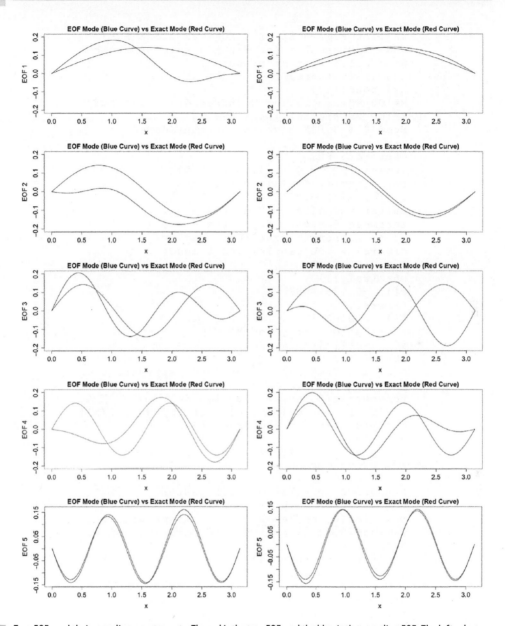

Figure 6.5 True EOFs and their sampling counterparts: The red is the true EOF, and the blue is the sampling EOF. The left column shows the first five EOFs with sample size $R = 100$, and the right for $R = 1,000$.

mode mixing. The two EOFs with closest eigenvalues with insufficient samples are mixed and are not clearly separate.

The upper left panel of Figure 6.5 can be generated by the follow computer code. The other panels of Figure 6.5 may be generated by entering corresponding parameters in the code. Because this is a simulation of a random field, your result may not be the same as what is shown here. You may repeat your numerical experiment many times and observe the result variations.

```
#R plot Fig. 6.5a: SVD EOF1 vs exact EOF1
#Ms= 100 samples for North's rule of thumb
#install.packages("mvtnorm")
library(mvtnorm)
set.seed(112)
M=100 #M samples or M independent time steps
N=100 #N spatial locations
Mode=5 # 5 EOF modes to be considered

#Generate PCs
lam=c(10.0, 9.0, 4.1, 4.0, 2.0)
round(sqrt(2/M)*lam, digits = 2)
#[1] 1.41 1.27 0.58 0.57 0.28
univar =diag(rep(1,Mode)) #SD = 1
zeromean = rep(0, Mode) #mean = 0
rs <- rmvnorm(M, zeromean, univar)
dim(rs) # 100 samples and 5 modes
#[1] 100     5
mean(rs[,1])
#[1] -0.02524037
var(rs[,1])
#[1] 0.9108917
t <- seq(0, 2*pi, len=M)
a51 <-rs[,1]*sqrt(1/M)
sum(a51^2)
#[1] 0.9026492 is the variance approximation

pcm=matrix(0, nrow=M, ncol=Mode)
for(m in 1:Mode){pcm[,m] = rs[,m]*sqrt(1/M)}
dim(pcm) #random and independent PCs
#[1] 100     5
sum(pcm[,1]^2) #verify the normality of a PC
#[1] 1.021924

#Generate EOFs for spatial patterns
eof <- function(n, x) sqrt(2)* sin(n*x)*sqrt((1/N))
x <- seq(0, pi,len=N) #N locations within [0,1]
sum(eof(3,x)^2) #verify the normality of an EOF
#[1] 0.99
sum(eof(1,x)*eof(2,x)) #verify the EOF orthogonality
#[1] 3.035766e-18

eofm <- matrix(0, nrow=N, ncol=Mode)
for (m in 1:Mode){eofm[,m]=eof(m,x)}
dim(eofm) #eofm are the 5 modes of EOF data
#[1] 100     5 #100 spatial locations and 5 modes

#Generate the random data with given EOFs
Lam = diag(sqrt(lam)) #eigenvalue matrix
da <- eofm%*%Lam%*%t(pcm) #spectral decomposition
dim(da) #random data at 100 spatial locations
#[1]  100 100
svdda <- svd(da)
round((svdda$d[1:5])^2, digits =2)
#[1] 10.14  8.12  3.83  3.26  2.10
```

```
png(file="fig0605a.png", width=600,height=300)
k=1
sum(svdda$u[,k]^2)
#[1] 1
par(mar=c(4.5,4.7,2,0.2))
plot(x,-svdda$u[,k], type="l", ylim=c(-0.2,0.2),
    xlab="x", ylab=paste("EOF", k),
    main="EOF␣Mode␣(Blue␣Curve)␣vs␣Exact␣Mode␣(Red␣Curve)",
    col="blue",
    cex.lab=1.4, cex.axis=1.4,cex.main=1.4)
lines(x, eof(k,x), col='red')
dev.off()
```

```
#Python plot Fig. 6.5a: SVD mode1 vs exact mode1
import random
random.seed(10)
M=100 #M samples or M independent time steps
N=100 #N spatial locations
Mode=5 # 5 EOF modes to be considered
#Generate PCs
lam = [10.0, 9.0, 4.1, 4.0, 2.0]
np.round(np.array(np.sqrt(2/M))*lam, 2)
#[1] 1.41 1.27 0.58 0.57 0.28
univar = np.diag([1]*Mode) #SD = 1
zeromean = [0]*Mode #mean = 0
rs = np.random.multivariate_normal(zeromean, univar, M)
print(rs.shape)
#(100, 5)  100 samples and 5 modes
pcm = rs*np.sqrt(1/M) #PCs
print(pcm.shape)
#(100, 5)
x = pd.DataFrame(np.linspace(0, np.pi, N))
n = pd.DataFrame(np.arange(1, Mode + 1))
nx = np.dot(x, n.transpose())
eofm = np.sqrt(2)*np.sin(nx)*np.sqrt((1/N))#EOFs
Lam = np.diag(np.sqrt(lam)) #eigenvalue matrix
da0 = np.matmul(eofm, Lam)
da = np.matmul(da0, pcm.transpose())
print(da.shape) #(100, 100) generated space-time data
u,d,v = np.linalg.svd(da) #SVD mode
plt.ylim(-0.2, 0.2)
plt.plot(x, u[:,1], color = 'b') #SVD model
plt.plot(x, eofm[:,0], color = 'r')#Exact mode
plt.ylabel('EOF1')
plt.xlabel('x')
plt.title('EOF␣mode␣(Blue␣curve)␣vs␣Exact␣mode␣(Red␣curve)')
```

The left column of Figure 6.5 shows the true EOFs (red) and the sampling EOFs (blue) with sample size equal to $R = 100$. The right column is for $R = 1,000$. The scree plot shows that $\lambda_1 = 10.0$ and $\lambda_2 = 9.0$ are well separated when $R = 1,000$, but not when $R = 100$. Thus, EOF1 and EOF2 are well separated for $R = 1,000$ and have little sampling error.

The sample EOF1 and EOF2 are close to the true EOF1 and EOF2, as shown in the first two panels of the right column. In contrast, EOF1 and EOF2 are mixed for $R = 100$ as shown in the first two panels of the left column. They have relatively large sampling errors. The sample EOF1 has two local maxima, but the true EOF1 has only one maximum. The sample EOF2 has only one clear local minimum and has an unclear local maximum, while the true EOF2 clearly has one local maximum and one local minimum. Thus, the sample EOF1 and EOF2 are mixed.

The scree plot Figure 6.4 shows that eigenvalues $\lambda_3 = 4.1$ and $\lambda_4 = 4.0$ are not well separated even for $R = 1,000$. Thus, EOF3 and EOF4 are mixed for both $R = 100$ and $R = 1,000$ and are expected to have large sampling errors. The sample EOF3 patterns in Figure 6.5 show four local extrema, while true EOF3 has only three extrema. The sample EOF4 patterns has three extrema while the true EOF4 has four extrema. Thus, the sample EOF3 and EOF4 are mixed.

The scree plot Figure 6.4 also shows that $\lambda_5 = 2.0$ is well separated from the other four eigenvalues for both 100 and 1,000 samples. The sample EOF5 is very close to true EOF5 as shown in Figure 6.5 for both 100 and 1,000 samples. EOF5 has a more complex pattern than the first four EOFs, but a smaller sampling error for a given sample size. This seemingly counterintuitive result implies the seriousness of the mode mixing problem and the high level of difficulty of obtaining correct EOFs when two or more eigenvalues are close to each other.

Thus, if an eigenvalue is well separated from other eigenvalues, the corresponding EOFs have a small sampling error even with a small sample size. The complexity of the EOF patterns is less important than the separation of the eigenvalues. When two neighboring EOFs are mixed, the sampling error for each EOF can be very large. The error size can be comparable to its neighbor EOF, according to North's rule-of-thumb formula Eq. (6.78). For example, the sampling error of EOF4 is comparable to that of EOF3.

Figure 6.6 shows a simulation example for the error of EOF4 $\delta \mathbf{u}_4$, which has a shape and magnitude comparable to the neighboring EOF3 $\mathbf{u}_3$ for $R = 100$. Figure 6.6 may be generated by the following computer code. Again, because this is a simulation of a random field, the result of your first try may not be the same as what is shown here. You can repeat your experiment many times and observe the result variations.

```
#R plot Fig. 6.6:  EOF4 error
k = 4 #Use the SVD result from 100 samples R = 100
plot(x,svdda$u[,k], type="l", ylim=c(-0.33,0.33),
     xlab="x", ylab="EOFs [dimensionless]",
     main="EOF4 error (black) vs Exact EOF3 (Orange)",
     col="blue",
     cex.lab=1.4, cex.axis=1.4, cex.main=1.4)
legend(0.2,0.37, bty = "n", cex=1.4, text.col = 'blue',
       lwd=1.2,legend="Sample EOF4",col="blue")
lines(x, eof(k,x), col='red')
legend(0.2,0.41, bty = "n", cex=1.4, text.col = 'red',
       lwd=1.2,legend="True EOF4",col="red")
lines(x,svdda$u[,k] - eof(k,x), lwd=2, col='black'  )
legend(0.2,0.33, bty = "n", cex=1.4, text.col = 'black',
       lwd=2,legend="Sample EOF4 - True EOF4",col="black")
lines(x, -eof(3,x), col='orange')
```

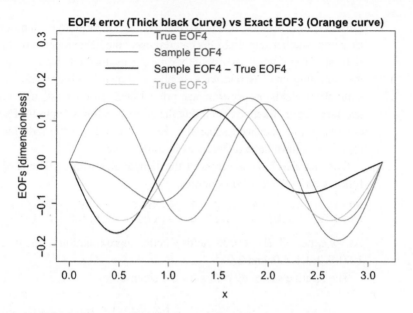

Figure 6.6 Sampling errors of EOF4 for $R = 100$: The red is the true EOF4, the blue is the sampling EOF4, the black is the difference of the true EOF4 minus the sample EOF4, and the orange is the true EOF3.

```
legend(0.2,0.29, bty = "n", cex=1.4, text.col = 'orange',
       lwd=1.2,legend="True␣EOF3",col="orange")
```

```
#Python plot Fig. 6.6: EOF4 error
plt.ylim(-0.26, 0.3)
plt.plot(x, eofm[:,3], color = 'r',
         label = 'True␣EOF4')#Exact mode 4
plt.plot(x, u[:,4], color = 'b',
         label = 'Sample␣EOF4') #SVD mode 4
plt.plot(x, eofm[:,3] - u[:,4], color = 'k',
         label = 'True␣EOF4␣-␣Sampel␣EOF4')
plt.plot(x, eofm[:,2], color = 'orange',
         label = 'True␣EOF3') #Exact mode 3
plt.ylabel('EOFs␣[dimensionless]')
plt.xlabel('x')
plt.title('EOF4␣error␣(black)␣vs␣exact␣EOF3␣(Orange)')
plt.legend(loc="upper␣left", fontsize = 17)
```

The thick black line in Figure 6.6 is the difference of the true EOF4 minus the sample EOF4 for $R = 100$, which has a shape and size similar to EOF3 shown by the green curve. Note that the sample EOF4 is Figure 6.6 is different from that in Figure 6.5, because the random fields are generated by the random PCs and are different for each generation. Therefore, although in general the sampling EOFs have large errors when two neighborhood eigenvalues are not well separated, there can be some individual cases that still produce accurate sample EOFs with not well-separated eigenvalues. However, the latter is rare. For these reasons, when two eigenvalues are close to each other, we need to carefully examine whether sample EOFs can represent the behavior of a real physical field.

In summary, when two neighboring eigenvalues are well separated, it is certain that the sampling error is relatively small, as shown in Figure 6.5 for EOF5. Almost every simulation yields an accurate sampling EOF5. In contrast, uncertainties, mode mixing, and large errors likely exist among the corresponding sample EOFs when the neighboring eigenvalues are not well separated. Keep in mind that the individual random simulation results may have large differences from one other.

One can also test the errors of the sample eigenvalues. The sample eigenvalues are as follows for 100 and 1,000 samples:

```
R=  100   10.8149   8.3288   4.3849   3.4097   1.8479
R=1000    9.9948   8.8308   4.1748   3.8353   1.9690
```

As we expected, $R = 1,000$ yields a better approximation to the true specified eigenvalues $10.0, 9.0, 4.1, 4.0,$ and 2.0.

The relative sampling errors for the eigenvalues

$$\left(\frac{\lambda_k - \hat{\lambda}_k}{\lambda_k} \right) \times 100 \tag{6.81}$$

are around 10%, comparable with $\sqrt{2/R} \times 100\% = 14\%$ when $R = 100$. This may be considered a validation to the simulation of the standard error formula (6.77) for sample eigenvalues.

6.5.5 EOF Errors and Mode Mixing: A 2D Example

We consider the following given eigenvalues and EOFs in a 2D square $[0, \pi] \times [0, \pi]$: $\lambda = 10.0, 9.0, 4.1, 4.0,$ and 2.0, and

$$\psi_k(x, y) = \frac{2}{\pi} \sin(kx) \sin(ky). \tag{6.82}$$

On the $N \times N$ grid, the discretized EOFs are

$$E_k(i, j) = \frac{2}{N} \sin(kx_i) \sin(ky_j), \tag{6.83}$$

where

$$x_i = i\pi/N, \quad y_j = j\pi/N, \quad i, j = 1, 2, \ldots, N \tag{6.84}$$

are the coordinates of the $N \times N$ sample grid points over the 2D square domain. The discrete functions $E_k(i, j)$ are considered to be the true EOFs and are shown in the left column of Figure 6.7.

We then generate a random field using $R = 100$ samples based on the these given eigenvalues and EOFs, and perform the SVD on the generated field to recover the EOFs. These are the sample EOFs, which are shown in the middle column of Figure 6.7. Compared with the left column of the true EOFs, the middle column displays the obvious distortion of the spatial patterns for the first four EOFs, but not for EOF5 whose eigenvalue $\lambda_5 = 2$ is well separated from other eigenvalues.

True EOFs Compared with Sample EOFs

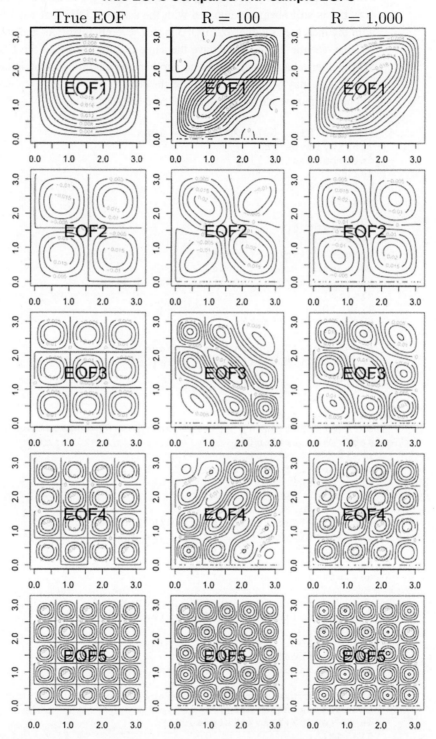

True EOFs and their sampling counterparts: The left column are the true EOFs, the middle column are the sample EOFs with sample size 100, and the right column are the sample EOFs with sample size 1,000.

For sample size $R = 1,000$, the distortion of EOF1 and EOF2 is less than that for $R = 100$, but is still obvious compared with the sample EOF5 that has little distortion. Even with the increased sample size, EOF3 is still severely distorted, and so is EOF4.

Thus, for the 2D case, we have the same conclusion about the EOF sampling errors: When eigenvalues are close to one another, the sampling error is large. Increasing the sample size may not effectively reduce the error.

The computer code for generating Figure 6.7 is as follows.

```
#R plot Fig. 6.7: 2D sample EOFs
library(mvtnorm)
dmvnorm(x=c(0,0))
M=100
N=100
Mode=5

#Generate PCs
lam=c(10.0, 9.0, 4.1, 4.0, 2.0)
univar =diag(rep(1,Mode))
zeromean = rep(0, Mode)
rs <- rmvnorm(M, zeromean, univar)
dim(rs)
#[1] 100    5
t <- seq(0, len=M)
a51 <-rs[,1]*sqrt(1/M)
pcm=matrix(0, nrow=M, ncol=Mode)
for(m in 1:Mode){pcm[,m] = rs[,m]*sqrt(1/M)}
dim(pcm)
#[1] 100    5

#Generate true EOFs
eof <- function(n, x, y){(outer(sin(n*x),sin(n*y)))*(2/N)}
#eof <- function(n, x, y){outer(sin(n*x),sin(n*y))}
x = y <- seq(0,pi,len=100)
sum((eof(2,x, y))^2) #verify the normality
#[1] 0.9801

#Plot true EOF1 2D
png(file="fig0607a.png", width=200,height=200)
par(mar=c(2,2,0.5,0.5))
contour(x,y, eof(1,x,y))
text(1.5,1.5, 'EOF1', cex=2)
dev.off()

eofm <- matrix(0, nrow=N^2, ncol=Mode)
for (m in 1:Mode){eofm[,m]=c(eof(n=m,x,y))}
dim(eofm)
#[1] 10000    5 #10000 = 100*100
t(eofm)%*%eofm
#approximately equal to I5 an identify matrix

#Generate the random data with given EOFs
Lam = diag(sqrt(lam))
da <- eofm%*%Lam%*%t(pcm)
dim(da)
#[1] 10000    100
svdda <- svd(da)
```

```
#Plot sample EOF1-5 2D: R=100
png(file="fig0607n.png", width=200,height=200)
par(mar=c(2,2,0.5,0.5))
k=5
contour(x,y, matrix(svdda$u[,k], ncol=100),
         cex.lab=1.4, cex.main=1.4)
text(1.5,1.5, 'EOF5', cex=2)
dev.off()
```

```
#Python Fig. 6.7a: 2D true EOF1
from scipy.stats import multivariate_normal as mvn
d = mvn([0,0])
x = d.rvs(size = 1)
dist = d.pdf(x)
M = 100
N = 100
Mode = 6
# generate PCs
lam = [10, 9, 4.1, 4, 2, 1]
univar = np.diag(np.repeat(1,Mode))
zeromean = np.repeat(0,Mode)
rs = np.asmatrix(np.random.multivariate_normal(zeromean,
                                        univar, M))
print(rs.shape)
t = np.linspace(0,99,100)
a51 = rs[:,1]*np.sqrt(1/M)
pcm = np.asmatrix(np.zeros([M, Mode]))
for m in range(Mode):
    pcm[:,m] = rs[:,m]*np.sqrt(1/M)
print(pcm.shape)
# generate EOFs
def eof(n,x,y):
    return((np.outer(np.sin(n*x),np.sin(n*y)))*(2/N))
x = y = np.linspace(0,np.pi,100)
print(np.sum((eof(2,x,y))**2))
# plot true EOF 1 2D
fig,ax = plt.subplots()
cs = ax.contour(x,y, eof(1,x,y))
ax.clabel(cs, fontsize = 15)
plt.text(1.35,1.5, "EOF1", size = 35)
# save figure
plt.savefig("fig0607.png")
plt.show()

#Python Fig. 6.7: 2D sample EOF5
eofm = (np.zeros([N**2, Mode]))
# potential problem with for-loop
for m in range(Mode):
    eofm[:,m] = (eof(m, x, y)).flatten()
print(eofm.shape)
np.dot(eofm.T, eofm)
```

```
# generate the random data with given EOFs
Lam = np.diag(np.sqrt(lam))
da = multi_dot([eofm, Lam, pcm.T])
print(da.shape)
from scipy import linalg
svdda, s, Vh = linalg.svd(da,full_matrices=False)

# plot sample EOF5 2D: R = 100
k = 4
plt.contour(x, y, (svdda[:,k]).reshape(100,100))
# save figure
plt.savefig("fig0607EOF5.eps")

plt.show()
```

6.5.6 The Original Paper of North's Rule-of-Thumb

North et al. (1982) provided a numerical example of a case similar to a meteorological field, a 500 hPa geopotential height field in midlatitudes. They showed estimates of the EOFs for sample size $R = 300$ and 1,000. Notice that the $R = 300$ sampling errors for the shapes are very large and the shapes are unrecognizably distorted. When $R = 1,000$, the first two EOFs are recognizable, but the last two are not. The figures were reproduced in von Storch and Zwiers (1998) and Wilks (2011).

If the sampling error in the eigenvalue is comparable to the distance to a nearby eigenvalue, then the sampling errors in the EOF will be comparable to the "nearby" EOF. This is North's rule-of-thumb. A physical interpretation of this result can be made. If the sampling error in an eigenvalue is comparable to the spacing, a kind of "effective degeneracy" occurs. We have already seen that a degeneracy leads to an intrinsic ambiguity in defining the EOF, since any linear combination of the possible eigenvectors is also an eigenvector. Even if no degeneracy actually exists, some eigenvalues may be close enough to each other that sampling errors lead to an effective degeneracy and EOF mixing occurs. That is, a particular sample will lead to one linear combination and another sample may pick out a drastically different linear combination of the nearby eigenvectors. The result is wildly differing patterns from one sample to the next.

6.5.7 When There Is Serial Correlation

Practical estimates of the (true) EOFs are usually computed from space-time data, which might be serially correlated. The serial correlation can expand the estimates of the confidence interval for an eigenvalue. Since the EOFs are sensitive to the nearness of neighboring eigenvalues, the serial correlation may impact the stability of EOF shapes.

The number of degrees of freedom for a time series at a given spatial location is usually smaller than the number of steps in the time series. The PC time series may be less serially correlated, but not totally uncorrelated. The relationship between the eigenvalue and the variance of its corresponding PC time series may require reevaluation.

6.6 Chapter Summary

This chapter describes covariance, EOFs, and PCs. The spatial covariance matrix may be computed as the product of a space-time anomaly matrix and its transpose. The eigenvectors of the spatial covariance matrix are called EOFs. The orthonormal projection of the anomaly data matrix on the EOFs are the PCs, each being the time series associated with an individual EOF. When the number of columns of the space-time anomaly data matrix is much smaller than the number of rows, you may wish to compute the temporal covariance matrix that is equal to the transposed space-time anomaly data matrix times the space-time anomaly data matrix. The order of this temporal covariance matrix is smaller than that of the spatial covariance matrix. The eigenvectors of the temporal covariance matrix are PCs. The orthonormal projection of the anomaly data on PCs lead to EOFs. The idea of computing a temporal covariance matrix is useful when dealing with a big anomaly dataset with a relatively small number of columns. When you are dealing with an anomaly data matrix of less than 1.0 GB, you can obtain the EOFs and PCs more directly using SVD, and you do not need to compute the covariance matrix. However, the covariance matrix itself may have climate science implications, as shown in Figure 6.1.

The first of a few EOFs, particularly the first two or three, often have clear climate science interpretations. The higher order of EOFs and PCs may be noise in the sense of climate physics, although they may help with a better mathematical approximation in climate data analysis and modeling. You always have to face a decision about how many EOFs to use in your data analysis. There is no specific rule for the decision, but you may use the total variance explained, such as 80%, to justify your number of EOFs to be used. See Section 6.2.2 and its scree plot for details. Note that the sample EOFs form a complete set of basis vectors, and can be used to represent the field as a set of statistical modes.

A climate field is often continuous in both space and time, yet our data analysis, observations, and modeling are often in grid boxes in 2D or 3D space and at discrete steps in time. The concepts of covariance and its eigenvector for the continuous climate field are expressed by integrals, which must be discretized for numerical calculations. The factors of area, volume, and time must be used in this discretization process. See Section 6.3.

In general, climate fields can be taken to be random for statistical modeling. Sections 6.4 discusses the method of generating a random climate field with given standard deviations and EOFs. Section 6.5 discusses the sampling errors of sample EOFs and sample eigenvalues. We presented North's rule-of-thumb: the sampling error of an EOF becomes large when its eigenvalue is close to another EOF's eigenvalue. Examples for EOF sampling errors were provided for both 1D and 2D domains.

References and Further Reading

[1] T. W. Anderson, 1963: Asymptotic theory for principal component analysis. *Annals of Mathematical Statistics*, **34**, 122–148.

> This paper used the perturbation method to estimate the errors of sample EOFs. The original derivation of the North's rule-of-thumb followed this paper and the perturbation theory in quantum mechanics.

[2] D. N. Lawley, 1956: Tests of significance for the latent roots of covariance and correlation matrices. *Biometrika*, **43**, 128–136.

> This was an early paper that studies the errors of eigenvalues of a covariance matrix using the perturbation method.

[3] G. R. North, T. L. Bell, R. F. Cahalan, and F. J. Moeng, 1982: Sampling errors in the estimation of empirical orthogonal functions. *Mon. Weather. Rev*, **110**, 699–706. https://journals.ametsoc.org/view/journals/mwre/110/7/1520/0493/1982/110/0699/seiteo/2/0/co/2.xml?tab/body=pdf

> This is the original paper of North's rule-of-thumb. Numerous papers had been motivated by this theory that quantifies the sample errors of EOFs.

[4] T. M. Smith, R. W. Reynolds, T. C. Peterson, and J. Lawrimore, 2008: Improvements to NOAA's historical merged land-ocean surface temperatures analysis (1880–2006). *Journal of Climate*, **21**, 2283–2296.

> This was an earlier paper that described the merged land-ocean monthly SAT anomaly data from 1880.

[5] H. Von Storch and F. W. Zwiers, 2001: *Statistical Analysis in Climate Research.* Cambridge University Press.

This is a comprehensive climate statistics book that helps climate scientists apply statistical methods correctly. The book has both rigorous mathematical theory and practical applications. Numerous published research results are discussed.

[6] R. S. Vose, D. Arndt, V. F. Banzon et al., 2012: NOAA's merged land-ocean surface temperature analysis. *Bulletin of the American Meteorological Society*, **93**, 1677–1685.

This paper provided a comprehensive summary of the quality control and other scientific procedures that were used to produce the dataset.

[7] D. S. Wilks, 2011: *Statistical Methods in the Atmospheric Sciences*. 3rd ed., Academic Press.

This excellent textbook is easy to read and contains many simple examples of analyzing real climate data. It is not only a good reference manual for climate scientists, but also a guide tool that helps scientists in other fields make sense of the data analysis results in climate literature.

Exercises

6.1 (a) Instead of the 2.5° S zonal band in Section 6.1, compute the covariance matrix for the January surface air temperature anomalies of the NOAAGlobalTemp data on a mid-latitude zonal band from 1961 to 1990. Take the *area-factor* $\sqrt{\cos(\text{latitude})}$ into account. You may choose any single latitude band between 25° N and 65° N. *Hint: Leave out the rows and columns in the space-time data matrix when the row or column has at least one missing datum.*

(b) Plot the covariance matrix similar to Figure 6.1a for the result of (a).

6.2 (a) Compute the eigenvalues and eigenvectors of the covariance matrix in Exercise 6.1.

(b) Plot the scree plot similar to Figure 6.2.

(c) Each EOF is a function of longitude and is thus a curve. Plot the first four EOFs on the same figure using different color. The horizontal axis is longitude and the vertical axis is the EOF values.

6.3 (a) Develop a space-time data matrix for the NOAAGlobalTemp surface air temperature anomalies on the same zonal band as in Exercise 6.1(a). Remove the rows and columns with missing data. The entire row or column is removed if it contains at least one missing datum.

(b) Use SVD to calculate the eigenvalues, EOFs, and PCs of the space-time data matrix.

6.4   (a)   Do the same as Exercise 6.3(a) but including a space-time data matrix with *area-factor* $\sqrt{\cos(\text{latitude})}$.

(b)   Use SVD to calculate the eigenvalues, EOFs, and PCs of the space-time data matrix with area-factor.

6.5   (a)   Comparing the EOFs and eigenvalues computed in Exercises 6.2–6.4, what can you conclude from the comparison? You may use text, numerical computing, and graphics to answer.

(b)   Does the area-factor make any difference in these three problems? Why?

6.6   Do the same as Exercise 6.1, but for a meridional band across the Pacific. Use Figure 6.1(b) as a reference. Leave out the rows and columns which have even a single missing datum.

6.7   (a)   Compute the December climatology of the monthly 1,000 mb (i.e., 1,000 hPa) air temperature data of a reanalysis product, such as NCEP/NCAR Reanalysis, using the 1951–2010 mean for the entire globe.

(b)   Compute the standard deviation of the same data for every grid box.

(c)   Plot the maps of climatology and standard deviation.

6.8   (a)   For the previous problem, form a space-time data matrix with area-factor for the anomalies of December air temperature at 1,000 mb level.

(b)   Apply SVD to the space-time matrix to compute eigenvalues, EOFs, and PCs.

(c)   Plot the scree plot.

(d)   Plot the maps of the first four EOFs.

(e)   Plot the line graphs of the first four PCs on the same figure using different colors.

6.9   (a)   Repeat the previous problem but for standardized anomalies, which are equal to the anomalies divided by the standard deviation for each grid box.

(b)   Comparing the EOFs and PCs of Step (a) of this problem with the EOF and PC results of the previous problem, discuss the differences and the common features.

6.10   (a)   Compute the December climatology of the monthly 1,000 mb *geopotential height* data of a reanalysis product, such as NCEP/NCAR Reanalysis, using the 1951–2010 mean for the entire globe.

(b)   Compute the standard deviation of the same data for every grid box.

(c)   Plot the maps of climatology and standard deviation.

6.11   (a)   For the previous problem, form a space-time data matrix with area-factor for the December standardized anomalies of the geopotential height data at the monthly 1,000 mb level.

(b)   Apply SVD to the space-time matrix to compute eigenvalues, EOFs, and PCs.

(c)   Plot the scree plot.

(d)   Plot the maps of the first four EOFs.

(e)   Plot the line graphs of the first four PCs on the same figure using different colors.

6.12   Develop a list of area factors for the 48 states of the contiguous United States, based on the theory of Section 6.3 and data you can find from the Internet.

6.13   (a)   The Global Precipitation Climatology Project (GPCP) monthly precipitation dataset from January 1979 is the observed precipitation data on a $2.5° \times 2.5°$ global grid. It has $144 \times 72 = 10368$ grid boxes, with latitude and longitude ranges in 88.75° N–88.75° S, and 1.25° E–358.75° E. You can download this dataset from the Internet, such as, https://psl.noaa.gov/data/gridded/data.gpcp.html. Compute the January climatology for every grid box in the time period from 1981 to 2020.

(b) Plot the map of the January climatology computed in (a).

(c) Compute the standard deviation of the same data for every grid box.

(d) Plot the map of the January standard deviation computed in (c).

6.14 (a) Compute the area-weighted spatial covariance matrix for the January standardized anomalies of the GPCP data using the 1981–2020 data computed in the previous problem. The result should be a 10368×10368 matrix.

(b) Compute the eigenvectors and eigenvalues of the above spatial covariance matrix.

(c) Plot the scree plot.

(d) Plot the maps of the first four EOFs.

(e) PCs are the anomaly data projections on the EOFs using matrix multiplication. Compute the PCs. Plot the line graphs of the first four PCs on the same figure using different colors.

6.15 (a) Following the temporal covariance theory in Sections 6.2 and 6.3, compute the area-weighted temporal covariance matrix for the January standardized anomalies of the GPCP data using the 1981–2020 climatology and standard deviation. The result should be a 40×40 matrix.

(b) Compute the eigenvectors and eigenvalues of the above temporal covariance matrix.

(c) Plot the scree plot.

(d) The eigenvectors here are PCs. Plot the line graphs of the first four PCs on the same figure using different colors.

(e) EOFs can be computed by the anomaly data projection on the PCs. Compute the EOFs. Plot the maps of the first four EOFs.

6.16 (a) Generate the space-time area-weighted standardized anomaly matrix from the January GPCP data from 1981 to 2020. The result should be a 10368×40 matrix.

(b) Compute the SVD for this space-time data matrix.

(c) The columns of the U-matrix from the SVD are the EOFs, and V-matrix from the SVD are the PCs. Plot the maps of the first four EOFs from this computing, and plot the line graphs of the first four PCs from the same computing. Arrange the figure into a 4×2 matrix, and place the EOFs on the left column and the corresponding PCs on the right column.

6.17 (a) Square the diagonal elements of the D-matrix from the SVD procedure in Exercise 6.16. Regard the squared diagonal elements as eigenvalues and plot the corresponding scree plot.

(b) Compare the scree plot from this SVD procedure and that from the spatial or temporal covariance matrix. What can you conclude?

6.18 (a) Compute the physical EOFs of the January GPCP precipitation data from 1981 to 2020 based on the theory in Section 6.3.

(b) Plot the maps of the first four physical EOFs in (a).

6.19 Stating from the perturbation expansion equations in Section 6.5.3, write down the detailed steps of the derivation of the formula for the EOF sampling error (Eq. (6.78)):

$$\delta \mathbf{u}_\ell \approx \frac{\delta \lambda_\ell}{\Delta \lambda_\ell} \mathbf{u}_{\ell'},\qquad\qquad (6.85)$$

6.20 Use both mathematical formulas and physical intuition to justify why area-factors are needed for computing the EOFs for a climate field across a large range of latitude, e.g., from $0° \, \mathrm{N}$ to $75° \, \mathrm{N}$.

6.21 (a) Following the theory of Sections 6.4 and 6.5, use a computer to generate random PCs and then to generate a random field with the prescribed 1D EOFs over a spatial line segment $[-\pi, \pi]$,

$$\psi_m(x) = \frac{1}{\sqrt{\pi}} \cos(mx), m = 1, 2, \ldots, M,\qquad\qquad (6.86)$$

and a set of eigenvalues of your choice. Work on two cases of spatial sampling: 100 and 2,000 samples. Do not forget the line segment factor.

(b) Plot the random field similar to Figure 6.3.

6.22 (a) Apply SVD to the random field in Exercise 6.21 to recover the EOFs.

(b) Plot the recovered EOFs and compare them with the originally prescribed EOFs in the interval $[-\pi, \pi]$.

6.23 Repeat the 2D example in Section 6.5 but use a new set of eigenvalues: $\lambda = 12.0$, 11.0, $6.1.1$, 6.0, and 1.0 with the spatial sample sizes: 200 and 3,000 samples.

## 7

Introduction to Time Series

Roughly speaking, a time series is a string of data indexed according to time, such as a series of daily air temperature data of San Diego in the last 1,000 days, and a series of the daily closing values of the Dow Jones Industrial Average in the previous three months. For these time series data, we often would like to know the following: What is their trend? Is there evidence of cyclic behavior? Is there randomness behavior? Eventually can we use these properties to make predictions. For example, a bride may want to plan for a wedding on the second Saturday of July of next year. She may use the temperature seasonal cycle to plan some logistics, such as clothes and food. She also needs to consider randomness of rain, snow, or a cold front, although she might choose to ignore the climate trend in her approximation.

Mathematically, a time series is defined as a sequence of random variables, indexed by time t, and is denoted by X_t. This means that for each time index t, the RV X_t has a probability distribution, ensemble mean (i.e., expected value), variance, skewness, etc. The time series as a whole may show trend, cycle, and random noise. A given string of data indexed by time is a realization of a discrete time series, and is denoted by x_t. A time series may be regarded as a collection of infinitely many realizations, which makes it different from a deterministic function of time that has a unique value for a given time. A stream of data ordered by time is an individual case drawn from the collection. This chapter will describe methods for the time series analysis, including the methods to quantify trends, cycles, and properties of randomness. In practice, a time series data analysis is for a given dataset, and the randomness property is understood to be part of the analysis but may not be included explicitly at the beginning.

When t takes discrete values, X_t is a discrete time series. When t takes continuous values, X_t is a continuous time series. If not specified, the time series dealt with in this book are discrete. This chapter begins with the time series data of CO_2, and covers the basic time series terminologies and methods, including white noise, random walk, stochastic processes, stationarity, moving average, autoregressive processes, Brownian motion, and data-model fitting.

7.1 Examples of Time Series Data

This section describes two examples of time series data: the monthly data of carbon dioxide concentration at Mauna Loa, Hawaii, USA, since 1958, and the daily minimum temperature data of St. Paul, Minnesota, USA, from 1941 to 1950. We will briefly show the features

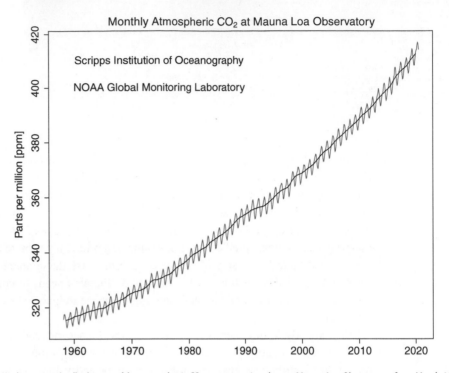

Figure 7.1 The Keeling curve (red): the monthly atmospheric CO_2 concentration data at Mauna Loa Observatory from March 1958 to July 2020. The black curve is the nonlinear trend of the Keeling curve. The trend was the result of the removal of seasonal oscillation and the computing of a moving average. The data and documents are from the NOAA Global Monitoring Laboratory (GML) website. `https://gml.noaa.gov/ccgg/trends/mlo.html`. Access date: August 2020.

of trend, seasonal cycle, random residual, and forecast. These results will help illustrate the time series methods to be presented in the subsequent sections.

7.1.1 The Keeling Curve: Carbon Dioxide Data of Mauna Loa

Figure 7.1 shows the monthly atmospheric carbon dioxide (CO_2) concentration data at Mauna Loa Observatory ($19.5362°$ N, $155.5763°$ W), elevation 3,394 meters, at Hilo Island, Hawaii, USA. The figure shows the data from March 1958 to July 2020. C. David Keeling of the Scripps Institution of Oceanography (SIO) started the observation at a NOAA facility. From May 1974, NOAA has been making its own CO_2 measurements, running in parallel with the SIO observations.

The red curve in Figure 7.1 displays the monthly CO_2 data. The up and down fluctuations of the red curve, in an approximately ± 3 ppm (parts per million) range, are due to the seasonal cycle of CO_2. The black curve was plotted according to the monthly trend data included in the same data table downloaded from NOAA GML. The trend curve demonstrates a steady increase of the CO_2 concentration. According to the trend data, the CO_2

concentration increased from approximately 314 ppm in 1958 to 413 ppm in 2020; an alarming 30% increase in 63 years!

The CO_2 seasonal cycle is due to the growth of plants and their leaves, and to the decomposition of the dead plants and fallen leaves. The seasonal cycle of the CO_2 concentration peaks in May, and reaches the bottom in October. From November, the dead plants and leaves break down. In this decomposition process, microbes respire and produce CO_2, and hence the CO_2 concentration begins to increase from November until May. In the spring, tree leaves grow, photosynthesis increases, and the plants and leaves absorb much CO_2 in the atmosphere. The photosynthesis burst in June and the following summer months makes the CO_2 concentration decrease from June until reaching the lowest level in October. Although the phase of the biological cycle in the Southern Hemisphere is exactly opposite, the smaller land area does not alter the CO_2 seasonal cycle dominated by the Northern Hemisphere.

Figure 7.1 can be plotted by the following computer code.

```
#R plot Fig. 7.1: Mauna Loa CO2 March 1958-July 2020
setwd("/Users/sshen/climstats")
co2m = read.table("data/co2m.txt", header = TRUE)
dim(co2m)
# [1] 749    7
co2m[1:3,]
#year mon      date average interpolated  trend days
#1 1958    3 1958.208  315.71       315.71 314.62    -1
#2 1958    4 1958.292  317.45       317.45 315.29    -1
#3 1958    5 1958.375  317.50       317.50 314.71    -1
mon = co2m[,3]
co2 = co2m[,5]
setEPS() # save the .eps figure file to the working directory
postscript("fig0701.eps", height = 8, width = 10)
par(mar=c(2.2,4.5,2,0.5))
plot(mon, co2, type="l", col="red",
     main =
        expression(paste("Monthly␣Atmospheric␣",
                       CO[2],"␣at␣Mauna␣Loa␣Observatory")),
     xlab ="",
     ylab="Parts␣Per␣Million␣[ppm]",
     cex.axis =1.5, cex.lab=1.5, cex.main= 1.6)
text(1975, 410, "Scripps␣Institution␣of␣Oceanography",
     cex=1.5)
text(1975, 400, "NOAA␣Global␣Monitoring␣Laboratory",
     cex=1.5)
lines(mon, co2m[,6]) #plot the trend data
dev.off()
```

```
#Python plot Fig. 7.1: The Keeling curve
os.chdir("/users/sshen/climstats")
co2m = pd.read_csv('data/co2_mm_mlo.csv', skiprows = 51)
print(co2m.head(5))
df = pd.DataFrame(co2m)
print(df.shape)
```

```
plt.plot(co2m['decimal␣date'], co2m['average'], color = 'r')
plt.plot(co2m['decimal␣date'], co2m['interpolated'],
        color = 'k')
plt.title('Monthly␣Atmospheric␣CO\N{SUBSCRIPT␣TWO}␣\
at␣Mauna␣Loa␣Observatory', pad = 15)
plt.ylabel('Parts␣Per␣Million␣[ppm]',
        labelpad = 15)
# text inside the figure
plt.text(1958,410,
      s='Scripps␣Institution␣of␣Oceanography', size = 18)
plt.text(1958,400,
         s='NOAA␣Global␣Monitoring␣Laboratory', size = 18)
# save figure
plt.savefig("fig0701.eps")
```

7.1.2 ETS Decomposition of the CO$_2$ Time Series Data

The CO$_2$ data has typical components of a time series: a trend due to the global greenhouse gas increase, a seasonal cycle due to the seasonal cycle of plants, and random residuals due to the random fluctuations of the plant's growth and climate conditions. This may be expressed by the following formula:

$$x_t = T_t + S_t + N_t,$$ (7.1)

where x_t stands for the time series data, T_t for trend, S_t for seasonal cycle, and N_t for random residuals (also called noise, or random error). We call this the ETS (i.e., error, trend, and season) decomposition of a time series.[1] A computer code can be used to decompose the data into the three components, as shown in Figure 7.2. This figure shows an additive model, as expressed by Eq. (7.1). You can also make a multiplicative decomposition, which is not discussed in this book.

The computer code for plotting Figure 7.2 is as follows.

```
#R plot Fig. 7.2: Time series decomposition
co2.ts = ts(co2, start=c(1958,3), end=c(2020,7),
        frequency =12)
co2.ts #Display time series with month and year
#      Jan    Feb    Mar    Apr    May    Jun    Jul    Aug
#1958               315.71 317.45 317.50 317.10 315.86 314.93
#1959 315.62 316.38 316.71 317.72 318.29 318.15 316.54 314.80
#1960 316.43 316.97 317.58 319.02 320.03 319.59 318.18 315.91

#Decompose a time series into components of trend,
#seasonal cycle, and random residual
co2.decompose <- decompose(co2.ts)
#Plot the time series and its three components
plot(co2.decompose, xlab ="")
```

[1] Although ETS is a commonly used term for error, trend, and season in statistics literatures, it might be more convenient to call Eq. (7.1) a TSN decomposition, standing for trend, season, and noise. The noise can be due to random errors or to the natural variability of the variable.

Monthly Atmospheric CO$_2$ [ppm] at Mauna Loa Observatory

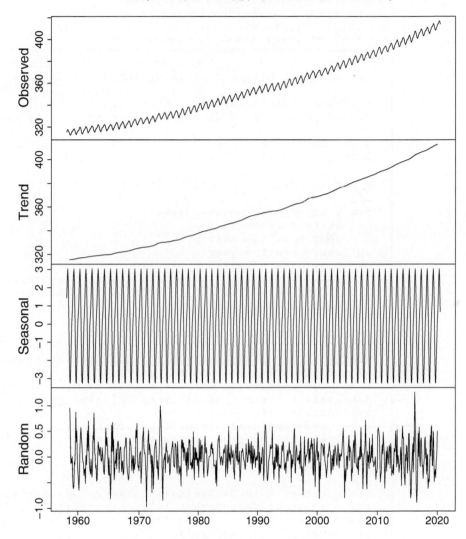

Figure 7.2 The Keeling curve as time series data and its components of trend, seasonal cycle, and random residual.

```
#Change the title of the plot
my_plot.decomposed.ts = function(x, title="", ...) {
  xx <- x$x
  if (is.null(xx))
    xx <- with(x, if (type == "additive")
      random + trend + seasonal
      else random * trend * seasonal)
  plot(cbind(observed = xx, trend = x$trend,
             seasonal = x$seasonal, random = x$random),
       main=title, ...)
}
par(mar=c(2,4.5,2.5,0.2))
```

```
my_plot.decomposed.ts(co2.decompose,
    expression(paste("Monthly␣Atmospheric␣",
    CO[2],"␣[ppm]␣at␣Mauna␣Loa␣Observatory")),
    xlab ="", cex.lab=1.5, cex.axis=1.8)
```

```
#Python plot Fig. 7.2: Time series decomposition

os.chdir("/users/sshen/climstats")
co2m = pd.read_csv('data/co2_mm_mlo.csv', skiprows = 51)

# get date into correct format in order to plot time series
date_index = []

# convert decimal date into datetime object, and store
df_co2 = df
for i in df_co2['decimal␣date']:
    start = float(i)
    year = int(float(start))
    rem = start - year
    base = datetime(year, 1, 1)
    result = base + timedelta(seconds=(base.replace(year=
        base.year + 1) - base).total_seconds() * rem)
    date_index.append(result)

date_index = pd.to_datetime(date_index)

from statsmodels.tsa.seasonal import seasonal_decompose
os.chdir("/users/sshen/climstats")
co2m = pd.read_csv('data/co2_mm_mlo.csv', skiprows = 51)
data_values = ser = pd.Series(co2m['average'])
ser.index = date_index
# get observed data in correct format
# convert values from string to float):
data_values = [float(i) for i in ser]
data_dates = pd.date_range('1958-03-17', '2020-08-17',
                            freq = 'M')
train_series = pd.Series(data_values, index=data_dates)
res = seasonal_decompose(ser, model='additive', freq = 12)
res.plot()
plt.savefig("fig0702.eps")
plt.show()
```

The R command decompose(co2.ts) for decomposing time series data co2.ts is for those data that have a known major seasonal cycle. Many climate variables do, such as the midlatitude surface air temperature and precipitation. The trend is computed by a moving average with its window length equal to a year, i.e., 12 months for the monthly data, and 365 days for the daily data. R has different moving average commands based on different algorithms. One of them is included in the R package itsmr, which has a moving average smoothing function smooth.ma(x, q), where x is the time series data, and q is the length of the moving average window, thus 12 in the case of our monthly CO_2 data.

After the trend is removed by y = x - smooth.ma(x, q), we have the de-trended data y. We then compute the climatology of the de-trended data for the entire data history.

This January to December climatology cycle is extended to the entire data history for every year to form the seasonal cycle of the time series. The random residuals are then equal to the de-trended data y minus the seasonal cycle.

Note that this time series decomposition procedure is for data that have a clear cycle with a fixed period. When the cycle is quasiperiodic with a variable period, such as the data dominated by the El Niño signal, this procedure may not yield meaningful results.

Let us further examine the results shown in Figure 7.2. The trend here is largely monotonic and increases from 315.41 ppm in 1958 to 412.81 ppm in 2020.

```
summary(co2.decompose$trend, na.rm=TRUE)
#   Min. 1st Qu.  Median   Mean 3rd Qu.    Max.    NA's
#315.4   330.1   352.9   355.2   377.9   412.8     12
```

Here, the trend line value 315.41 ppm in Figure 7.2 is for September 1958, and is slightly different from the trend value 314.62 for March 1958 in the original CO_2 data shown in Figure 7.1. The trend line value 412.81 ppm is for January 2020 in Figure 7.2, and is also slightly different from the trend value 414.04 for July 2020 in the original CO_2 data shown in Figure 7.1. The small difference is due to (i) the different ways to compute the trend for Figures 7.1 and 7.2 and (ii) the different beginning month and end month.

The seasonal cycle has an amplitude approximately equal to 3 ppm.

```
round(summary(co2.decompose$seasonal, na.rm=TRUE),
      digits = 3)
#Min. 1st Qu.  Median   Mean 3rd Qu.    Max.
#-3.256  -1.484   0.678   0.013   2.318   3.031
```

The mean is 0.013 ppm, not zero. The seasonal oscillation trough is -3.26 ppm. The seasonal oscillation crest is 3.03 ppm. Thus, the seasonal oscillation is not symmetric around zero, which may be a consequence of nonlinear properties of CO_2 variation. You may plot a few cycles to see the asymmetry and the nonlinear nature using R:

```
plot(co2.decompose$seasonal[1:50], type = 'l')
```

The random residuals are generally bounded between -1.0 ppm and 1.0 ppm, with two exceptions: August 1973 and April 2016, as found by the following R code.

```
round(summary(co2.decompose$random, na.rm=TRUE),
      digits = 4)
# Min. 1st Qu.  Median    Mean 3rd Qu.    Max.    NA's
#-0.9628 -0.1817  0.0034 -0.0009  0.1818  1.2659     12
sd(co2.decompose$random, na.rm=TRUE)
#[1] 0.2953659
which (co2.decompose$random > 1.2659369)
#[1] 698
co2m[698,]
#year mon     date average interpolated trend days
#698 2016   4 2016.292  407.45       407.45 404.6   25
```

The mean of the random residuals is approximately zero. The standard deviation is approximately 0.3 ppm. The random component is relatively small compared with the seasonal cycle and trend. This makes the CO_2 forecast easier since the trend and seasonal cycle can provide excellent forecasting skills.

7.1.3 Forecasting the CO_2 Data Time Series

The strong trend and seasonal cycle together with the weak random residuals make the forecast for CO_2 relatively reliable. Figure 7.3 shows the forecast of 48 months ahead (the blue line). The upper and lower bounds of the blue forecasting line are indicated by the shades. The forecast was made based on the aforementioned error, trend, and seasonal cycle (ETS) theory. The detailed theory of time series forecasting is beyond the scope of this book. Readers are referred to online materials and specialized books and papers.

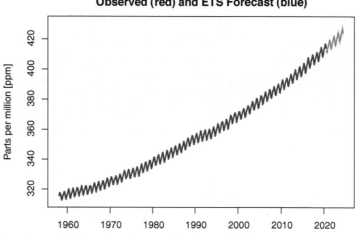

Observed (red) and ETS Forecast (blue)

Figure 7.3   The blue line is the forecast of the Keeling curve with a lead of 48 months based on the ETS method. The red line is the observed CO_2 data, and the black line almost covered by the red line is the ETS fitting result.

The gap between the red and blue lines is due to a drop of 1.8 ppm CO_2 concentration value from July 2020 as the last observed value to August 2020 as the first forecasting value. The drop in the range 1.5–2.6 ppm from July to August has occurred every year in the past.

Figure 7.3 can be reproduced by the following computer code.

```
#R plot Fig. 7.3: Time series forecast for CO2
#by ETS(Error, Trend, and Seasonal) smoothing
library(forecast)
co2forecasts <- ets(co2.ts,model="AAA")
#ETS model AAA means exponential smoothing with
#season and trend
#Forecast 48 months to the future
co2forecasts <- forecast(co2forecasts, h=48)
plot(co2forecasts,
     ylab="Parts␣Per␣Million␣[ppm]",
     main = "Observed␣(red)␣and␣ETS␣Forecast␣(blue)",
     cex.axis =1.5, cex.lab=1.5, cex.main= 1.2)
lines(co2.ts, col='red')
```

```
#Python plot Fig. 7.3: Time series forecast
from statsmodels.tsa.api import SimpleExpSmoothing
#Use the data "train_series" and "ser" from
#  the code for Fig. 7.2
# Fit Holt Winter's Exponential Smoothing model,
# with additive trend and season, then make
# a forecast
model = ExponentialSmoothing(train_series,
          seasonal_periods = 12, trend='add',
          seasonal='add', freq='M')
model_fit = model.fit()
# forecast 48 months
pred = model_fit.forecast(48)

s = seasonal_decompose(ser, model='additive', freq=12)

# set up plot
fig, ax = plt.subplots()
# labels
plt.title('Observed␣(red)␣and␣ETS␣Forecast␣(blue)', pad=15)
plt.ylabel('Parts␣Per␣Million␣[ppm]', labelpad = 15)

# plot observed and forecast lines
plt.plot(s.observed, color='red', linewidth=2)
plt.plot(pred, linewidth = 2)
# save figure
plt.savefig("fig0703.eps")
plt.show()
```

7.1.4 Ten Years of Daily Minimum Temperature Data of St. Paul, Minnesota, USA

The monthly CO_2 time series has a strong trend, a clear seasonal cycle, and a weak random component. However, many climate data time series have a large random component, particularly the time series of short time scales, such as the daily data and hourly data. Although the seasonal cycles are usually clear, the trend can be weak, nonexistent, or nonlinear. Figure 7.4 shows the data of ten years (January 1, 1941 to December 31, 1950) of daily minimum temperature at St. Paul, Minnesota, USA (44.8831° N, 93.2289° W) with an elevation of 265.8 meters.

Figure 7.4 shows that (i) a clear and strong seasonal cycle of about $-20°$C in winter and $20°$C in summer; (ii) the random residual data are bounded in $(-20, 20)°$C, the fluctuations are often ± 3 or $4°$C, and the variations are obviously larger in winter than summer; and (iii) the trend is bounded in by $(1, 4)°$C, it is relatively weak compared to the amplitude of seasonal cycle and random residuals, and further the trend component is nonlinear, varies slowly up and down in time, and is not monotonic. Some extreme values in the ETS components may correspond to abnormal weather phenomena and may deserve detailed scrutiny. Thus, the ETS decomposition is a helpful diagnostic tool for climate series data.

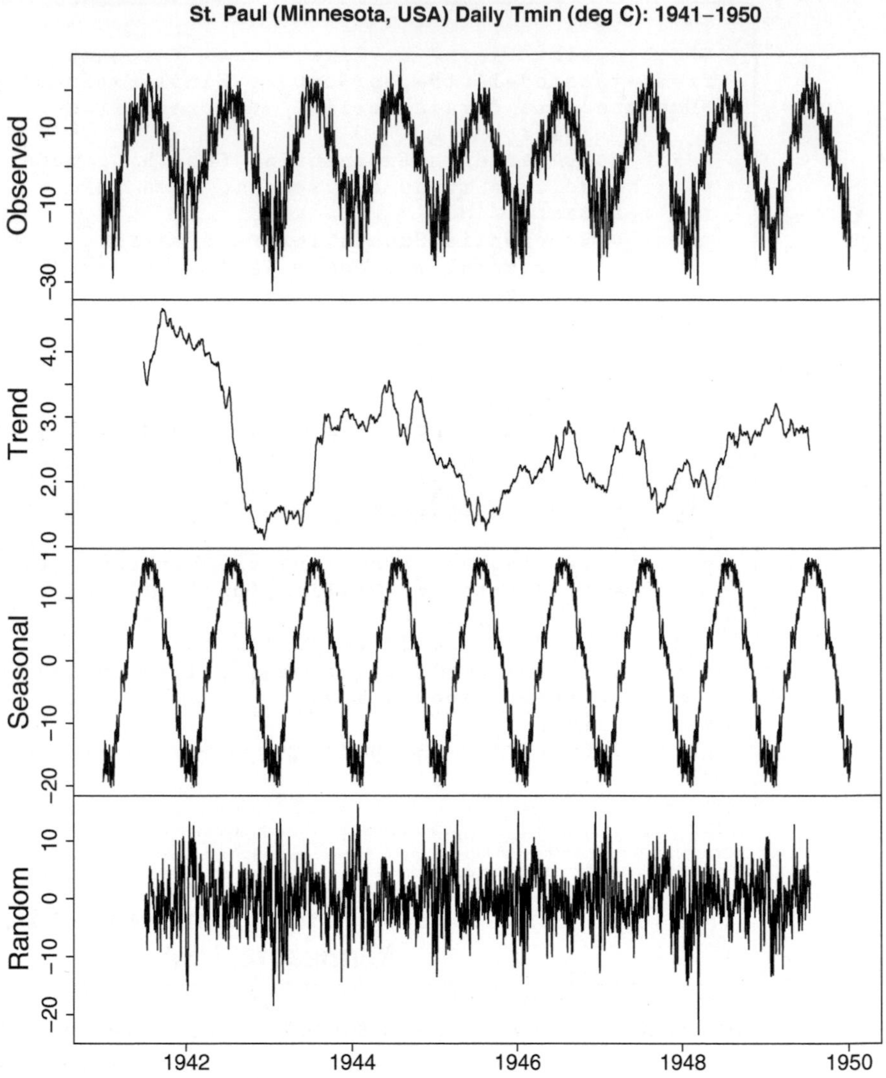

St. Paul (Minnesota, USA) Daily Tmin (deg C): 1941–1950

Figure 7.4 The daily minimum temperature observed data of St. Paul station, MA, USA, from January 1, 1941 to December 31, 1950, and their ETS components of trend, seasonal cycle, and random residual.

The large random component and the non-monotonic and nonlinear trend in Figure 7.4 make the forecast of the daily minimum temperature of St. Paul have large errors if the forecast is based solely on the ETS time series method. This is very different from the Mauna Loa CO_2 forecast shown in Figure 7.3. The practical daily temperature and precipitation forecast often uses many other constraints, such as atmospheric pressure, precipitable water in the air, and more atmospheric dynamical properties. Three-dimensional observations based on stations, radar, satellite and other instruments, together with numerical models are often combined to make modern weather forecasts.

Figure 7.4 can be reproduced by the following computer code.

```r
#R plot Fig. 7.4: Daily Tmin at ST PAUL Station, MN
#1941-1950 MINNEAPOLIS/ST PAUL
#GHCND:USW00014927  44.8831N  93.2289W  265.8Elev
setwd("/Users/sshen/climstats")
tda = read.table("data/StPaulStn.txt", header = F)
dim(tda)
#[1] 7589    13
#Col12 = Tmin, col9 = PRCP, col8 = Date
t1 = tda[,8] #date
tmin = tda[, 12] #Tmin data
t1[1462]
#[1] 19410101
t1[5112]
#[1] 19501231
da = tmin[1462:5112]
ta = t1[1462:5112]
dav = as.vector(t(da))
Tmin.ts = ts(dav, start=c(1941,1,1),
             end=c(1950,12,31),
             frequency = 365)
#ETS time series decomposition
Tmin.decompose <- decompose(Tmin.ts)
plot(Tmin.decompose)
#Change the title of the ETS decomposition figure
my_plot.decomposed.ts = function(x, title="", ...) {
  xx <- x$x
  if (is.null(xx))
    xx <- with(x, if (type == "additive")
      random + trend + seasonal
      else random * trend * seasonal)
  plot(cbind(observed = xx, trend = x$trend,
             seasonal = x$seasonal, random = x$random),
       main=title, ...)
}
par(mar=c(2,4.5,2.5,0.2))
my_plot.decomposed.ts(Tmin.decompose,
 title =
"St. Paul (Minnesota, USA) Daily Tmin (deg C): 1941-1950",
             xlab ="", cex.lab=1.5, cex.axis=1.8)
```

```python
#Python plot Fig. 7.4: Daily Tmin at ST PAUL Station,
MN
os.chdir("/users/sshen/climstats")
#Read .txt data into Pandas DataFrame
dat = pd.read_fwf('data/StPaulStn.txt', header=None)
print(dat[1461:1462][:])
#GHCND:USW00014927 ST. PAUL AIRPORT MN US
19410101
print(dat[5111:5112][:])
#GHCND:USW00014927 ST. PAUL AIRPORT MN US
19501213
print(dat.shape)
Tmin = dat[1461:5111][11] #Extract Tmin data from column 12
```

```
ets = seasonal_decompose(Tmin, model='additive', freq = 365)
ets.plot() #plot the decomposition
#Make correct time ticks
x_ticks = np.linspace(1461+365+365, 5111, num = 5)
labels = ['1942', '1944', '1946', '1948', '1950']
plt.xticks(x_ticks, labels)#plot the time ticks
plt.savefig("fig0704.eps") #save the figure
plt.show()
```

Because a leap year has 366 days, strictly speaking, using `frequency = 365` in the time series decomposition is only an approximate period of the annual cycle. However, since our data string covers only 10 years, the error caused by the leap year is small.

Compared with the monthly CO_2 data in Figure 7.2, the daily temperature data at St. Paul in Figure 7.4 has a much stronger noise component, and a nonlinear and non-monotonic trend. The seasonal signals are clear and stable in both the monthly CO_2 and the daily Tmin data strings. Most climate data have a clear seasonal signal in the ETS decomposition, because of the intrinsic seasonality of weather.

7.2 White Noise

A time series W_t is called *white noise* (WN) if each random variable W_t, for a given discrete time t, follows a zero-mean normal distribution, and the random variables at different times t and t' are uncorrelated. Namely,

$$W_t \sim \mathcal{N}(0, \sigma^2) \text{ (zero-mean normal distribution)}, \tag{7.2}$$

$$\mathrm{E}[W_t] = 0 \text{ (zero mean)}, \tag{7.3}$$

$$\mathrm{E}[W_t W_{t'}] = \sigma^2 \delta_{tt'} \text{ (zero autocorrelation for a nonzero lag)}, \tag{7.4}$$

where $\delta_{tt'}$ is the Kronecker delta. That is, each member in a white noise time series is uncorrelated with all past and future members. This can also be expressed as the autocorrelation function (ACF) with a time lag τ:

$$\rho(\tau) = \mathrm{Cor}[W_t W_{t+\tau}] = \delta_{0\tau}. \tag{7.5}$$

The name "white noise" is composed of (i) "noise," because W_t has zero autocorrelation with both its past and future, and (ii) "white," because the time series has neither trend nor seasonal cycle, i.e., the spectral power of the entire time series is evenly distributed among all frequencies as white light waves. The white light has uniform spectra across all frequencies in nature. The spectral theory of a time series will be described in Chapter 8 of this book.

Figure 7.5 shows a realization of white noise with 500 time steps, generated by the following computer code.

```
#R plot Fig. 7.5: Generate white noise
set.seed(119) # set seed to ensure the same simulation result
```

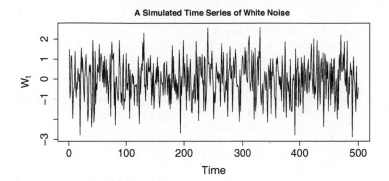

Figure 7.5 A simulated time series of white noise of 500 time steps.

```
wa <- rnorm(500, mean = 0, sd = 1)
par(mar=c(4.5,4.5,2.5, 0.3))
plot(wa, type = 'l',
     xlab = 'Time', ylab = expression(W[t]),
     main = 'A␣Simulated␣Time␣Series␣of␣White␣Noise␣',
     cex.lab=1.5, cex.axis=1.5)
```

```
#Python plot Fig. 7.5: White noise time series
seed(1)# set seed to ensure the same simulation result
# create the white noise series
white_noise = [gauss(0.0, 1.0) for i in range(500)]
white_noise_ser = pd.Series(white_noise)
# set up plot
fig, ax = plt.subplots(figsize=(12,6))
# white noise
plt.plot(white_noise_ser, 'k')
#plt.plot(white_noise_ser, color='black')
# add labels
plt.title('A␣Simulated␣Time␣Series␣of␣White␣Noise',
          pad = 15)
plt.ylabel(r'$W_{t}$', labelpad = 15)
plt.xlabel('Discrete␣Time␣Steps', labelpad = 15)
plt.savefig("fig0705.eps") # save figure
plt.show()
```

Figure 7.5 shows complete randomness without an obvious trend or a seasonal cycle.
Figure 7.6 shows a histogram and ACF $\rho(\tau)$ of the time series. The blue line over the histogram is the pdf of the standard normal distribution, which is a good fit to the histogram.
This implies that the simulated time series is approximately normally distributed with zero mean and standard deviation equal to one. The ACF figure shows that the autocorrelations are mostly close to zero and bounded in $[-0.1, 0.1]$ for the time lag $\tau = 1, 2, 3, \ldots$. We regard the ACF approximately zero for any time lag of a positive integer.

Figure 7.6 shows the histogram, its PDF fit, and the ACF of a realization of white noise with 500 time steps, generated by the following computer code.

```
#R plot Fig. 7.6: Histogram and ACF of white noise
par(mfrow=c(1,2))
par(mar=c(4.5,4.5,2.5, 0))
hist(wa, freq=F, breaks=16,
     xlim =c(-4,4), ylim=c(0,0.4),
     xlab = expression(W[t]),
     main='Histogram of a White Noise Time Series',
     cex.lab=1.5, cex.axis=1.5)
x=seq(-4,4, by=0.1)
lines(x,dnorm(x, mean=0, sd=1), col='blue',
      type = 'l', lwd=1.5)
par(mar=c(4.5,4.5,3, 0.3))
acf(wa, main='Auto-correlation of White Noise',
    cex.lab=1.5, cex.axis=1.5)
dev.off()#go back to R's default figure setting
```

```
#Python plot Fig. 7.6: Histogram and ACF of white noise
#set up the plot layout: two panels in a row
fig, ax = plt.subplots(1,2, figsize = (12, 5))
#histogram set up
ax[0].set_xlim(-4,4)
ax[0].set_ylim(0,.5)
ax[0].set_title('Histogram of a White Noise Time Series',
                pad = 20, size = 18)
ax[0].set_xlabel(r'$W_{t}$', labelpad = 15, size = 20)
ax[0].set_ylabel('Density', labelpad = 15, size = 20)

#Plot the histogram of white noise
ax[0].hist(white_noise_ser, density=True,
           edgecolor='black', fc='None')
#Fit the histogram to the standard normal distribution
mu = 0
variance = 1
sigma = math.sqrt(variance)
x = np.linspace(-4,4,100)
ax[0].plot(x, stats.norm.pdf(x, mu, sigma), color='blue')

#Plot ACF
sm.graphics.tsa.plot_acf(white_noise_ser,
                         ax=ax[1], color = 'k')
#Add labels and ticks for ACF
ax[1].set_title('Auto-correlation of White Noise',
                pad = 20, size = 20)
ax[1].set_xlabel('Lag', labelpad = 15, size = 20)
ax[1].set_ylabel('ACF', labelpad = 15, size = 20)
ax[1].set_xticks([0,5,10,15,20,25])

plt.tight_layout()
plt.savefig("fig0706.pdf") # save figure
plt.show()
```

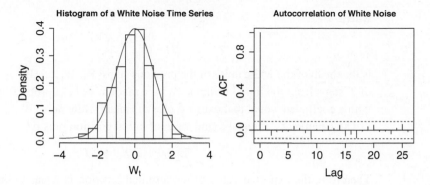

Figure 7.6 Left: Histogram of the simulated time series of white noise. Right: The autocorrelation function of the white noise.

White noise is the simplest discrete time series. It provides building blocks of almost all time series models. However, most time series in nature are not white noise, because the elements are correlated. Next, you will learn how to use the white noise to build time series models.

7.3 Random Walk

Random walk (RW) literally means a random next step, forward or backward from the present position, when walking on a line. Both the size and the direction of the next step are random. Let X_t be a random walk time series, or represent the stochastic process of a random walk. By definition, the next step X_{t+1} is equal to the present position X_t plus a completely random step modeled by a white noise W_t:

$$X_{t+1} = \delta + X_t + W_t, \tag{7.6}$$

where δ is a fixed constant called *drift*, $W_t \sim N(0, \sigma^2)$, and σ^2 is the variance of the white noise.

From the initial position X_0 at time $t = 0$, we can derive the position X_T at time $t = T$ as follows:

$$X_1 = X_0 + \delta + W_0, \tag{7.7}$$
$$X_2 = X_1 + \delta + W_1 = X_0 + 2\delta + (W_0 + W_1), \tag{7.8}$$
$$\ldots\ldots \tag{7.9}$$
$$X_T = X_0 + T\delta + (W_0 + W_1 + \cdots + W_{T-1}). \tag{7.10}$$

Thus, the final result of a random walk from the initial position is the sum of white noise of all the previous steps.

We can also write it as

$$X_T - X_0 = (\delta + \bar{W}_{T-1})T, \tag{7.11}$$

where

$$\bar{W}_{T-1} = \frac{\sum_{t=0}^{T-1} W_t}{T} \tag{7.12}$$

is the mean of the white noise of the previous steps: $W_0, W_1, \ldots, W_{T-1}$. Thus, the departure of T steps from the initial position X_0 by a random walk is proportional to time length T with a coefficient equal to the drift δ plus the mean white noise $\bar{W}_{T-1}$.

The mean of X_T is equal to the mean of initial position plus δT:

$$\mathrm{E}[X_T] = \mathrm{E}[X_0] + \delta T. \tag{7.13}$$

Therefore, the expected value of the terminal position is equal to the expected value of the initial position plus an integrated drift δT. This is reasonable since the white noise is completely random and does not have a preferred direction. This drift formulation may be used to model a systematic drift of satellite remote sensing measurements, or a drift due to ambient wind or ocean current.

The variance of X_T grows linearly with T, as shown:

$$\mathrm{Var}[X_T] = \mathrm{Var}[X_0] + \sum_{t=0}^{T-1} \mathrm{Var}[W_t] = \mathrm{Var}[X_0] + \sigma^2 T. \tag{7.14}$$

This result is important and can be used to develop stochastic models for science and engineering, such as Brownian motion, which describes a random motion of fine particles suspended in a liquid or gas. Brownian motion is named after Scottish botanist Robert Brown (1773–1858) who, in 1827, described the grains of pollen of a plant suspended in water under a microscope. Of course, the proper mathematical model of the Brownian motion should use (x, y) coordinates as a function of t. Both $x(t)$ and $y(t)$ are generated by a sum of many random walk steps simulated by white noise according to Eq. (7.10). This is a 2-dimensional random walk. Three-dimensional Brownian motion can be used to model a diffusive process in climate science, such as heat diffusion in both water and air, and aerosol particle diffusion in atmosphere.

Figure 7.7 shows five realizations of random walk time series of 100 steps with drift $\delta = 0.05$ and standard deviation of the white noise $\sigma = 0.1, 0.4, 0.7, 1.0,$ and 1.5 and $x_0 = 0$. When the standard deviation is relatively small compared to the drift, the drift plays the dominant role in leading the walk directions. When the standard deviation is large, the large variance in the white noise part imposes much uncertainty and makes the final position highly uncertain, even being uncertain whether the final terminal position is on the positive or negative side.

Figure 7.7 can be generated by the following computer code.

```
#R plot Fig. 7.7: Random walk time series
#(a) Five realizations with different sigma values
n = 1000 #Total number of time steps
m = 5 #Number of time series realizations
a = 0.05 #Drift delta = 0.05
b = c(0.1, 0.4, 0.7, 1.0, 1.5) #SD sigma
#Generate the random walk time series data
x = matrix(rep(0, m*n), nrow=m)
for(i in 1:m){
```

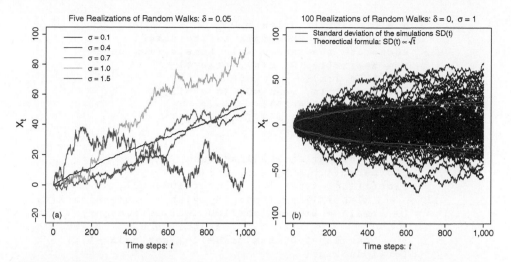

Figure 7.7 Left: The first 1,000 steps of five realizations of random walk for various values of standard deviation and a fixed drift value. Right: One hundred realizations of random walk to show that the standard deviation $SD[X_t]$ of the 100 realizations, indicated by the red line, is proportional to $\sqrt{t}$, indicated by the blue line.

```
  w = rnorm(n, mean =0, sd = b[i])
  for(j in 1:(n-1)){
    x[i,j+1] = a + x[i,j] + w[j]
  }
}
#Plot the five time series realizations
par(mfrow=c(1,2))
par(mar=c(4.5,4.5, 2.5, 0.5))
plot(x[1,], type='l', ylim=c(-20,100),
     xlab ="Time steps: t", ylab = expression(X[t]),
     main = expression('Five realizations of random walks:'
                       ~ delta ~'= 0.05'),
     cex.lab=1.5, cex.axis=1.5, cex.main =1.2)
lines(x[2,], type='l', col='blue')
lines(x[3,], type='l', col='red')
lines(x[4,], type='l', col='orange')
lines(x[5,], type='l', col='brown')
legend(-100, 110,
       legend=c(expression(sigma ~ '= 0.1'),
                expression(sigma ~ '= 0.4'),
                expression(sigma ~ '= 0.7'),
                expression(sigma ~ '= 1.0'),
                expression(sigma ~ '= 1.5')),
       col=c('black','blue','red','orange','brown'),
       x.intersp = 0.2, y.intersp = 0.4,
       seg.len = 0.6, lty =1, bty ='n', cex = 1.3)
text(20,-20, "(a)", cex =1.4)

#(b) 100 realizations with fixed parameters
#Random walk to show variance increasing with time
n = 1000 #Total number of time steps
m = 100 #Number of time series realizations
```

```r
a = 0.0 #Drift delta = 0
b = rep(1, m) #SD sigma is the same
#Generate the random walk time series data
x = matrix(rep(0, m*n), nrow=m)
for(i in 1:m){
  w = rnorm(n, mean =0, sd = b[i])
  for(j in 1:(n-1)){
    x[i,j+1] = a + x[i,j] + w[j]
  }
}
#Plot the series realizations
par(mar=c(4.5,4.5, 2.5, 0.8))
plot(x[1,], type='l', ylim=c(-100,100),
     xlab ="Time steps: t", ylab = expression(X[t]),
     main = expression('100 realizations of random walks:'
                       ~ delta ~'= 0,'~~ sigma ~'= 1'),
     cex.lab=1.5, cex.axis=1.5, cex.main =1.2)
for(i in 2:m){
  lines(x[i,], type='l')
}
library(matrixStats)
y = colSds(x)
lines(y,type='l', lwd=2, col='red')
lines(-y,type='l', lwd=2, col='red')
z = sqrt(1:n)
lines(z,type='l', lwd=2, col='blue')
lines(-z,type='l', lwd=2, col='blue')
legend(-150, 120,
    legend=c('Standard deviation of the simulations SD(t)',
  expression('Theorectical formula: SD(t)'%prop% sqrt(t))),
       col=c('red','blue'), cex = 1.3,
       x.intersp = 0.2, y.intersp = 0.3,
       seg.len = 0.4, lty=1, bty='n', lwd =2)
text(20,-100, "(b)", cex =1.4)
dev.off() #go back to R's default figure setting
```

```python
#Python plot Fig. 7.7(a): Random walk time series
#set seed to ensure the same simulation result
seed(100)
# Fig. 7.7(a): Five realizations with different sigma values
n = 1000  #Total number of time steps
m = 5  #Number of time series realizations
drift = 0.05  #Drift delta = 0.05
sd = [0.1, 0.4, 0.7, 1.0, 1.5]  #SD sigma

#generate the random walk time series data
X = np.zeros((m,n))
for i in range(m):
    white_noise = [gauss(0.0, sd[i]) for k in range(n)]

    for j in range(n-1):
        X[i, j+1] = drift + X[i,j] + white_noise[j]
```

```
# plot the five time series realizations
seed(100) #set seed
fig, ax = plt.subplots() #set up plot
#plot five random walks
plt.plot(X[0,:], color='black', linewidth=2,
         label=r'$\sigma$ = 0.1')
plt.plot(X[1,:], color='blue', linewidth=2,
         label=r'$\sigma$ = 0.4')
plt.plot(X[2,:], color='red', linewidth=2,
         label=r'$\sigma$ = 0.7')
plt.plot(X[3,:], color='orange', linewidth=2,
         label=r'$\sigma$ = 1.0')
plt.plot(X[4,:], color='brown', linewidth=2,
         label=r'$\sigma$ = 1.5')
#add legend
fig.legend(loc=(.15,.65), fontsize = 18)
blue_series = X[1,:]
#add labels
plt.title(r'Five RW realizations: $\delta = 0.05$',
          pad = 15)
plt.ylabel(r'$X_{t}$', labelpad=15)
plt.xlabel('Time steps: t', labelpad=15)
plt.savefig("fig0707a.eps") # save figure
plt.show()
```

```
# Fig. 7.7(b): 100 realizations
# random walk to show variance increasing with time
seed(100)# set seed
n = 1000  #Total number of time steps
m = 100  #Number of time series realizations
drift = 0  #Drift delta = 0
sd = np.ones(m)  #SD sigma is the same for all realizations
#Generate the random walk time series data
X = np.zeros((m,n))
for i in range(m):
    white_noise = [gauss(0.0, sd[i]) for k in range(n)]

    for j in range(n-1):
        X[i, j+1] = drift + X[i,j] + white_noise[j]

# plot the time series realizations
seed(100)# set seed
# set up plot
fig, ax = plt.subplots()
# random walks
for i in range(m):
    plt.plot(X[i,:], color='black', linewidth=1.5)

#sd of cols of X
col_sd = np.std(X,axis=0)
plt.plot(col_sd, color='red', linewidth=2,
```

```
     label='Standard deviation of the simulations: sd(t)')
plt.plot(-col_sd, color='red', linewidth=2)
#theoretical sd
theo_sd = np.sqrt(np.arange(1,n+1))
plt.plot(theo_sd, color='blue', linewidth=2,
         label=r'Theoretical formula: SD=$\sqrt{t}$')
plt.plot(-theo_sd, color='blue', linewidth=2)
#Add legend
fig.legend(loc=(.16,.79), fontsize = 18)

#Add labels
plt.title(r'100 RW Realizations: $\delta = 0, \sigma = 1$',
          pad=15)
plt.ylabel(r'$X_{t}$', labelpad=10)
plt.xlabel('Time steps: t', labelpad=15)
ax.set_ylim(-100,110)
ax.set_yticks([-100,-50,0,50,100])

plt.savefig("fig0708b.eps") # save figure
plt.show()
```

7.4 Stochastic Processes and Stationarity

7.4.1 Stochastic Processes

A stochastic process, also called a random process, is a collection of random variables, denoted by $\{X_t\}$. The simplest stochastic process is the Bernoulli process, in which the random variable takes only two possible values, 0 or 1. The probability of taking 0 is p and that of 1 is $1 - p$. Tossing a fair coin is a Bernoulli process, with the head being 0 and $p = 0.5$ and the tail being 1 for $1 - p = 0.5$.

A time series may be regarded as a sample of a stochastic process, indexed by time t. The time may take either discrete or continuous values. The coin toss is a discrete random process. The process of heat diffusion is a continuous random process.

When discussing time series, stochastic processes are mentioned, since they are related by probabilistic properties of random variables. We often specify a time series from a particular stochastic process, such as the global average annual mean temperature time series from the complex stochastic climate process of the Earth.

7.4.2 Stationarity

Stationary literally means "not moving" or "not changing in condition or quantity." A time series X_t is stationary if none of the probability characteristics of each random variable X_t change with respect to t. This basically means that the mean, variance, skewness, kurtosis, and all the higher statistical moments do not vary with time. This is called *strict-sense stationarity*. A milder form of stationarity for a time series is that only the mean and auto-covariance do not vary with time. In this case, we call it *wide-sense stationarity*, or *weak stationarity*. In climate science, we often deal with weakly stationary time series.

The two conditions for wide-sense stationarity can be expressed by mathematical formulas as follows. The expected value does not change:

$$\mathrm{E}[X_t] = \mathrm{E}[X_{t'}], \tag{7.15}$$

and the autocovariance depends only on the time lag $\tau = |t - t'|$:

$$\mathrm{Cov}[(X_t, X_{t'}] = h(\tau), \tag{7.16}$$

where the auto-covariance h is a function determined by the particular stochastic process being studied. Because $h(0) = \mathrm{Var}[X_t^2] = \mathrm{Var}[X_{t'}^2]$, the variance of a weakly stationary time series does not change with time. Because the variance is constant, ACF can be calculated as follows:

$$\mathrm{Cor}[X_t, X_{t+\tau}] = \frac{\mathrm{Cov}[X_t, X_{t'}]}{\sqrt{\mathrm{Var}[X_t]\mathrm{Var}[X_{t'}]}} = \frac{h(\tau)}{h(0)} \equiv \rho(\tau). \tag{7.17}$$

Namely, the ACF depends on only the time lag τ, not on the beginning time t.

Example 7.1 The standard white noise is a weakly stationary time series because $\mathrm{E}[W_t] = 0, \mathrm{Var}[W_t] = 1$, and the ACF

$$\rho(\tau) = \begin{cases} 1, & \text{if } \tau = 0 \\ 0, & \text{otherwise} \end{cases} \tag{7.18}$$

is independent of t.

Example 7.2 A random walk is nonstationary. If there is a nonzero drift, then the mean $\mathrm{E}[X_t] = \mathrm{E}[X_0] + \delta t$ varies with t, where δ is the drift. The first condition of stationarity is violated, and hence the random walk is nonstationary. Even in the absence of drift, the variance $\mathrm{Var}[X_t] = \mathrm{Var}[X_0] + \sigma^2 t$ also varies with t, where σ^2 is the variance of the white noise that builds the random walk. Hence the random walk without a drift is still nonstationary.

7.4.3 Test for Stationarity

Augmented Dickey–Fuller (ADF) test, named after American statisticians Wayne Dickey (1945–) and Wayne Fuller (1931–), may be used to check if a time series is stationary. R has a code for the ADF test. The null hypothesis is that the time series is nonstationary, and the alternative hypothesis is that it is stationary. For example, we can use it to test the blue time series in the left panel of Figure 7.7 is nonstationary.

```
library(tseries) #tseries is an R package
adf.test(x[2,])
#Dickey-Fuller = -1.4192, Lag order = 9, p-value = 0.8242
```

The large p-value implies that the null hypothesis is accepted, and hence the random walk time series is nonstationary.

Similarly, we can test that a white noise time series is stationary.

```
library(tseries) #tseries is an R package
adf.test(rnorm(500))
#Dickey-Fuller = -8.4851, Lag order = 7, p-value = 0.01
```

The small p-value implies that the alternative hypothesis is accepted and the white noise time series is stationary.

The corresponding Python code is as follows.

```
#Stationarity test for the blue TS in Fig. 7.7 by Python:
adf_test = adfuller(blue_series, autolag='AIC')
print('p-value␣=␣', adf_test[1])
#p-value =  0.5653243650129005

#Stationarity test for white noise generated by Python:
adf_test = adfuller(white_noise, autolag='AIC')
print('p-value␣=␣', adf_test[1])
#p-value =  0.0
```

7.5 Moving Average Time Series

A moving average (MA) stochastic process is a weighted average of white noise. MA(1) denotes a moving averaging process that smoothes the white noise at the present time step and the previous step by a weighted average:

$$X_t = aW_t + bW_{t-1}, \qquad (7.19)$$

where $W_t \sim N(0, \sigma^2)$ are Gaussian white noise; a, b, and σ are constants; and

$$a + b = 1. \qquad (7.20)$$

The three parameters a, b, and σ can be used to fit data or a model. The constraint Eq. (7.20) may not be necessary when modeling a stochastic process.

Figure 7.8 shows three realizations of moving average time series. The black line is an MA(1) time series with uniform weights. The blue is an MA(1) with uneven weights with a larger weight at time t and a smaller one for $t - 1$. The red is a MA(5) time series with uniform weights. In MA(5), X_t is the weighted average of the last five steps including the present step. One can of course extend this concept to MA(q) for any positive integer q.

Figure 7.8 can be generated by the following computer code.

```
#R plot Fig. 7.8: Moving average time series
set.seed(65)
n = 124
```

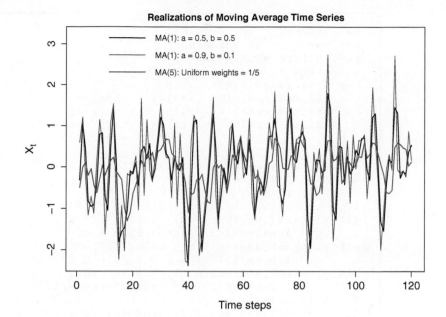

Figure 7.8 The first 120 steps of three realizations of moving average time series.

```
w = rnorm(n)
x1 = x2 = x3 = rep(0, n)
for (t in 5:n){
  x1[t] = 0.5*w[t] + 0.5*w[t-1]
  x2[t] = 0.9*w[t] + 0.1*w[t-1]
  x3[t] = (1/5)*(w[t] + w[t-1] + w[t-2] + w[t-3] + w[t-4])
}
par(mar=c(4.3, 4.5, 2, 0.2))
plot(x1[5:n], ylim= c(-2.3,3.2), type = 'l',
        main="Realizations␣of␣moving␣average␣time␣series",
        xlab ="Time␣steps", ylab = expression(X[t]),
        cex.lab=1.3, cex.axis=1.3, lwd=2)
lines(x2[5:n], col='blue', lwd = 1)
lines(x3[5:n], col='red', lwd=2)
legend(5, 3.8, bty='n', lty=1,
        legend=(c("MA(1):␣a␣=␣0.5,␣b␣=␣0.5",
                  "MA(1):␣a␣=␣0.9,␣b␣=␣0.1",
                  "MA(5):␣Uniform␣weights␣=␣1/5")),
        col=c('black','blue','red'),
        lwd = c(2,1,2))
```

```
#Python plot Fig. 7.8: MA time series
seed(101)# set seed
n = 124
# white noise
w = [gauss(0,1) for k in range(n)]
```

```
ma1_uniform = np.zeros(n)
ma1_uneven = np.zeros(n)
ma5_uniform = np.zeros(n)
#generate three MA time series
for t in range(4,n):
    ma1_uniform[t] = .5*w[t] + .5*w[t-1]
    ma1_uneven[t] = .9*w[t] + .1*w[t-1]
    ma5_uniform[t] = (w[t] + w[t-1] + w[t-2]
                      + w[t-3] + w[t-4]) / 5

# plot the time series realizations
seed(101)# set seed
# set up plot
fig, ax = plt.subplots(figsize = (12,6))
plt.plot(ma1_uniform[4:n], color='black',
         label='MA(1): a = 0.5, b = 0.5')
plt.plot(ma1_uneven[4:n], color='blue',
         label='MA(1): a = 0.9, b = 0.1')
plt.plot(ma5_uniform[4:n], color='red',
         label='MA(5): Uniform weight = 1/5')
# add legends
fig.legend(loc=(.16,.71), prop={"size":16})
# add labels
plt.title('Realizations of MA time series',
          pad = 15)
plt.ylabel(r'$X_{t}$', labelpad=15)
plt.xlabel('Time steps', labelpad=15)
ax.set_ylim(-2,4)
plt.savefig("fig0708.eps")# save figure
plt.show()
```

It is known that the white noise is a stationary process. We may claim that the smoothing of a white noise is also stationary. This claim is true. We use MA(1) time series as an example to justify this claim.

First, consider the mean of X_t:

$$\mathrm{E}[X_t] = 0, \tag{7.21}$$

since $\mathrm{E}[W_t] = 0$.

Next, we consider the auto-covariance of X_t:

$$\mathrm{Cov}[X_t X_{t+\tau}] = \sigma^2 \left[(a^2 + b^2)\delta_{0,\tau} + ab(\delta_{1,\tau} + \delta_{-1,\tau})\right], \tag{7.22}$$

where $\delta_{0,\tau}$ is the Kronecker delta function. The auto-covariance $\mathrm{Cov}[X_t X_{t+\tau}]$ is independent of t. The time independence of the mean and auto-covariance of X_t implies that the MA(1) time series is weakly stationary.

Because

$$\mathrm{Var}[X_t] = \mathrm{Cov}[X_t X_{t+0}] = \sigma^2(a^2 + b^2), \tag{7.23}$$

ACF of an MA(1) is

$$\rho(\tau) = \frac{\mathrm{Cov}[X_t X_{t+\tau}]}{\mathrm{Var}[X_t]} = \delta_{0,\tau} + \frac{ab(\delta_{1,\tau} + \delta_{-1,\tau})}{a^2 + b^2}. \tag{7.24}$$

This formula implies that

$$\rho(0) = 1, \tag{7.25}$$

$$\rho(1) = \rho(-1) = \frac{ab}{a^2 + b^2} < 1, \tag{7.26}$$

$$\rho(\tau) = 0 \ \text{otherwise}. \tag{7.27}$$

The autocorrelation depends only on the time lag τ, and is independent of the beginning time t.

7.6 Autoregressive Process

7.6.1 Brownian Motion and Autoregressive Model AR(1)

The motion of a Brownian particle mentioned in Section 7.3 may be modeled by the following differential equation:

$$m\frac{dv}{dt} = -bv + f(t), \tag{7.28}$$

where m is mass of the particle, v is velocity, t is time, dv/dt is the derivative of v with respect to t, b is a frictional damping coefficient, and $f(t)$ is white noise forcing due to random buffeting by molecules. The damping provides a stabilizing feedback to the stochastic system to prevent the variance of v from growing indefinitely. This problem was originally solved by Albert Einstein (1879–1955) as part of his contributions to statistical mechanics.

An equation that includes derivatives is called a *differential equation*. The solution of a deterministic differential equation without any random element is a function. When a random forcing is involved, the differential equation becomes stochastic and its solution is a time series. The study of differential equations belongs to a branch of mathematics and is not included in the scope of our book. Here we only discuss the finite difference solution of Eq. (7.28). By finite difference, we mean to discretize the stochastic differential equation (7.28) with a time step size Δt:

$$m\frac{v(t_n + \Delta t) - v(t_n)}{\Delta t} = -bv(t_n) + f(t_n), \tag{7.29}$$

where $t_n = n\Delta t$ is the time at the nth time step. Denote $X_n = v(t_n)$ and $X_{n+1} = v(t_n + \Delta t)$. Then

$$X_{n+1} = \lambda X_n + W_n, \tag{7.30}$$

where $\lambda = 1 - b\Delta t/m$ is a decay parameter with a bound $0 \leq \lambda < 1$, and $W_n = f(t_n)\Delta t/m$ is white noise. The condition $0 \leq \lambda$ implies that $\Delta t \leq m/b$, a restriction on the time step size.

This model implies that the time series one step ahead is a regression to the current step with white noise as its intercept. This is a first-order autoregressive model, denoted by AR(1).

We claim that the AR(1) process is an infinitely long moving average process MA(∞). This can be demonstrated by the following derivations.

Multiplying Eq. (7.30) by $(1/\lambda)^{n+1}$ yields

$$\frac{X_{n+1}}{\lambda^{n+1}} - \frac{X_n}{\lambda^n} = \frac{W_n}{\lambda^{n+1}}. \tag{7.31}$$

Summing this equation from 0 to $N-1$ results in

$$\frac{X_N}{\lambda^N} - \frac{X_0}{\lambda^0} = \frac{1}{\lambda}\sum_{n=0}^{N-1}\frac{W_n}{\lambda^n}. \tag{7.32}$$

This can be further written as

$$X_N = X_0\lambda^N + \lambda^{N-1}\sum_{n=0}^{N-1}\frac{W_n}{\lambda^n}. \tag{7.33}$$

This formula means X_N is the sum of the decaying initial condition $X_0\lambda^N$ plus the weighted average of white noise terms from time step 0 to $N-1$. Consequently, you may regard the AR(1) process as MA(∞) when N is large.

The first term of Eq. (7.33) represents the decay of the initial condition X_0 when $|\lambda| < 1$. When N is large enough, $\lambda^N < 1/e$ and hence the process forgets the initial condition after the time step N. This condition leads to

$$N > \frac{1}{|\ln\lambda|}. \tag{7.34}$$

If $\lambda = 0.9$, then $N > 10$, and if $\lambda = 0.7$, then $N > 3$.

The second term of Eq. (7.33) is an MA(N) process and is hence stationary. Thus, AR(1) process is stationary as time goes to infinity. This is called the asymptotic stationarity.

7.6.2 Simulations of AR(1) Time Series

Figure 7.9 shows four realizations of an AR(1) time series in color and their mean in a thick black line. The parameters are $x_0 = 4, \lambda = 0.9, \sigma = 0.25$. All four realizations start at the same point: $x_0 = 4$. The decay parameter $\lambda = 0.9$ leads to $1/|\ln\lambda| = 9.5$, implying that the initial condition is forgotten when time reaches a number larger than 10, say 20 or 30. The figure supports this claim. The simulations show that the decay is completed in the first 30 time steps, and the stationary random variations dominate from 30 time steps and on. The mean of the four time series realizations shows the decay even more clearly. The white noise has zero mean and 0.25 as its standard deviation, which determines the range of fluctuations of the time series.

Figure 7.9 can be generated by the following computer code.

```
#R plot Fig. 7.9: Autoregressive time series AR(1)
set.seed(791)
n = 121 #Number of time steps
m = 4 #Number of realizations
lam = 0.9 #Decay parameter
x = matrix(rep(4, n*m), nrow=m) #x0 = 4
#Simulate the time series data
```

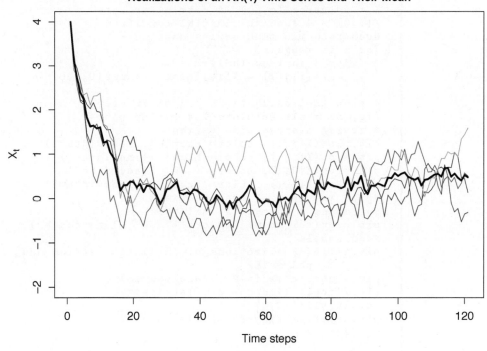

Figure 7.9   The first 120 steps of four realizations of an AR(1) time series (the four color lines) and their mean (the thick black line)

```
for(k in 1:m){
  for(i in 1:(n-1)){
    x[k,i+1] = x[k,i]*lam +
      rnorm(1, mean=0, sd=0.25)
  }
}
#Plot the realizations and their mean
plot.ts(x[1,], type = 'l', ylim=c(-2,4),
  main="Realizations␣of␣an␣AR(1)␣time␣series␣and␣their␣mean",
  xlab ="Time␣steps", ylab = expression(X[t]),
      cex.lab=1.5, cex.axis=1.5,
      col='purple')
lines(x[2,], col='blue')
lines(x[3,], col='red')
lines(x[4,], col='orange')
lines(colMeans(x),  lwd=3)
```

```
#Python plot Fig. 7.9: Autoregressive time series AR(1)
seed(791)# set seed
n = 121  # total number of time steps
m = 4  # number of time series realizations
lam = .9  # decay parameter
```

```
X = np.zeros((m,n))  # matrix for data
X[:,0] = 4 #set the initial condition at t=0
#generate the time series data
for i in range(m):
    for j in range(n-1):
        X[i,j+1] = X[i,j]*lam + gauss(0,.25)

# plot the realizations and their mean
fig, ax = plt.subplots() # set up plot
# moving average time series
plt.plot(X[0,:], color='purple', linewidth=2)
plt.plot(X[1,:], color='blue', linewidth=2)
plt.plot(X[2,:], color='red', linewidth=2)
plt.plot(X[3,:], color='orange', linewidth=2)
#plot the mean of above series
col_means = np.mean(X, axis=0)
plt.plot(col_means, color='black', linewidth=2)
#add labels
plt.title('Realizations of AR(1) time series and their mean',
         pad = 15)
plt.ylabel(r'$X_{t}$', labelpad=15)
plt.xlabel('Time steps', labelpad=15)
ax.set_ylim(-2,4)
plt.savefig("fig0709.eps")# save figure
plt.show()
```

This computer code is based on the definition of AR(1): Eq. (7.30). One can also do the AR(1) simulation using the moving average formula (7.33).

7.6.3 Autocovariance of AR(1) Time Series when $X_0 = 0$

Based on Eq. (7.33), when $X_0 = 0$, the AR(1) process in terms of white noise is

$$X_N = \lambda^{N-1} \sum_{n=0}^{N-1} \frac{W_n}{\lambda^n}. \tag{7.35}$$

Since $E[X_N] = 0$, the auto-covariance $\text{Cov}[X_N, X_{N+l}]$ can be calculated as follows:

$$E[X_N X_{N+l}] = \lambda^{2N+l-2} \sum_{n=0}^{N-1} \frac{\sigma^2}{\lambda^{2n}}, \tag{7.36}$$

because $E[W_m W_n] = \sigma^2 \delta_{m,n}$ where $\delta_{m,n}$ is the Kronecker delta.
Since

$$\sum_{n=0}^{N-1} \frac{1}{\lambda^{2n}} = \sum_{n=0}^{N-1} (\lambda^{-2})^n = \frac{1-\lambda^{-2N}}{1-\lambda^{-2}} = \lambda^{-2N+2} \frac{1-\lambda^{2N}}{1-\lambda^2}, \tag{7.37}$$

the auto-covariance here becomes

$$\text{Cov}[X_N, X_{N+l}] = \sigma^2 \lambda^l \frac{1-\lambda^{2N}}{1-\lambda^2}. \tag{7.38}$$

The variance of X_N is

$$\text{Var}[X_N] = \text{Cov}[X_N, X_N] = \sigma^2 \frac{1 - \lambda^{2N}}{1 - \lambda^2}. \tag{7.39}$$

This is an increasing function of N. When N is large, λ^{2N} is negligible when $|\lambda| < 1$. Hence,

$$\text{Var}[X_N] \to \frac{\sigma^2}{1 - \lambda^2} \quad \text{as} \quad N \to \infty. \tag{7.40}$$

So, the variance approaches a constant from below as N goes to ∞.

The autocorrelation function (ACF) of AR(1) with a lag l is thus

$$\rho_l = \frac{\text{Cov}[X_N, X_{N+l}]}{\text{Var}[X_N]} = \lambda^l. \tag{7.41}$$

The zero mean, the independence of the autocorrelation from N, and the existence of the limit of variance in Eq. (7.40) imply that the AR(1) process becomes approximately stationary when N is large. After this time, the initial condition is forgotten, and the stochastic process completes its transition to a stationary process. Therefore, the AR(1) process is asymptotically stationary.

Another perspective is the decaying property of ACF ρ_l. As the decay parameter $|\lambda| < 1$, λ^l becomes very small when the time lag l is large. Thus, ACF approaches zero as the time lag increases. The initial condition X_0 is forgotten after a certain number of steps in the AR(1) process.

The third property is the sum of the autocorrelation at all the time lags, denoted by τ. The sum is equal to the characteristic time scale of the AR(1) process, as justified as follows:

$$\tau = \sum_{l=0}^{\infty} \rho_l = \sum_{l=0}^{\infty} \lambda^l = \frac{1}{1 - \lambda}. \tag{7.42}$$

This agrees with the dynamical equation (7.28) for the Brownian motion, where

$$\gamma = \frac{b}{m} \Delta t = 1 - \lambda \tag{7.43}$$

is a parameter related to the damping coefficient b, mass m, and time step size Δt. Its inverse is the correlation interval or characteristic time scale of decay $\tau = 1/\gamma$ whose time unit is Δt. Namely, $\tau \Delta t$ is the characteristic time.

When $\lambda \to 1$, the time scale $\tau \to \infty$. This is because $\lambda \to 1$ implies the frictional damping coefficient b in Eq. (7.28) of motion of Brownian particle becomes zero. Thus, the damping mechanism disappears, which implies an infinitely long time scale. In the absence of damping with $b = 0$, the Brownian motion becomes a random walk $X_{n+1} = X_n + W_n$. The variance of the corresponding time series grows linearly with time T, in contrast to the case of Brownian motion whose variance goes to a constant $\sigma^2/(1 - \lambda^2)$ as described in Eq. (7.39).

Figure 7.10 shows the ACF for two different decay parameters: $\lambda = 0.9$ and $\lambda = 0.6$. The corresponding time scales are thus $\tau = 10$ and $\tau = 2.5$. The figure shows that ACF is less than $1/e = 0.37$ when the lag is larger than 11 for $\lambda = 0.9$, and 4 for $\lambda = 0.6$. Therefore, the simulation agrees well with the theoretical result based on mathematical derivations.

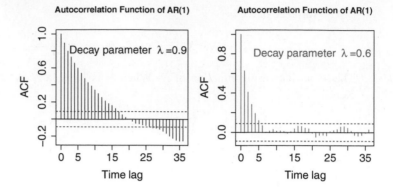

Figure 7.10 ACFs and the corresponding time scales and decay parameters.

The fluctuations of small ACF values (mostly bounded inside $[-0.1, 0.1]$ indicated by two blue dashed lines in Figure 7.10) for large time lags are likely due to the insufficient length of the sample data string and the noise in numerical simulations. Theoretically, these values should be almost zero, because $\rho_l = \lambda^l \to 0$ as $l \to \infty$.

Figure 7.10 may be generated by the following computer code.

```
#R plot Fig. 7.10: ACF of AR(1)
setwd('/Users/sshen/climstats')
n = 481 #Number of time steps
m = 2 #Number of realizations
lam = c(0.9, 0.6) #Decay parameter
x = matrix(rep(4, n*m), nrow=m) #x0 = 4
#Simulate the time series data
for(k in 1:m){
  for(i in 1:(n-1)){
    x[k,i+1] = x[k,i]*lam[k] +
      rnorm(1, mean=0, sd=0.25)
  }
}
#Plot the auto-correlation function
setEPS() #Automatically saves the .eps file
postscript("fig0710.eps", width=10, height=5)
par(mfrow=c(1,2))
par(mar=c(4.5, 4.5, 3, 1))
acf(x[1,], lag.max = 36,
    main = 'Auto-correlation function of AR(1)',
    xlab='Time lag',
    cex.lab=1.5, cex.axis=1.5)
text(20, 0.8, bquote('Decay parameter'~lambda == 0.9),
    cex=1.5)
par(mar=c(4.5,4.5,3, 0.3))
acf(x[2,], lag.max = 36,
    main = 'Auto-correlation function of AR(1)',
    xlab='Time lag', col='red',
    cex.lab=1.5, cex.axis=1.5)
text(20, 0.8, expression('Decay parameter'~lambda == 0.6),
    col='red', cex=1.5)
dev.off()
```

```
#Python plot Fig. 7.10: ACF of AR(1)
seed(82) #set seed to ensure the same result
n = 481  # total number of time steps
m = 2  # number of time series realizations
lam = [0.9, 0.6]  # decay parameter
X = np.zeros((m,n))  # placeholder for data matrix
X[:,0] = 4

# simulate the time series data
for k in range(m):
    for i in range(n-1):
        X[k,i+1] = X[k,i]*lam[k] + gauss(0,0.25)

 # plot the auto-correlation function
fig, ax = plt.subplots(1,2, figsize = (14,6))
# ACF
sm.graphics.tsa.plot_acf(X[0,:], ax=ax[0],
                          lags=36, color = 'k')
sm.graphics.tsa.plot_acf(X[1,:], ax=ax[1], lags=36,
    vlines_kwargs={'color':'red'}, color = 'red')

#add labels
ax[0].set_title('Auto-correlation function of AR(1)',
                pad = 15)
ax[0].set_xlabel('Time lag', labelpad=15)
ax[0].set_ylabel('ACF', labelpad = -5)
ax[0].set_xticks([0,5,15,25,35])
ax[0].set_yticks([-0.2, 0.2, 0.6,1.0])
ax[0].annotate(r'Decay parameter $\lambda$ = 0.9',
                xy=(5,.9), size = 18)
ax[1].set_title('Auto-correlation function of AR(1)',
                pad = 15)
ax[1].set_xlabel('Time lag', labelpad=15)
ax[1].set_ylabel('ACF', labelpad=10)
ax[1].set_xticks([0,5,15,25,35])
ax[1].set_yticks([0, 0.4,0.8])
ax[1].annotate(r'Decay parameter $\lambda$ = 0.6',
                xy=(5,.9), color='red', size = 18)
plt.tight_layout()
plt.savefig("fig0710.pdf")# save figure
plt.show()
```

7.7 Fit Time Series Models to Data

In the previous sections about time series, we described mathematical properties and simulations of MA and AR models. This section answers a question: For a given time series
dataset, can we find an appropriate time series model so that the model can best fit the
data? This is also called a model estimation problem.

7.7.1 Estimate the MA(1) Model Parameters

Let us find out how to use the given data $x_1, x_2, \ldots, x_N$ in an MA(1) process

$$X_n = aW_n + bW_{n-1}, \quad W_n \sim N(0, \sigma^2) \tag{7.44}$$

to estimate the model parameters a, b, and σ. The idea is to use the correlation coefficient ρ computed from the data.

Neither $a = 0$ nor $\sigma = 0$, because the former means white noise and the latter means a deterministic process. Our MA(1) is neither. Thus, $a\sigma \neq 0$. We can normalize the MA model Eq. (7.44) by dividing both sides of the equation by $a\sigma$. This leads to

$$Y_n = Z_n + \theta Z_{n-1}, \quad Z_n \sim N(0, 1) \tag{7.45}$$

where

$$Y_n = \frac{X_n}{a\sigma}, \quad Z_n = \frac{W_n}{\sigma}, \quad \theta = \frac{b}{a}. \tag{7.46}$$

Compute the ACF $\rho(\tau)$ as follows:

$$
\begin{aligned}
\rho(\tau) &= \frac{E[Y_{n+\tau}Y_n]}{E[Y_n^2]} \\
&= \frac{E[(Z_{n+\tau} + \theta Z_{n+\tau-1})(Z_n + \theta Z_{n-1})]}{E[(Z_n + \theta Z_{n-1})(Z_n + \theta Z_{n-1})]} \\
&= \begin{cases} 1, & \text{when } \tau = 0 \\ \frac{\theta}{1+\theta^2}, & \text{when } \tau = 1 \\ 0, & \text{when } \tau > 1 \end{cases}
\end{aligned} \tag{7.47}
$$

because

$$E[Z_{n+\tau}Z_n] = \delta_{\tau,0}. \tag{7.48}$$

This formula provides an equation between the model parameter θ and the data correlation $\rho(1)$:

$$\rho(1)(1 + \theta^2) = \theta. \tag{7.49}$$

The parameter θ has two solutions; hence the MA(1) model is not unique. However, the two models represent the same process.

Therefore, for a given data sequence $x_1, x_2, \ldots, x_N$, we may compute the lag-1 correlation $\rho(1)$ using the following formula:

$$\rho(1) \approx \frac{\sum_{n=1}^{N-1}(x_{n+1} - \bar{x})(x_n - \bar{x})}{\sum_{n=1}^{N-1}(x_n - \bar{x})^2}, \tag{7.50}$$

where

$$\bar{x} = \frac{\sum_{n=1}^{N} x_n}{N} \tag{7.51}$$

is the data mean. Then, the AR(1) parameter θ in the model (7.45) can be computed by solving Eq. (7.49).

7.7.2 Estimate the AR(1) Model Parameters

Similar to the model estimation for MA(1), we can also use the data sequence $x_1, x_2, \ldots, x_N$ to estimate the AR(1) model:

$$X_{n+1} = \lambda X_n + W_n. \tag{7.52}$$

Equation (7.41) gives the ACF of AR(1):

$$\rho(\tau) = \lambda^\tau. \tag{7.53}$$

For lag-1, we have

$$\lambda = \rho(1). \tag{7.54}$$

Thus, we can use the given data sequence and formulas (7.50) and (7.51) to estimate $\rho(1)$ that is λ in the AR(1) model (7.52). Consequently, the fitted AR(1) model is

$$X_{n+1} = \rho(1)X_n + W_n. \tag{7.55}$$

7.7.3 Difference Time Series

Random walk

$$X_t = X_{t-1} + W_t, \tag{7.56}$$

defined earlier, can be written as

$$X_t - X_{t-1} = W_t. \tag{7.57}$$

The right-hand side is white noise. The left-hand side is a difference, which may be regarded as a discretization of the first derivative, as explained for the motion equation of a Brownian particle. It is known that random walk is nonstationary, but its difference time series

$$X_t^{(1)} = X_t - X_{t-1} \tag{7.58}$$

is stationary.

It sometimes happens that the difference $X_t^{(1)}$, also called the first difference, from a nonstationary time series becomes stationary. For example, a monthly air temperature anomaly time series is often nonstationary, but its month-to-month changes may be stationary, although the stationarity still needs to be rigorously verified by a proper statistical hypothesis test, such as the ADF test discussed earlier in this chapter. The method of difference time series has been used in analyzing real climate data. Peterson et al. (1998) used the first difference $X_t^{(1)}$ method to include more long-term stations to calculate the global average temperature change. Smith et al. (1998) used the method for studying the variation of the historical SST data.

Sometimes, the first difference time series is still nonstationary, but the second or third difference time series is stationary. You may use $X_t^{(d)}$ to denote the difference time series of the dth order difference, which may be regarded as a formula from approximating the dth order derivative of X, e.g.,

$$X_t^{(2)} = X_t - 2X_{t-1} + X_{t-2} \tag{7.59}$$

for $d = 2$. Of course, there are cases where the difference time series of any order is still nonstationary, although the method of difference time series is a powerful tool for analyzing climate data sequences.

7.7.4 AR(p) Model Estimation Using Yule-Walker Equations

We may extend AR(1) model to an AR(p) model as

$$X_t = \phi_0 + \phi_1 X_{t-1} + \phi_2 X_{t-2} + \cdots + \phi_p X_{t-p} + W_t, \tag{7.60}$$

where $\phi_i (i = 0, 1, \ldots, p)$ are constant model parameters. Our task is to estimate these parameters from the given data sequences using ACF.

When $\phi_0 = 0$, calculating the autocorrelation of Eq. (7.60) with X_{t-k}, we obtain

$$\rho(k) = \phi_1 \rho(k-1) + \phi_2 \rho(k-2) + \cdots + \phi_p \rho(k-p), \quad k = 1, 2, \ldots, p. \tag{7.61}$$

These p equations are called Yule-Walker equations that can determine the p parameters $\phi_1, \phi_2, \ldots, \phi_p$, when the lag-$k$ autocorrelations $\rho(k)$ are known for all k, or estimated from data. A general principle is that, given the error size, the more unknowns we want to estimate, the longer the time series of data has to be.

Example 7.3   Use the Yule–Walker equations to estimate AR(2).

When $p = 2$, the Yule–Walker equations are

$$\rho(1) = \phi_1 + \phi_2 \rho(1), \tag{7.62}$$

$$\rho(2) = \phi_1 \rho(1) + \phi_2. \tag{7.63}$$

Solving these two equations yields

$$\phi_1 = \frac{\rho(1)(1 - \rho(2))}{1 - \rho^2(1)}, \tag{7.64}$$

$$\phi_2 = \frac{\rho(2) - \rho^2(1)}{1 - \rho^2(1)}. \tag{7.65}$$

7.7.5 ARIMA(p, d, q) Model and Data Fitting by R

We may also extend an MA(1) model to an MA(q) model as

$$X_t = \theta_0 + \theta_1 W_{t-1} + \theta_2 W_{t-2} + \cdots + \theta_q W_{t-q} + W_t, \tag{7.66}$$

for a positive integer q.

A further extension is to the sum of an AR(p) and an MA(q). The result model is called ARIMA(p, d, q):

$$X_t^{(d)} = c + \phi_1 X_{t-1} + \phi_2 X_{t-2} + \cdots + \phi_p X_{t-p} + \theta_1 W_{t-1} + \theta_2 W_{t-2} + \cdots + \theta_q W_{t-q} + W_t. \quad (7.67)$$

The letter "I" in ARIMA stands for "Integration." It means the ARIMA model development is as if integrating a dth order differential equation. R has a command to fit time series data to an ARIMA model.

The R command `arima(ts, order = c(p, d, q))` can fit the time series data `ts` to an ARIMA(p, d, q) model and calculate all the model parameters. Since Brownian motion is common in nature, climate science often uses AR(1) models, although the general ARIMA(p, d, q) model may be applied in some special cases. In the following, we show an example of fitting the Mauna Loa CO_2 data to an AR(1) model (see Fig. 7.11). We will also show that an MA(1) model is not a good fit to the data sequence.

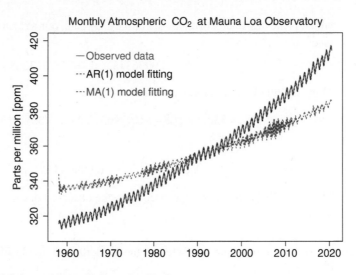

Monthly Atmospheric CO_2 at Mauna Loa Observatory

Figure 7.11 AR(1) and MA(1) fitting to the monthly Mauna Loa CO_2 data

Figure 7.11 shows a very good AR(1) fitting by R command `arima(co2.ts, order = c(1,0,0))`: The fitted AR(1) model data (in black dashed lines) are almost overlapping with the observed data.

The `arima` fitting is based on the least square principle, which may be illustrated by the AR(1) process as an example. The `arima` is designed for a time series of zero mean:

$$X_t - \mu = \phi(X_{t-1} - \mu) + W_t, \quad (7.68)$$

where μ is a constant and may be considered the ensemble mean of X_t. We regard W_t as residuals:

$$W_t = X_t - \mu - \phi(X_{t-1} - \mu). \quad (7.69)$$

The `arima` fitting algorithm minimizes the sum of the residual squares:

$$S_c(\phi,\mu) = \sum_{t=2}^{n} W_t^2. \tag{7.70}$$

This is also called the conditional sum-of-squares (CSS) function. Given the time series data x_t of n time steps, CSS is a function of ϕ and μ:

$$S_c(\phi,\mu) = \sum_{t=2}^{n} [x_t - \mu - \phi(x_{t-1}-\mu)]^2. \tag{7.71}$$

We use e_t to denote the term inside the square bracket for time step t:

$$e_t = x_t - \mu - \phi(x_{t-1}-\mu). \tag{7.72}$$

Minimization of $S_c(\phi,\mu)$ yields the parameter estimation of $\hat{\phi}$ and $\hat{\mu}$. It can be proven that

$$\hat{\phi} \approx \rho(1), \tag{7.73}$$

which is the lag-1 autocorrelation. For the CO_2 data, our `arima` fitting for the AR(1) model leads to

$$\phi = 0.9996343, \quad \mu = 358.3138084. \tag{7.74}$$

Thus, the fitted AR(1) model is

$$X_t = 358.3138084 + 0.9996343(X_{t-1} - 358.3138084) + W_t, \tag{7.75}$$

or

$$X_t = 0.1310354 + 0.9996343 X_{t-1} + W_t. \tag{7.76}$$

The fitted AR(1) model data $\hat{x}_t$ is equal to the observed data minus the residuals

$$\hat{x}_t = x_t - e_t. \tag{7.77}$$

These are the fitted data depicted by the black dashed line in Figure 7.11. We note that the CO_2 data

$$x_t = \hat{x}_t + e_t, \tag{7.78}$$

are only a single realization of the stochastic process represented by (7.76), and we cannot recover the CO_2 data by simulation based on (7.76).

Figure 7.11 also shows that MA(1) model (the blue dotted line) is not a good fit to the CO_2 data. The model fitting is also based on the principle of least squares.

We have tested other ARIMA model fittings, such as ARIMA(2, 0, 0) and ARIMA(0, 1, 1). Both models also fit the data well.

ACF and partial ACF (PACF) are often used to make a calculated guess of the best ARIMA model. When ACF $\rho(\tau)$ decreases as a power function of the time lag τ, and PACF(p+1) value drops suddenly to a small value, AR(p) may be a good model. In the case of the CO_2 data, PACF(1) is almost one, but PACF(2) is around -0.15. This suggests that AR(1) may be a good model for the CO_2 data. The details about PACF referred to modern time series books, such as Shumway and Stoffer (2011).

Figure 7.11 can be reproduced by the following computer code.

```r
#R plot Fig. 7.11: ARIMA model fitting
setwd("/Users/sshen/climstats")
co2m = read.table("data/co2m.txt", header = TRUE)
mon = co2m[,3]
co2 = co2m[,5]
co2.ts = ts(co2, start=c(1958,3), end=c(2020,7),
            frequency =12)
#Then fit an AR(1) model
co2.AR1 <- arima(co2.ts, order = c(1,0,0))
#Obtain the AR(1) model fit data
AR1_fit <- co2.ts - residuals(co2.AR1)
#Fit an MA(1) model
co2.MA1 <- arima(co2.ts, order = c(0,0,1))
#Obtain the MA(1) model fit data
MA1_fit <- co2.ts - residuals(co2.MA1)

setEPS() #Automatically saves the .eps file
postscript("fig0711.eps", width=8, height=6)
#Plot the CO2 time series
par(mar=c(3,4.5,2.5, 0.5))
plot(mon, co2, ylim=c(310,420),
     type='l', lwd=2, col = 'red',
     main = bquote("Monthly Atmospheric "~
                    CO[2]~" at Mauna Loa Observatory"),
     xlab ="",
     ylab="Parts Per Million [ppm]",
     cex.main=1.5,
     cex.lab =1.5, cex.axis=1.5)
#Plot the AR(1) model data on the CO2 data
points(AR1_fit, type = "l", lwd=1,
       col = 'black', lty = 2)
points(MA1_fit, type = "l", lwd=1,
       col = 'blue', lty = 2)
legend(1960, 420, lty=c(1, 2, 2),
       col=c('red','black', 'blue'),
       legend=c("Observed data",
                "AR(1) model fitting",
                "MA(1) model fitting"),
       text.col = c('red','black', 'blue'),
       x.intersp = 0.2, y.intersp = 1.4,
       seg.len = 0.8, cex=1.5, bty = 'n')
dev.off()
```

```python
#Python plot Fig. 7.11: ARIMA model fitting
## NOTE: Run the code for Figs. 7.1 and 7.2 first ##
# set up series data variable
data_values = [float(i) for i in ser]
data_dates = pd.date_range('1958-03-17',
                '2020-08-17', freq = 'M')
```

```
co2_series = pd.Series(data_values,
                       index=data_dates)

# Fit AR(1) model
AR1_model = ARIMA(co2_series, order=(1,0,0))
AR1_model_fit = AR1_model.fit()
AR1_residuals = AR1_model_fit.resid
AR1_fit = pd.Series(co2_series.values -
            AR1_residuals.values, data_dates)

# Fit MA(1) model
MA1_model = ARIMA(co2_series, order=(0,0,1))
MA1_model_fit = MA1_model.fit()
MA1_residuals = MA1_model_fit.resid
MA1_fit = pd.Series(co2_series.values -
            MA1_residuals.values, data_dates)

# drop first values of models
AR1_fit.drop(AR1_fit.index[0], inplace=True)
MA1_fit.drop(MA1_fit.index[0], inplace=True)

fig, ax = plt.subplots()#plot setup
#prepare labels
plt.title('Monthly Atmospheric CO\N{SUBSCRIPT TWO} \
           at Mauna Loa Observatory',
        pad = 15)
plt.ylabel('Parts Per Million [ppm]', labelpad = 15)
#plot observed line
plt.plot(ser, color='red', linewidth=2,
        linestyle='-', label='Observed data')
#plot AR(1) model
plt.plot(AR1_fit, color='black', linestyle='--',
        linewidth=1, label='AR(1) model fitting')
#plot MA(1) model
plt.plot(MA1_fit, color='blue', linestyle='--',
        linewidth=2, label='MA(1) model fitting')
#add legend
plt1 = plt.legend(loc=(.10,.7), fontsize = 18)
plt1.get_texts()[0].set_color('r')
plt1.get_texts()[2].set_color('b')
plt.savefig("fig0711.pdf")# save figure
plt.show()
```

The residuals for the AR(1) model fitting can be can be verified as follows for $t = 2$.

```
mu = 358.3138084
phi = 0.9996343
co2[2] - mu - phi*(co2[1]-mu)
#[1] 1.72442
residuals(co2.AR1)[2]
#[1] 1.724418
```

7.8 Chapter Summary

This chapter has included the basic theory and computer code of time series for climate data analysis. Time series is an important branch of statistics. Many excellent books are available, such as the modern textbook with R and real climate data by Shumway and Stoffer, and the classical textbook by Box and Jenkins. Our chapter is different from the comprehensive treatment of time series in these books. Instead, we present a few carefully selected time series methods, concepts, datasets, R code, and Python code that are useful in climate data analysis. These methods and concepts are summarized as follows.

(i) ETS decomposition of a time series data sequence: Many time series can be decomposed into three ETS components: seasonal cycle (S), trend (T), and random error (E). R code and Python code are included to make the ETS decomposition and to generate corresponding graphics. The monthly atmospheric carbon dioxide data and the daily minimum temperature data are used as examples.

(ii) White noise time series: A white noise W_t at a given time t is normally distributed: $W_t \sim N(0, \sigma^2)$, where the zero mean and standard deviation σ do not vary in time. The autocorrelation with a nonzero lag is zero. We use white noise as building blocks for the commonly used time series: random walk (RW), autoregression (AR), and moving average (MA).

(iii) Random walk: The difference time series from a random walk is white noise, i.e., the next step of a RW is W_t. This concept can be extended from 1-dimensional to n-dimensional random variables at different time steps. An important result of the random walk is that its variance grows linearly with time.

(iv) Stationary versus nonstationary: Many methods used in published research papers use statistical or mathematical methods that require the assumption of stationarity. Section 7.4 suggests you pay careful attention to the concept of stationarity and provides you a method to test the stationarity of a time series. When the original data sequence is not stationary, you may consider its difference time series which may become stationary. The difference time series method is a very important tool to homogenize the historical climate data from observational stations (see Peterson et al. 1998 and Smith et al. 1998).

(v) AR(1) process: The AR(1) model is often applicable to fit a climate data sequence because its difference time series corresponds to the commonly used first derivative of a climate variable with respect to time. The AR(1) model provides a very good fit to the Mauna Loa CO_2 data. Mathematically, an AR(1) time series may be regarded as MA(∞).

(vi) When does a time series forget its initial condition? This is an important problem in climate science, and can be explored via the ACF method.

(vii) Concise theory of the RW, AR, MA, and ARIMA models: We have presented short derivations of the mathematical theory on white noise, random walk, autoregression time series, moving average time series, and ARIMA time series. Our presentation approach of concise derivation was designed to facilitate a climate scientist readily learning the method, theory, and coding for analyzing real data, and was not intended for a professional statistician to make a mathematical investigation. A proper choice of the statistical model to analyze a given set of climate time series data is often motivated by common sense, prior evidence of physics, or climate science theories, in addition to the statistical model assumptions and hypothesis tests.

References and Further Reading

[1] T. C. Peterson, T. R. Karl, P. F. Jamason, R. Knight, and D. R. Easterling, 1998: First difference method: Maximizing station density for the calculation of long-term global temperature change. *Journal of Geophysical Research: Atmospheres*, **103**(D20), 25967–25974.

> This paper explains the detailed procedures of the first difference method to homogenize the station data for the monthly surface air temperature.

[2] M. Romer, R. Heckard, and J. Fricks, 2020: *Applied Time Series Analysis*, `https://online.stat.psu.edu/stat510/lesson/5/5.1`, PennState Statistics Online. Access date: September 2020.

> This time series analysis text with R has many examples of real data. It has excellent materials on time series decomposition.

[3] Scripps Institution of Oceanography, 2022: *The Keeling Curve*, `https://keeling-curve.ucsd.edu`. Access date: March 2022.

> The Keeling curve is named after Charles David Keeling (1928–2005), an American scientist who pioneered in the accurate measurement of the atmospheric carbon dioxide concentration. This website has many kinds of information on the atmospheric carbon dioxide measurement, including data of different time scales, ranging from hourly to monthly to annual and to longer time scales. The time coverage ranges from the last week, back to 1958, and all the way back to 800K years ago.

[4] R. H. Shumway and D. S. Stoffer, 2016: *Time Series Analysis and Its Applications: With R Examples*. 4th ed., Springer.

> This book includes many inspiring examples. R code is provided for each example. The book contains some examples of climate datasets, such as global average annual mean temperature, precipitation, soil surface temperature, fish population, El Niño, wind speed, and dew point.

[5] T. M. Smith, R. E. Livezey, and S. S. P. Shen, 1998: An improved method for interpolating sparse and irregularly distributed data onto a regular grid. *Journal of Climate*, **11**, 1717–1729.

> This paper includes the concept of first guess, which can form a difference time series.

Exercises

7.1 Write a computer code to make ETS decomposition for the monthly total precipitation at Omaha, Nebraska, USA (Station code: USW00014942, 41.3102° N, 95.8991° W) from January 1948 to December 2017. Plot the observed data sequence and its ETS components. Use 100–300 words to comment on the seasonal, trend, and error components you have obtained. The data can be downloaded from internet sites such as www.ncdc.noaa.gov/cdo-web.

You can also use OmahaP.csv data file from data.zip for this book. The file data.zip can be downloaded from the book website www.climatestatistics.org.

7.2 Plot the histogram of the random error component from the previous problem on Omaha precipitation. Comment on the probabilistic distribution of the error data based on this histogram.

7.3 Write a computer code to make ETS decomposition for the monthly minimum surface air temperature data for a station of your choice from January 1948 to December 2017. Plot the observed data sequence and its ETS components. Use 100–300 words to comment on the seasonal, trend, and error components you have obtained.

7.4 Following Figure 7.3, use the data from the previous problem and the ETS forecasting method to forecast the monthly minimum surface air temperature for the next 12 months: January to December 2018. Compare your forecast data with the observed data of the same station in the forecasting period January to December 2018.

7.5 Write a computer code to generate two realizations of 1,000 time steps for a white noise time series with zero mean and standard deviation equal to 1.0. Plot the two realizations on the same figure. See Figure 7.5 and its computer code for reference.

7.6 Plot the two histograms and two ACF functions for the two realizations of the white noise in the previous problem. See Figure 7.6 and its computer code for reference.

7.7 From the random walk model defined by:

$$X_{t+1} = \delta + X_t + W_t,$$

(7.79)

where δ is a fixed constant called *drift*, $W_t \sim N(0, \sigma^2)$, and σ^2 is the variance of the white noise, show that the mean of X_T is equal to the mean of initial position plus δT:

$$\mathrm{E}[X_T] = \mathrm{E}[X_0] + \delta T. \tag{7.80}$$

Please include all the details in your mathematical derivation.

7.8 From the random walk X_t defined in the previous problem, show that the variance of X_T grows linearly with T in the following way:

$$\mathrm{Var}[X_T] = \mathrm{Var}[X_0] + \sum_{t=0}^{T-1} \mathrm{Var}[W_t] = \mathrm{Var}[X_0] + \sigma^2 T. \tag{7.81}$$

Please include all the details in this mathematical derivation.

7.9 Use a computer to reproduce Figure 7.7. Because of randomness, the time series realizations will not be exactly the same as those in Figure 7.7. Choose a set of simulation results that are a reasonably good mimic of Figure 7.7.

7.10 Generate a 2-dimensional random walk (x_t, y_t) of 20 time steps with mean zero and standard deviation equal to 1.0. Plot the track of the random particle motion on an xy-plane.

7.11 Generate a 3-dimensional random walk (x_t, y_t, z_t) of 20 time steps with mean zero and standard deviation equal to 1.0. Plot the projection of the track of the random particle motion on the yz-plane.

7.12 Using the method similar to that for deriving formula (7.33) from (7.30), derive formula (7.11) from the random walk definition (7.6).

7.13 Use the ADF test to show whether the monthly total precipitation at Omaha, Nebraska, from January 1948 to December 2017 in Exercise 7.1 is stationary or not.

7.14 Write a computer code to generate a difference time series data from the monthly Omaha precipitation data in Exercise 7.1. Plot the result difference time series. Use the ADF test to check whether the difference time series is stationary.

7.15 Generate a realization of 2,000 time steps for a white noise time series with zero mean and standard deviation equal to 2.0. Use the ADF test to verify that this time series is stationary.

7.16 Simulate the AR(1) process using the formula of the moving average process (7.33) and reproduce a figure similar to Figure 7.9.

7.17 Derive formula (7.33) from the AR(1) definition (7.30) using the following iterative approach:

$$X_1 = \lambda X_0 + W_0, \tag{7.82}$$
$$X_2 = \lambda X_1 + W_1 = \lambda(\lambda X_0 + W_0) + W_1, \tag{7.83}$$
$$\cdots \tag{7.84}$$

Finally, express the AR(1) term X_N in terms of the initial condition X_0 and white noise, as shown in Eq. (7.33).

7.18 For an AR(3) model, explicitly write the three Yule–Walker equations. You may reference Eqs. (7.61)–(7.63).

7.19 Use the ARIMA fitting method to fit the observed CO_2 data in Figure 7.1 to an AR(2) model. Plot the observed CO_2 data and the fitted model data on the same figure. You may reference Figure 7.11 and its relevant computer code, formulas, and numerical results.

7.20 Use the ARIMA fitting method to fit the Omaha monthly precipitation data in Exercise 7.1 to an AR(1) model. Plot the observed data and the fitted model data on the same figure. You may reference Figure 7.11 and its relevant computer code, formulas, and numerical results.

8

Spectral Analysis of Time Series

Climate has many cyclic properties, such as seasonal and diurnal cycles. Some cycles are more definite, e.g., sunset time of London, the UK. Others are less certain, e.g., the monsoon cycles and rainy seasons of India. Still others are quasiperiodic with cycles of variable periods, e.g., El Niño Southern Oscillation and Pacific Decadal Oscillation. In general, properties of a cyclic phenomenon critically depend on the frequency of the cycles. For example, the color of light depends on the frequency of electromagnetic waves: red corresponding to the energy in the range of relatively lower frequencies around 400 THz (1 THz = 10^{12} Hz, 1 Hz = 1 cycle per second), and violet to higher frequencies around 700 THz. Light is generally a superposition of many colors (frequencies). The brightness of each color is the spectral power of the corresponding frequency. Spectra can also be used to diagnose sound waves. We can tell if a voice is from a man or a woman, because women's voices usually have more energy in higher frequencies while men's have more energy in relatively lower frequencies. The spectra of temperature, precipitation, atmospheric pressure, and wind speed often are distributed in frequencies far lower than light and sound. Spectral analysis, by name, is to quantify the frequencies and their corresponding energies. Climate spectra can help characterize the properties of climate dynamics. This chapter will describe the basic spectral analysis of climate data time series. Both R and Python codes are provided to facilitate readers to reproduce the figures and numerical results in the book.

8.1 The Sine Oscillation

The word "spectral" was derived from Latin "spectrum," meaning appearance or look. Further, "spec" means to look at or regard. In science, spectrum, or spectra in its plural form, means the set of colors into which a beam of light can be separated. The colors are lined up according to their wavelength. For example, the seven colors isolated by a prism from white sunlight are red, orange, yellow, green, blue, indigo, and violet, lined up according to the wavelength of each color, from red in longer wavelengths around 700 nm (1 nm = 10^{-9} m) to violet in shorter wavelengths around 400 nm. Besides color, light has another important property: brightness, i.e., the energy in a given color. Spectral analysis studies both color and brightness. In general, it examines wavelength and its associated energy for any kind of variations, such as climate.

As shown in Figure 8.1, a deterministic cyclic phenomenon may be described by three parameters: period or frequency, amplitude, and phase. The period is the time length of a full oscillation. The period in a spatial oscillation, such as lightwaves, is the aforementioned wavelength. The frequency is the number of full oscillations in a unit time, e.g., a second, and is equal to one divided by period. Thus, the period or frequency shows how rapid the oscillation is. The amplitude shows the oscillation strength and is an indicator for the amount of energy. The phase determines the beginning position of an oscillation. In theory, there are infinitely many ways to mathematically express an oscillation with given period, amplitude, and phase. In practice, we often use the simple sine or cosine functions, also known as harmonic functions, to conveniently express a deterministic oscillation. A single harmonic function is

$$T(t) = A \sin(2\pi t/\tau + \phi), \qquad (8.1)$$

where A is the amplitude, τ is the period, and ϕ is the phase. Figure 8.1 shows curves of this function $T(t)$ with different values of A, τ, and ϕ. The first row of Figure 8.1 shows two harmonic functions of the same amplitude, but different periods (or frequencies). The first column of Figure 8.1 compares different amplitudes and phases while keeping the same period.

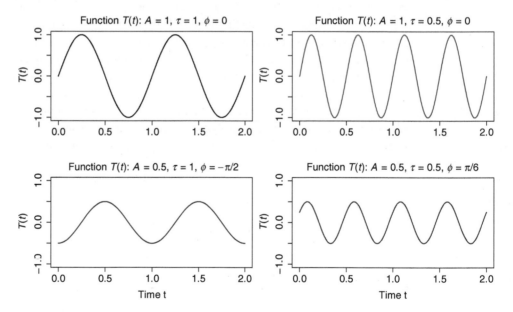

Figure 8.1 Sine function of different amplitude A, period τ, and phase ϕ.

Figure 8.1 can be generated by the following computer code.

```
#R plot Fig. 8.1: Waves of different amplitudes,
#                          periods and phases
setEPS()
postscript("fig0801.eps", height=6, width=10)
par(mar = c(4.5, 4.5, 2.5, 0.5))
par(mfrow=c(2,2))
```

```
t = seq(0, 2, len = 1000)
y1 = sin(2*pi*t) #A=1, tau=1, phi=0
y2 = sin(2*pi*t/0.5) #A=1, tau=0.5, phi=0
y3 = 0.5*sin(2*pi*t  - pi/2) #A=0.5, tau=1, phi=0
y4 = 0.5*sin(2*pi*t/0.5 + pi/6) #A=0.5, tau=1, phi=pi/6
plot(t, y1,
     type = 'l', lwd = 2,
     xlab = "␣", ylab = "T(t)",
     cex.lab =1.5, cex.axis = 1.5,
     cex.main = 1.5,
main=expression(paste(
  'Function␣T(t):␣A=1,␣', tau, '=1,␣', phi,'=0')))
plot(t, y2,
     type = 'l', lwd = 2,
     xlab = "␣", ylab = "T(t)",
     cex.lab =1.5, cex.axis = 1.5,
     cex.main = 1.5, col = 'red',
     main=expression(paste(
       'Function␣T(t):␣A=1,␣', tau, '=0.5,␣', phi,'=0')))
plot(t, y3,
     type = 'l', lwd = 2,
     xlab = "Time␣t", ylab = "T(t)",
     cex.lab =1.5, cex.axis = 1.5,
     cex.main = 1.5, col = 'blue',
     main=expression(paste(
       'Function␣T(t):␣A=0.5,␣', tau, '=1,␣', phi,'=', - pi/2)))
plot(t, y4,
     type = 'l', lwd = 2,
     xlab = "Time␣t", ylab = "T(t)",
     cex.lab =1.5, cex.axis = 1.5,
     cex.main = 1.5, col = 'purple',
     main=expression(paste(
  'Function␣T(t):␣A=0.5,␣', tau, '=0.5,␣', phi,'=', pi/6)))
dev.off()
```

```
#Python plot Fig. 8.1: Waves of different amplitudes,
#                            periods and phases
t = np.linspace(0,2,2000)
# A=1, tau=1, phi=0
y1 = np.sin(2*np.pi*t)
# A=1, tau=0.5, phi=0
y2 = np.sin(2*np.pi*t/0.5)
# A=0.5, tau=1, phi=pi/2
y3 = 0.5 * np.sin(2*np.pi*t - np.pi/2)
# A=0.5, tau=1, phi=pi/6
y4 = 0.5 * np.sin(2*np.pi*t/0.5 + np.pi/6)

fig, ax = plt.subplots(2, 2, figsize=(13, 8))
# plot t,y1
ax[0, 0].plot(t, y1, color = 'k', linewidth = 2)
ax[0, 0].set_title("Function␣T(t):␣\
A=1,␣$\u03C4$=1,␣$\phi$=0",
                    pad=15, size = 18)
ax[0,0].set_ylabel("T(t)", labelpad=10, size = 18)
```

```
ax[0,0].tick_params(labelsize=15)
ax[0,0].set_yticks([-1.0,0.0,1.0])
# plot t,y2
ax[0, 1].plot(t, y2, color = 'red', linewidth = 2)
ax[0, 1].set_title("Function␣T(t):␣A=1,␣\
$\u03C4$=0.5,␣$\phi$=0",
                   pad=16, size = 18)
ax[0,1].set_ylabel("T(t)", labelpad=10, size = 18)
ax[0,1].tick_params(labelsize=15)
ax[0,1].set_yticks([-1.0,0.0,1.0])
# plot t,y3
ax[1, 0].plot(t, y3, color = 'blue', linewidth = 2)
ax[1, 0].set_title("Function␣T(t):␣A=0.5,␣\
$\u03C4$=1,␣$\phi=\dfrac{\pi}{2}$",
                   pad=18, size = 18)
ax[1,0].set_xlabel("Time␣t", labelpad=15, size = 18)
ax[1,0].set_ylabel("T(t)", labelpad=10, size = 18)
ax[1,0].tick_params(labelsize=15)
ax[1,0].set_yticks([-1.0,0.0,1.0])
# plot t,y4
ax[1, 1].plot(t, y4, color = 'purple', linewidth = 2)
ax[1, 1].set_title('Function␣T(t):␣A=0.5,␣\
$\u03C4$=0.5,␣$\phi=\dfrac{\pi}{6}$',
                   pad=18, size = 18)
ax[1,1].set_xlabel("Time␣t", labelpad=15, size = 18)
ax[1,1].set_ylabel("T(t)", labelpad=10, size = 18)
ax[1,1].tick_params(labelsize=15)
ax[1,1].set_yticks([-1.0,0.0,1.0])
fig.tight_layout(pad=3.0)
plt.savefig("fig0801.eps") # save the figure
```

Why do we choose sine or cosine? First, sine and cosine functions have important properties of orthogonality, expressed as follows:

$$\int_0^1 \sin(2\pi m x) \sin(2\pi n x)\, \mathrm{d}x = 0, \qquad (8.2)$$

if m and n are different integers. This is like the zero dot product of two orthogonal vectors. The orthogonality property turns out to be extremely useful in the analysis of all kinds of signals, whether climate, electrical, or acoustical. Second, sine and cosine functions are the simplest periodic orthogonal functions and can be modeled by the x and y coordinates of a point on a unit circle, or by the harmonics of a simple pendulum oscillation. These two properties make sine and cosine functions convenient to use in many signal analysis problems. The signal analysis based on sine or cosine functions and their summations is known as the Fourier analysis, named after the French mathematician and physicist Jean-Baptiste Joseph Fourier (1768–1830). This chapter is mainly about the Fourier analysis of time series.

Figure 8.2 shows a periodic function, whose oscillation is not as regular as a simple sine function. It is a linear combination of the four sine functions shown in Figure 8.1.

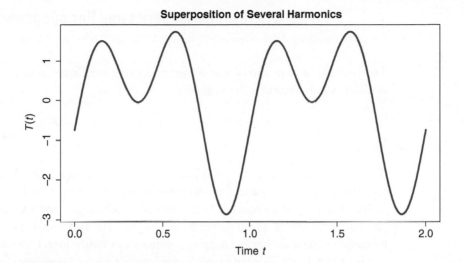

Superposition of Several Harmonics

Figure 8.2   Superposition of the four harmonics from Figure 8.1.

The linear combination is also called superposition. In fact, almost all of the periodic functions in geoscience can be expressed as a superposition of simple harmonics. While the precise statement and proof of this claim are not the aim of this book, we show the intuition, numerical computing, and physical meaning of Fourier analysis, i.e., the spectral analysis.

Figure 8.2 can be generated by the following computer code.

```
#R plot Fig. 8.2: Wave superposition
setEPS()
postscript("fig0802.eps", height=6, width=10)
par(mar = c(4.5, 4.8, 2.5, 0.5))
plot(t, y1 + y2 + 2*y3 + y4,
     type = 'l', lwd = 4,
     xlab = "Time␣t", ylab = "T(t)",
     cex.lab =1.5, cex.axis = 1.5,
     cex.main = 1.5, col = 'blue',
     main='Superposition␣of␣several␣harmonics')
dev.off()
```

```
#Python plot Fig. 8.2: Wave superposition
fig, ax = plt.subplots(figsize=(12, 6))
plt.plot(t, y1 + y2 + 2*y3 + 2*y4, color = 'b', linewidth=2)
plt.title("Superposition␣of␣several␣harmonics", pad=15)
plt.xlabel("Time␣t", labelpad = 15)
plt.ylabel("T(t)", labelpad = 15)
plt.yticks([-3,-2,-1,0,1], rotation = 90)
plt.xticks([0,0.5,1.0,1.5,2.0])
plt.savefig("fig0802.eps") # save figure
plt.show()
```

8.2 Discrete Fourier Series and Periodograms

The previous section shows that a harmonic oscillation gives a signal with a specified period τ, whose inverse is the frequency f:

$$f = \frac{1}{\tau}. \tag{8.3}$$

If $\tau = 0.005$ [sec], then $f = 200$ [Hz], meaning 200 cycles per second. The unit Hz, pronounced as Hertz, is after German physicist Heinrich Hertz (1857–1894).

The frequency of an average adult woman's voice ranges approximately from 165 to 255 Hz, while that of a man ranges from 85 to 155 Hz. Because women and men's voice frequency ranges are not overlapping, our ears can easily detect whether the voice is a woman's or a man's. This is a simple signal detection method: separate signals according to frequency. This kind of detection is called filtering in engineering. The filtering algorithm or device is called a filter.

In climate science, we often deal with slower oscillations, such as annual or semiannual cycles. Thus, the unit Hz is rarely used for the climate data of temperature, precipitation, or atmospheric pressure. Nonetheless, the meteorological observations often involve high-frequency instruments, such as meteorology radiosonde operating at a radio frequency around 403 or 1680 MHz (megahertz) (1 MHz = 10^6 Hz), weather radar waves in a frequency range of 1–40 GHz (gigahertz) (1 GHz = 10^9 Hz), and infrared satellite remote sensing in a range of 10–400 THz (terahertz) (1 THz = 10^{12} Hz).

If we regard all the signals as superpositions of harmonic oscillations expressed in Eq. (8.1) as indicated by the Fourier analysis theory, then the frequencies $f = 1/\tau$ and the corresponding amplitudes A can determine the signal up to a phase shift ϕ, which determines when the signal begins in the cycle. The square of the amplitude of a given frequency is called a spectral component, which varies according to frequency. The plot of the spectral component as a function of frequency is called *power spectrum*, or *periodogram*. Conventionally, engineering often uses periodogram to mean discrete power spectrum. This chapter focuses on the discrete case. We will use the term periodogram.

We can use these ideas to analyze climate data for signals, such as the surface air temperature variations of Tokyo, Japan. The mathematical procedure is to transform climate data in the time domain into the amplitudes squared in the frequency domain, i.e., a frequency–spectrum relationship.

8.2.1 Discrete Sine Transform

Consider time series data of length M:

$$X = (x_1, x_2, \cdots, x_M).$$

An $M \times M$ orthogonal matrix of sine transform can be defined as follows:

$$\Phi = \left[\sqrt{\frac{2}{M}} \sin\left(\frac{\pi \left(t - \frac{1}{2}\right)\left(k - \frac{1}{2}\right)}{M} \right) \right]_{k,t=1,\cdots,M}. \tag{8.4}$$

It can be proven that matrix Φ is orthogonal, i.e.,

$$\Phi\Phi^t = I, \tag{8.5}$$

where Φ^t is the transpose matrix of Φ, and I is an M-dimensional identity matrix.

The vector

$$S = \Phi^t X \tag{8.6}$$

is called the discrete sine transform (DST) of the time series data X. The word "discrete" is in the name, because we are considering discrete time series data $X = (x_1, x_2, \ldots, x_M)$. Of course, one can also do a sine transform for a continuous function.

From the spectral data $S = (s_1, s_2, \ldots, s_M)$, one can recover the original time series data X, by simply multiplying Eq. (8.6) by the sine transform matrix Φ:

$$X = \Phi S. \tag{8.7}$$

This step is called the inverse DST, denoted by iDST.

The computer code for a DST example is as follows.

```
#R code for the discrete sine transform (DST)
M = 5
time = 1:M - 1/2
freq = 1:M - 1/2
time_freq = outer(time, freq)
#Construct a sine transform matrix
Phi = sqrt(2/M)*sin((pi/M)*time_freq)
#Verify Phi is an orthogonal matrix
round((t(Phi)%*%Phi), digits = 2)
#The output is an identity matrix

ts = time^2  #Given time series data
ts
#[1]  0.25  2.25  6.25 12.25 20.25
ts_dst = t(Phi)%*%ts #DST transform
Recon = Phi%*%ts_dst #inverse DST
t(Recon) #Verify Recon = ts
#[1,] 0.25 2.25 6.25 12.25 20.25
```

```
#Python code for the discrete sine transform (DST)
import numpy as np
import math
M = 5
time = np.linspace(1, M, M)-1/2
freq = np.linspace(1, M, M)-1/2
time_freq = np.outer(time, freq)
# Construct a sine transform matrix
Phi = np.sqrt(2/M)*np.sin((math.pi/M)*time_freq)
```

```
# Verify Phi is an orthogonal matrix
abs(np.dot(Phi,Phi.transpose()).round(2))
# The output is an identity matrix

# Given time series data
ts = time**2
# DST transform
ts_dst = np.dot(Phi.transpose(),ts)
# Inverse DST
Recon = np.dot(Phi, ts_dst)
# Verify Recon = ts
Recon.transpose()
#array([ 0.25,   2.25,   6.25, 12.25, 20.25])
```

8.2.2 Discrete Fourier Transform

Similarly, one can define discrete cosine transform (DCT). Although DCT has been used frequently in image compression, climate scientists often use the discrete transform involving both sine and cosine, which are connected to a complex-valued function through Euler's formula:

$$e^{i\theta} = \cos\theta + i\sin\theta, \tag{8.8}$$

where $i = \sqrt{-1}$ is the imaginary unit, and θ is an angle in geometry or a phase lag in waves. Leonhard Euler (1707–1783) was a Swiss mathematician and physicist.

8.2.2.1 A Review of Basics of Complex Numbers

Before formulating the discrete Fourier transform (DFT), we briefly recapitulate the basics of complex numbers. Complex numbers are an essential mathematical tool for electrical engineering and signal analysis.

A complex number z is written as

$$z = x + iy, \tag{8.9}$$

with the real number $x = \text{Re}(z)$ called the real part of z, and the real number $y = \text{Im}(z)$ the imaginary part of z. The modulus of z is defined as

$$|z| = \sqrt{x^2 + y^2}, \tag{8.10}$$

and is often denoted by r. Tradition refers to the modulus of z as the absolute value, or magnitude.

The polar expression of z is

$$z = re^{i\theta} = r\cos\theta + i\,r\sin\theta, \tag{8.11}$$

where θ is called the argument of z. See Figure 8.3 for the geometric meaning of the polar expression on a complex plane spanned by a real axis x and an imaginary axis y. Figure 8.3 may be generated by the following computer code.

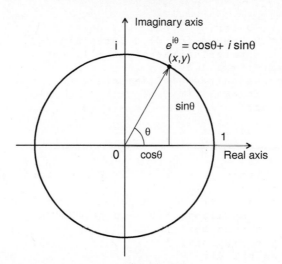

Figure 8.3 Polar expression of a complex number $z = x + iy$ on a complex plane.

```
setwd("~/climstats")
setEPS()
postscript("fig0803.eps", height=5.5, width=5.5)
par(mar = c(0,0,0,0))
r=10
bb=1.5*r
t=seq(0, 2*pi, length=200)
x=r*cos(t)
y=r*sin(t)
plot(x,y,type="l", lwd=3, asp=1,
    xlim=c(-bb + 3, bb),ylim=c(-bb + 3, bb),
    xaxt="n",yaxt="n",ann=FALSE,bty="n")
aa=1.4*r
x1=r*cos(pi/3)
y1=r*sin(pi/3)
arrows(c(-aa + 2, 0), c(0, -aa + 2), c(aa,0), c(0, aa),
        length=0.1, code=2, lwd=2)
arrows(0,0,0.98*x1,0.98*y1,
        length=0.15,code=2, lwd=1, angle=15)
t2=seq(0,pi/3,length=10)
x2=0.22*r*cos(t2)
y2=0.22*r*sin(t2)
lines(x2,y2, type="l")
points(x1,y1,pch=19,cex=1)
segments(x1,0, x1,y1)
text(1.1*r,0.1*r, "1", cex=1.3)
text(-0.1*r, 1.1*r, "i", cex=1.5)
text(1.2*x1,1.1*y1, "(x,y)", cex=1.3)
text(0.3*r,-0.1*r, expression(paste(cos,theta)),
    cex=1.3)
text(1.35*x1,0.5*y1, expression(paste(sin,theta)),
    cex=1.3)
text(1.35*r,-0.1*r, "Real␣Axis", cex=1.3)
text(0.5*r, 1.35*r, "Imaginary␣Axis", cex=1.3)
```

```
text(-0.1*r, -0.1*r, "0", cex=1.5)
text(0.3*r*cos(pi/6),0.3*r*sin(pi/6),
     expression(paste(theta)), cex=1.3)
text(1.9*x1,1.3*y1,
expression(paste(e^{i*theta},"=", "cos",theta,"+ i sin",theta)),
     cex=1.5)
dev.off()
```

```python
#Python plot Fig. 8.3: Polar expression of a complex number
r = 10
bb = 1.4*r
t = np.linspace(0,2*np.pi,200)
x = r*np.cos(t)
y = r*np.sin(t)
aa = 1.4*r
x1 = r*np.cos(np.pi/3)
y1 = r*np.sin(np.pi/3)
t2 = np.linspace(0,np.pi/3,10)
x2 = 0.22*r * np.cos(t2)
y2 = 0.22*r * np.sin(t2)

import pylab as pl
from matplotlib import collections  as mc
fig, ax = plt.subplots(figsize=(10, 10))
ax.plot(x,y,color='k')# plot circle
ax.set_xlim(-14,14)
ax.set_ylim(-14,14)
ax.axis('off')# hide axis
# add arrows
ax.annotate("", xy=(14,0), xytext=(-13, 0),
    arrowprops=dict(width=0.5, color = 'k'))
ax.annotate("", xy=(0,14), xytext=(0, -14),
    arrowprops=dict(width=0.5, color = 'k'))
ax.plot(x2,y2,color='k')# plot angle
# plot line segments
segments = [[(x1, 0),(x1,y1)], [(0,0),(x1,y1)]]
ls = mc.LineCollection(segments, colors='k', linewidths=2)
ax.add_collection(ls)
# add text annotations
ax.text(10.3,-1.5, "Real Axis", size = 18)
ax.text(0.5,12.3, "Imaginary Axis", size = 18)
ax.text(-1,-1, "0", size = 18)
ax.text(1,-1.2, r"x=rcos$\theta$", size = 18)
ax.text(5.2,3, r"y=rsin$\theta$", size = 18)
ax.text(2, 4.7, "r", size = 18)
ax.text(2, 1.3, r"$\theta$", size = 18)
ax.text(5, 9, "(x,y)", size = 18)
ax.text(4.5, 10.5,
 r"z =re$^{i\theta}$ = rcos$\theta$ + $i$ rsin$\theta$",
        size = 18)
plt.savefig("fig0803.eps") # save figure
plt.show()
```

The complex conjugate of z is defined as

$$z^* = x - iy. \tag{8.12}$$

This can also be expressed as

$$z^* = re^{-i\theta}, \tag{8.13}$$

because

$$re^{-i\theta} = r\cos(-\theta) + i\sin(-\theta) = r\cos(\theta) - i\sin(\theta) = x - iy. \tag{8.14}$$

The following identify is often useful for the wave amplitude calculation:

$$z^*z = r^2, \tag{8.15}$$

because

$$z^*z = (x - iy)(x + iy) = x^2 + ixy - iyx - (iy)^2 = x^2 + y^2 = r^2. \tag{8.16}$$

This equation can also be shown using the polar expression as follows:

$$z^*z = re^{-i\theta} \times re^{i\theta} = r^2 e^{-i\theta+i\theta} = r^2 e^0 = r^2. \tag{8.17}$$

8.2.2.2 A Formulation of DFT

An orthogonal matrix of complex numbers is called the *unitary matrix U*, whose conjugate transpose, denoted by $U^H = (U^*)^t$ and called the Hermitian matrix of U, is orthogonal to the original matrix, i.e.,

$$UU^H = U^H U = I, \tag{8.18}$$

where I is the identity matrix. If u_{ij} are used to denote the elements of U, then the complex conjugate of U is defined as

$$U^* = [u_{ij}^*], \tag{8.19}$$

i.e., a matrix of the complex conjugates of all the corresponding elements of the original matrix.

The $M \times M$-order unitary matrix for DFT is

$$U = \left[\frac{1}{\sqrt{M}} e^{2\pi ikt/M} \right]_{k,t=0,1,...,M-1}, \tag{8.20}$$

where t is an integer time step, and k is an integer frequency. It can be proven that this matrix U is a unitary matrix satisfying

$$U^H U = I. \tag{8.21}$$

The complex unitary matrix U can be decomposed into a real part A plus an imaginary part B as follows:

$$U = A + iB, \tag{8.22}$$

where

$$A = \text{Re}(U), \quad B = \text{Im}(U). \tag{8.23}$$

Figures 8.4(a) and (b) show the real part A and the imaginary part B of the unitary DFT transformation matrix U when $M = 200$.

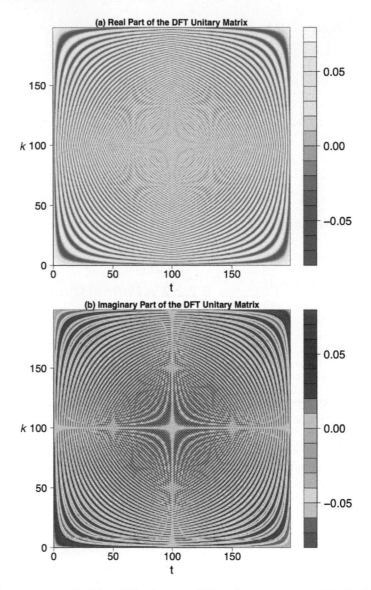

Figure 8.4 Real part and imaginary part of a 200×200-order unitary DFT transformation matrix U defined by (8.20).

The real and imaginary parts have many symmetry and other properties, such as $AB^t = 0$, $BA^t = 0$, and AA^t and BB^t are diagonal matrices with

$$AA^t + BB^t = I. \tag{8.24}$$

These properties and patterns help decompose a time series into many components at different frequencies and hence help with signal analysis of a climate data time series. Also see the patterns of DFT matrices A and B for $M = 32$ produced by Zalkow and Müller (2022).

Figure 8.4 may be generated by the following computer code.

```R
#R plot Fig. 8.4: The unitary DFT matrix
M = 200
i  = complex(real = 0, imaginary = 1)
time_freq = outer(0:(M-1), 0:(M-1))
U = exp(i*2*pi*time_freq/M) / sqrt(M)
Ure = Re(U) #Real part of U
Uim = Im(U) #Imaginary part of U

setEPS()
postscript("fig0804a.eps", height=8.1, width=10)
par(mar=c(4.2, 5.5, 1.5, 0))
#Plot the real part
filled.contour(0:(M-1), 0:(M-1), Ure,
                color.palette = heat.colors,
                #      xlab = 't', ylab ='k',
                plot.axes={
                   axis(1, cex.axis=1.8)
                   axis(2, cex.axis=1.8)},
                plot.title={
    title(main = '(a) Real Part of the DFT Unitary Matrix',
                    xlab="t", cex.lab=2, cex.main = 1.5)
                mtext("k",2,cex=2,line=4,las=1)
                },
                key.axes = axis(4, cex.axis = 2)
)
dev.off()
#Plot the imaginary part
setEPS()
postscript("fig0804b.eps", height=8.1, width=10)
par(mar=c(4.2, 5.5, 1.5, 0))
#Plot the real part
filled.contour(0:(M-1), 0:(M-1), Uim,
                color.palette = rainbow,
                #      xlab = 't', ylab ='k',
                plot.axes={
                   axis(1,cex.axis=1.8)
                   axis(2,cex.axis=1.8)},
                plot.title={
    title(main = '(b) Imaginary Part of the DFT Unitary Matrix',
                    xlab="t", cex.lab=2, cex.main = 1.5)
                mtext("k",2,cex=2,line=4,las=1)
                },
                key.axes = axis(4, cex.axis = 2)
)
dev.off()
```

```python
#Python plot Fig. 8.4: The unitary DFT matrix
M = 200
i = complex(0,1)#construct the imaginary unit
time_freq = np.outer(np.linspace(0,M-1,M),
```

```
                        np.linspace(0,M-1,M))
#Construct the unitary DFT matrix U
U = np.exp(i*2*np.pi*time_freq/M) / np.sqrt(M)
Ure = np.real(U)# get real part of U
Uim = np.imag(U)# get imaginary part of U

# plot the real part of U
fig, ax = plt.subplots(figsize=(10, 9))
plt.contourf(np.linspace(0,M-1,M),np.linspace(0,M-1,M),
             Ure, cmap = "autumn")
plt.title("Real Part of the DFT Matrix", pad = 10)
plt.xlabel("t", labelpad = 10)
plt.ylabel("k", labelpad = 10)
plt.yticks([0,50,100,150])
# add color bar
plt.colorbar(ticks = [-0.08,-0.04,0,0.04,0.08])
plt.savefig("fig0804a.png") # save figure
plt.show()

# plot the imaginary part of U
fig, ax = plt.subplots(figsize=(10,9))
plt.contourf(np.linspace(0,M-1,M),np.linspace(0,M-1,M),
             Uim, cmap = "rainbow")
plt.title("Imaginary Part of the DFT Matrix", pad = 10)
plt.xlabel("t", labelpad = 10)
plt.ylabel("k", labelpad = 10)
plt.yticks([0,50,100,150])
# add color bar
plt.colorbar(ticks = [-0.08,-0.04,0,0.04,0.08])
plt.savefig("fig0804b.png")# save figure
plt.show()
```

The vector

$$\tilde{X} = U^H X \tag{8.25}$$

is called the DFT of the time series data vector X. Each entry in the complex vector $\tilde{X}$ corresponds to a harmonic oscillation of a fixed frequency. Thus, the DFT transform means that a time series X can be decomposed into M harmonic oscillations. Based on this idea, you can reconstruct the original time series from $\tilde{X}$ by multiplying Eq. (8.25) by U from left:

$$X = U\tilde{X}. \tag{8.26}$$

This step is called the inverse DFT, denoted by iDFT.

The computer code for an example of the DFT and iDFT of a data sequence is as follows.

```
#R code for DFT and iDFT
M = 9
ts = (1:M)^(1/2) #X is ts here
round(ts, digits = 2)
#[1]  1.00 1.41 1.73 2.00 2.24 ...
i  = complex(real = 0, imaginary = 1)
time_freq = outer(0:(M-1), 0:(M-1))
```

```
U = exp(i*2*pi*time_freq/M) / sqrt(M)
ts_dft = t(Conj(U))%*%ts #DFT of ts
ts_rec = U%*%ts_dft #Inverse DFT for reconstruction
#Verify the reconstruction
round(t(ts_rec), digits = 2)
#[1,] 1+0i 1.41+0i 1.73+0i 2+0i 2.24+0i ...
```

```
#Python code for an example of DFT and iDFT
M = 9
ts = (np.arange(1, M+1))**(1/2)
np.set_printoptions(precision=2)
print(ts)
#[1.   1.41 1.73 2.   2.24 ...
i = complex(0,1)#construct the imaginary unit
time_freq = np.outer(np.linspace(0,M-1,M),
                     np.linspace(0,M-1,M))
#Construct the unitary DFT matrix U
U = np.exp(i*2*np.pi*time_freq/M) / np.sqrt(M)
#DFT of ts
ts_dft = np.dot(U.conj().transpose(), ts)
#Inverse DFT for reconstruction
ts_rec = np.dot(U, ts_dft)
np.set_printoptions(precision=2)
print(ts_rec)
#[1.   -5.83e-15j 1.41-2.43e-15j 1.73-3.05e-16j ...
```

In signal analysis, the recovery of the original signal X from spectra $\tilde{X}$ is referred to as reconstruction or decoding. Although the inverse DFT reconstruction $U\tilde{X}$ is exact, we will see later that some noise-filtered reconstruction is only an approximation. The approximate reconstruction is often used in image compression or transmission, since only the important parts of the spectra are used and other parts are filtered out. So we may regard $U\tilde{X}$ as a reconstruction, denoted by X_{recon}.

When working on a very long data stream, say $M = 10^7$, the DFT $\tilde{X} = U^H X_t$ computing by a direction multiplication of the unitary matrix U takes too much time and computer memory resources, measured in the order of $O(M^2)$. There is a faster algorithm to do the computing for DFT, called the fast Fourier transform (FFT). The FFT algorithm commonly used today was invented by Cooley and Tukey (1965) and continues to be improved. The computational complexity of the Cooley–Tukey FFT algorithm is $O(M\log(M))$. For a data sequence of length equal to a million, data size approximately being 4 MB, the FFT's computational complexity is in the order of 6×10^6, while the direct DFT's is 10^{12}, a huge difference in terms of computing time and memory. This complexity of order 10^{12} means that even when the 128 GB (i.e., in the order of 10^{11} bytes) memory of your computer is completely exhausted, you still cannot perform the discrete Fourier transform for a data sequence of length $M = 10^6$, but the FFT computation can be performed easily by the same computer. If it takes a computer 1 second to compute the FFT for $M = 10^6$, then it will take about 50 hours to compute the corresponding DFT assuming that the computer has enough memory. In this case, FFT is 0.2 million times faster than DFT! In fact, since the 1960s, people have almost exclusively used FFT for practical data analysis. The DFT formulation is only used for mathematical proofs and the FFT result interpretation. Numerous software packages are available for various kinds of FFT algorithms. Almost all the computer languages have a FFT command. The R command for FFT is simply `fft(ts)`, and so is Python `fft(ts)` in the package `from numpy.fft import fft, ifft`.

Different FFT algorithms may use different normalization conventions in DFT and iDFT. The FFT result of one FFT algorithm may be equal to another FFT result multiplied or divided by a normalization factor, which is often $\sqrt{M}$. For example, the FFT result of the R version 3.6.3 is equal to the DFT result computed from Eq. (8.25) multiplied by $\sqrt{M}$. In the previous DFT R code example, this claim can be verified as follows:

```
round(fft(ts), digits = 2)
#[1] 19.31+0.00i   -1.55+2.73i    -1.25+1.24i
round(sqrt(M)*ts_dft, digits = 2)
#[1,] 19.31+0.00i
#[2,] -1.55+2.73i
#[3,] -1.25+1.24i
#       ......
```

The detailed FFT algorithm is beyond the scope of this book. Interested readers can find the relevant materials from numerical analysis books, e.g., Press et al. (2007).

8.2.2.3 Period and Frequency Calculation with Proper Units

Figure 8.5 shows the oscillation patterns of the real and complex parts of the first four harmonics in the DFT unitary matrix

$$U(t,k) = \frac{1}{\sqrt{M}}e^{2\pi ikt/M}, \quad k = 0,1,2,3,$$ (8.27)

for $M = 200$. Please note that the time t here takes discrete values from 0 to M. These curves are the horizontal cross sections of the color maps of Figure 8.4 for $k = 0,1,2,3$.

These curves show 0, 1, 2, and 3 cycles in the time interval $[0,M]$ with $M = 200$. When $U(k,t)$ is regarded as a function of t, it has k cycles when t goes from 0 to $M-1$. In climate application, we need to attached time units to time t and M, such as month, day, or hour. For monthly data, we have $U(k,t)$ oscillation k times in M months. Thus, the oscillation period is

$$p_k = \frac{M}{k} \text{ [month]}.$$ (8.28)

The frequency is

$$f_k = \frac{1}{p_k} \text{ [cycle/month]}.$$ (8.29)

Climate scientists frequently use the units of cycles per year as the frequency units, which can be computed as follows:

$$f_k = 12 \times \frac{1}{p_k} \text{ [cycle/year]}.$$ (8.30)

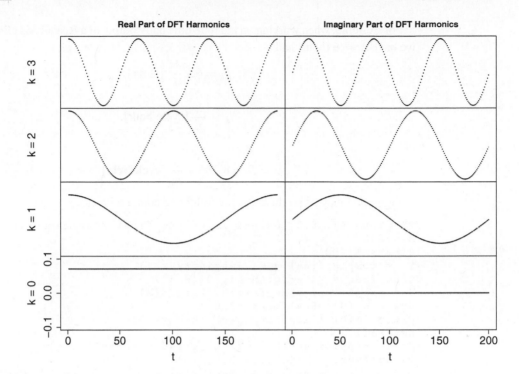

Figure 8.5 Real part and imaginary part of the first four DFT harmonics functions $\frac{1}{\sqrt{M}}e^{2\pi i k t/M}$, $k = 0, 1, 2, 3$ with $M = 200$.

The annual cycle corresponds to the period equal to $p_k = 12$ [month], which means a frequency $f_k = 1$ [cycle/year]. The semiannual cycle means $p_k = 6$ [month] and $f_k = 2$ [cycle/year].

If it is daily data, then the period unit is [day]

$$p_k = \frac{M}{k} \text{ [day]}, \tag{8.31}$$

the frequency unit is

$$f_k = \frac{1}{p_k} \text{ [cycle/day]}, \tag{8.32}$$

and the annual cycle is

$$f_k = 365.25 \times \frac{1}{p_k} \text{ [cycle/year]}. \tag{8.33}$$

Both climate science and mathematics allow noninteger frequencies.

For the general time step Δt [unit], the period and frequency can similarly be computed:

$$p_k = \frac{M}{k} \text{ [}\Delta t\text{]}, \tag{8.34}$$

$$f_k = \frac{1}{p_k} \text{ [cycle/}\Delta t\text{]}. \tag{8.35}$$

For example, when working on the four-time daily output of a Reanalysis climate model, we have $\Delta t = 6$ [hour] and

$$p_k = 6 \times \frac{M}{k} \text{ [Hour]}, \tag{8.36}$$

$$f_k = \frac{1}{p_k} \text{ [cycle/hour]}, \tag{8.37}$$

or

$$f_k = 8,766 \times \frac{1}{p_k} \text{ [cycle/year]}. \tag{8.38}$$

Figure 8.5 may be produced by the following computer code.

```
#R plot Fig. 8.5: Re and Im of the first four harmonics in DFT
M = 200
time = 1:200
i  = complex(real = 0, imaginary = 1)
time_freq = outer(0:(M-1), 0:(M-1))
U = exp(i*2*pi*time_freq/M) / sqrt(M)
Ure = Re(U) #Real part of U
Uim = Im(U) #Imaginary part of U
setEPS()
postscript("fig0805.eps", height=6, width=8)
layout(matrix(c(1,2,3,4,5,6,7,8),
              nrow = 4, ncol = 2),
       heights = c(0.92, 0.7, 0.7, 1.16,
                   0.92, 0.7, 0.7, 1.16),
       widths  = c(4, 3.3) # Widths of the 2 columns
)
par(mar=c(0,5,3,0)) #Zero space between (a) and (b)
plot(time, Ure[4,], pch =16, cex =0.3, xaxt="n",
     yaxt="n", xlab="", ylab="k=3",
     cex.axis = 1.6,  cex.lab = 1.6,
     main ='Real part of DFT harmonics')
par(mar=c(0,5,0,0))
plot(time,Ure[3,],pch =16, cex =0.3, xaxt="n",
     yaxt="n", xlab="", ylab="k=2",
     cex.axis = 1.6,  cex.lab = 1.6)
par(mar=c(0,5,0,0))
plot(time,Ure[2,], pch =16, cex =0.3, xaxt="n",
     yaxt="n", ylim = c(-0.1, 0.1),
     xlab="", ylab="k=1",
     cex.axis = 1.6, cex.lab = 1.6)
par(mar=c(6,5,0,0))
plot(time,Ure[1,], pch =16, cex =0.3,
     xaxt="n", yaxt="n",
     ylim = c(-0.1, 0.1),
     xlab="t", ylab="k=0",
     cex.axis = 1.6,  cex.lab = 1.6)
axis(1, at = c(0, 50, 100, 150), cex.axis=1.6)
axis(2, at = c(-0.1, 0, 0.1), cex.axis=1.6)

par(mar=c(0,0,3,1)) #Zero space between (a) and (b)
plot(time,Uim[4,], pch = 16, cex =0.3, xaxt="n",
     yaxt="n", xlab="", ylab="",
```

```
          cex.axis = 1.6,   cex.lab = 1.6,
          main ='Imaginary␣part␣of␣DFT␣harmonics')
par(mar=c(0,0,0,1))
plot(time,Uim[3,], pch = 16, cex = 0.3, xaxt="n",
     yaxt="n", xlab="", ylab="",
          cex.axis = 1.6,   cex.lab = 1.6)
par(mar=c(0,0,0,1))
plot(time,Uim[2,], pch = 16, cex = 0.3, xaxt="n",
     yaxt="n", ylim = c(-0.1, 0.1),
     xlab="", ylab="",
          cex.axis = 1.6, cex.lab = 1.6)
par(mar=c(6,0,0,1))
plot(time,Uim[1,], pch = 16, cex = 0.3,
     yaxt="n", ylim = c(-0.1, 0.1),
     xlab="t", ylab="",
          cex.axis = 1.6, cex.lab = 1.6)
dev.off()
```

```
#Python plot Fig. 8.5: The first four DFT harmonics
M = 200
i = complex(0,1)#construct the imaginary unit
time_freq = np.outer(np.linspace(0,M-1,M),
                     np.linspace(0,M-1,M))
#Construct the unitary DFT matrix U
U = np.exp(i*2*np.pi*time_freq/M) / np.sqrt(M)
Ure = np.real(U)# get real part of U
Uim = np.imag(U)# get imaginary part of U
time = np.linspace(0,M-1,M)
heights = [0.92,0.7, 0.7, 1.16,
           0.92,0.7, 0.7, 1.16]
widths = [4,3.3]

#Python plot Fig. 8.5: Continued
fig, ax = plt.subplots(4, 2, figsize=(13, 8), sharex=True)
fig.subplots_adjust(hspace=0)
fig.subplots_adjust(wspace=0)
# real k=3
ax[0,0].plot(time, Ure[3,:], 'k-')
ax[0,0].set_title("Real␣part␣of␣DFT␣harmonics",
                  pad=10, size = 17)
ax[0,0].set_ylabel("k=3", labelpad = 40, size = 17)
ax[0,0].axes.yaxis.set_ticks([])
# imaginary k=3
ax[0,1].plot(time, Uim[3,:], 'k-')
ax[0,1].set_title("Imaginary␣part␣of␣DFT␣harmonics",
                  pad=10, size = 17)
ax[0,1].axes.yaxis.set_ticks([])
# real k=2
ax[1,0].plot(time, Ure[2,:], 'k-')
ax[1,0].set_ylabel("k=2", labelpad = 40, size = 17)
ax[1,0].axes.yaxis.set_ticks([])
# imaginary k=2
ax[1,1].plot(time, Uim[2,:], 'k-')
ax[1,1].axes.yaxis.set_ticks([])
```

```
# real k=1
ax[2,0].plot(time, Ure[1,:], 'k-')
ax[2,0].set_ylabel("k=1", labelpad = 40, size = 17)
ax[2,0].axes.yaxis.set_ticks([])
# imaginary k=1
ax[2,1].plot(time, Uim[1,:], 'k-')
ax[2,1].axes.yaxis.set_ticks([])
# real k=0
ax[3,0].plot(time, Ure[0,:], 'k-')
ax[3,0].set_xlabel("t", size = 17, labelpad = 10)
ax[3,0].set_ylabel("k=0", labelpad = 15, size = 17)
ax[3,0].set_ylim(-0.1,0.1)
ax[3,0].tick_params(axis = 'y',labelsize = 15,
                    labelrotation=90)
ax[3,0].tick_params(axis = 'x',labelsize = 15)
ax[3,0].set_yticks([-0.1,0,0.1])
# imaginary k=0
ax[3,1].plot(time, Uim[0,:], 'k-')
ax[3,1].set_xlabel("t", size = 17, labelpad = 10)
ax[3,1].axes.yaxis.set_ticks([])
ax[3,1].tick_params(axis = 'x',labelsize = 15)
plt.savefig("fig0805.eps") # save figure
plt.show()
```

8.2.2.4 Sample Periodogram

The magnitude squares of all components of the complex DFT spectra vector $\tilde{X}$ form the so-called *sample periodogram*, or simply periodogram, of the sample data sequence X. Namely, periodogram is the graph of the discrete function

$$|\tilde{X}(k)|^2 \ (k = 1, 2, \ldots, M). \tag{8.39}$$

The periodogram is also called *spectrum*, or *power spectra*, since the periodogram magnitude $|\tilde{X}(k)|^2$ is an indicator of energy, or variance, of the oscillation at the kth harmonics:

$$\hat{X}(k,t) = \tilde{X}(k) \times \frac{1}{\sqrt{M}} e^{2\pi i k t / M}. \tag{8.40}$$

In signal analysis, the word "power" in the "power spectra" often corresponds to energy or variance, not power in the sense of physics, where it is energy divided by time.

A large value of $|\tilde{X}(k)|^2$ indicates that the time series has significant energy or variance at frequency k. Most climate signals have an annual cycle. Thus, the corresponding periodogram should have a relatively large value at a proper number k so that the frequency k corresponds to a year. A time series can have several obvious large values as peaks in its periodogram. The frequencies k for the peak values of the periodogram of a woman's voice have larger k values than those for a man.

Figure 8.6 shows the monthly temperature time series for years 2011–20 from NCEP/N-CAR Reanalysis data over the grid box that covers Tokyo (35.67° N, 139.65° E), Japan.[1] The Reanalysis data have a spatial resolution of 2.5° latitude–longitude. The grid box that covers Tokyo is centered at 35° N, 140° E. The Tokyo temperature time series shows a clear annual cycle: There were 10 cycles from 2011 to 2020. A small semiannual cycle can also be seen, but seems not to appear in every year. In fact, there even exists a three-month cycle, which cannot be seen from Figure 8.6, but can be seen from its periodogram, which is Figure 8.7.

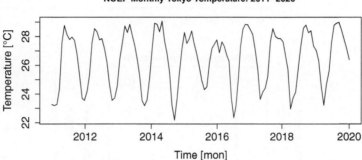

NCEP Monthly Tokyo Temperature: 2011–2020

Figure 8.6   Sample time series of the monthly surface temperature data of Tokyo, Japan, from the NCEP/NCAR Reanalysis data in the period of January 2011–December 2020.

For the periodogram of the Tokyo monthly temperature data from 2011 to 2020, we have $M = 120$ [months], and the time step equal to $\Delta t = 1$ [month]. Before applying the FFT, the mean of the Tokyo monthly temperature data is computed to be 26.43°C and is removed. Otherwise, the large mean signal can obscure the oscillatory signals that are more interesting to us. Thus, FFT is applied to the anomaly data with respect to the ten-year mean 26.43°C.

In the periodogram formula $|\tilde{X}(k)|^2$, k represents the number of cycles in the M months (i.e., 120 months or 10 years). Figure 8.7(a) is the periodogram $|\tilde{X}(k)|^2$. Although this figure is mathematically sound, its interpretation of k may be confusing. Based on the frequency and period interpretation in the previous section, the periodogram may be plotted versus cycles per year as shown in Figure 8.7(b), cycles per month as shown in Figure 8.7(c), or period [unit: month] in Figure 8.7(d). Figures 8.7(b) and (d) are often the preferred options, which clearly shows two peaks: The larger one corresponds to an annual cycle that has one cycle per year, and the smaller one corresponds to the semiannual cycle that has two cycles per year. The larger peak's period is 12 months and the smaller peak's period is 6 months, as shown in Figure 8.7(d). The annual cycle is due to the Earth's inclination angle when orbiting around the sun. The semiannual cycle is mainly due to the fact that the Sun crosses the equator twice a year. Both sea surface temperature and sea level of the Japan Sea have semiannual cycles.

---

[1] NCEP stands for the NOAA National Centers for Environmental Prediction. NCAR stands for the U.S. National Center for Atmospheric Research.

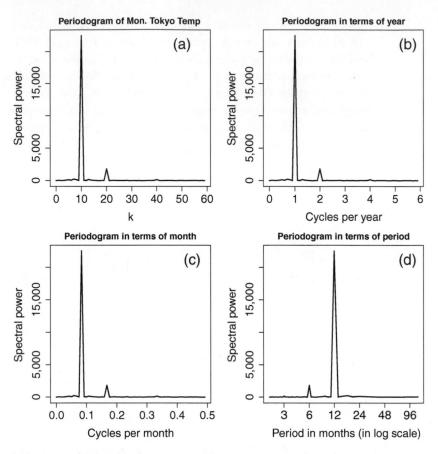

Figure 8.7 Peridogram of the monthly surface temperature data of Tokyo, Japan, from the NCEP/NCAR Reanalysis data in the period of January 2011–December 2020.

Figures 8.6 and 8.7 may be plotted by the following computer code.

```
#R plot of Figs. 8.6 and 8.7: Tokyo temperature and spectra
library(ncdf4)
nc=ncdf4::nc_open("/Users/sshen/climstats/data/air.mon.mean.nc")
Lon <- ncvar_get(nc, "lon")
Lat1 <- ncvar_get(nc, "lat")
Time <- ncvar_get(nc, "time")
library(chron)
month.day.year(1297320/24,
  c(month = 1, day = 1, year = 1800))
#1948-01-01
NcepT <- ncvar_get(nc, "air")
dim(NcepT)
#[1] 144 73 878,
#i.e., 878 months=1948-01 to 2021-02, 73 years 2 mons
#Tokyo (35.67N, 139.65E) monthly temperature data 2011-2020
Lat1[23]
#[1] 35oN
Lon[57]
#[1] 140
```

```r
m1 = month.day.year(Time/24,
      c(month = 1, day = 1, year = 1800))
m1$year[757]
#[1] 2011
m1$month[757]
#[1] 1  i.e., January
TokyoT = NcepT[56,29, 757:876] #2011-2020
M = length(TokyoT)
t1 = seq(2011, 2020, len = M)
#Plot Fig. 8.6: Tokyo temperature 2011-2020
plot(t1, TokyoT,
      type = 'l', lwd=1.5,
      xlab = "Time [mon]",
      ylab = 'Temperature [deg C]',
main = 'NCEP Monthly Tokyo Temperature: 2011-2020',
      cex.lab=1.4, cex.axis=1.4)
#Compute FFT
TokyoT_FFT = fft(TokyoT-mean(TokyoT))

#R plot Fig. 8.7: Peoriodogram of Tokyo temperature
setEPS()
postscript("fig0807.eps", height=8, width=8)
layout(matrix(c(1,2,3,4),
                nrow = 2, ncol = 2),
        heights = c(1, 1, 1, 1),
        widths  = c(1, 1) # Widths of the 2 columns
)
par(mar=c(4.5, 5, 2, 1))
kk = 0:59
plot(kk, Mod(TokyoT_FFT)[1:60]^2,
      type = 'l', lwd=2,
      xlab = 'k', ylab = 'Spectral Power',
      cex.lab = 1.5, cex.axis = 1.5,
main = 'Periodogram of Mon. Tokyo Temp')
text(50,21000, '(a)', cex = 2)

f_mon = kk/M
plot(f_mon, Mod(TokyoT_FFT)[1:60]^2,
      type = 'l', lwd=2,
      xlab = 'Cycles per month', ylab = 'Spectral Power',
      cex.lab = 1.5, cex.axis = 1.5,
      main = 'Periodogram in terms of month')
text(0.45,21000, '(c)', cex = 2)
#axis(1, at = c(0.08, 0.17, 0.33, 0.50), cex.axis=1.6)

f_year = 12*kk/M
plot(f_year, Mod(TokyoT_FFT)[1:60]^2,
      type = 'l', lwd=2,
      xlab = 'Cycles per year', ylab = 'Spectral Power',
      cex.lab = 1.5, cex.axis = 1.5,
      main = 'Periodogram in terms of year')
text(5.5,21000, '(b)', cex = 2)

tau = 120/kk[0:60]
plot(tau, Mod(TokyoT_FFT)[1:60]^2,
      log = 'x', xaxt="n",
```

```
       type = 'l', lwd=2,
       xlab = 'Period in Months (in log scale)',
       ylab = 'Spectral Power',
       cex.lab = 1.5, cex.axis = 1.5,
       main = 'Periodogram in terms of period')
text(90,21000, '(d)', cex = 2)
axis(1, at = c(3, 6, 12, 24, 48, 96), cex.axis=1.6)

dev.off()
```

```python
#Python plot Fig. 8.6: Tokyo temperature 2011-2020
import netCDF4 as nc
# read the data file from the folder named "data"
ncep ='data/air.mon.mean.nc' # file name and path
nc = nc.Dataset(ncep)
# get detailed description of dataset
print(nc)
# define variables
Lon = nc.variables['lon'][:]
Lat1 = nc.variables['lat'][:]
Time = nc.variables['time'][:]
NcepT = nc.variables['air'][:]
# 1684 months from Jan 1880-June 2019
print(np.shape(NcepT))
print(Lat1[22])
print(Lon[56])

# change dates from Julian to standard
m1 = []
for t in ((Time/24).filled()):
    day = t
    start = datetime.date(1800,1,1)
    delta = datetime.timedelta(t)
    offset = start + delta
    m = offset.strftime('%Y-%m-%d')
    m1.append(m)
print(m1[756][0:4]) # print out year
print(m1[756][5:7]) # print out month
# 2011, 01
# crop NcepT for years 2011-2020
TokyoT = NcepT[756:876, 28, 55]
M = len(TokyoT)
t1 = np.linspace(2011, 2020, M)

#plotting
fig, ax = plt.subplots()
ax.plot(t1, TokyoT, 'k-', linewidth = 2)
ax.set_title("NCEP Monthly Tokyo Temperature: 2011-2020",
            pad = 15)
ax.set_xlabel("Time [month]", labelpad = 15)
ax.set_ylabel("Temperature [$\degree$C]", labelpad = 15)
ax.set_xticks([2012, 2014, 2016, 2018, 2020])
ax.set_yticks([22,24,26,28])
# save figure
```

```
plt.savefig("fig0806.eps")
plt.show()
```

```
#Python plot Fig. 8.7: Periodogram of Tokyo temperature
from numpy.fft import fft, ifft
Tokyo_FFT = np.fft.fft(TokyoT - np.mean(TokyoT))
kk = np.linspace(0,59,60)
f_mon = kk/M
f_year = 12*kk/M
tau = 120/kk
Mod_squared = ((np.sqrt(np.real(Tokyo_FFT)**2 +
                np.imag(Tokyo_FFT)**2))[0:60]**2)/100
#plotting
fig, ax = plt.subplots(2, 2, figsize = (13,12))
fig.subplots_adjust(hspace=0.35)
fig.subplots_adjust(wspace=0.35)
# top left
ax[0,0].plot(kk, Mod_squared, 'k-')
ax[0,0].set_title("Periodogram of monthly Tokyo temp",
                pad = 10, size = 15)
ax[0,0].set_xlabel("k", size = 15)
ax[0,0].set_ylabel("Spectral Power", labelpad = 10,
                size = 15)
ax[0,0].tick_params(labelsize = 15)
# top right
ax[0,1].plot(f_year, Mod_squared, 'k-')
ax[0,1].set_title("Periodogram in terms of year",
                pad = 10, size = 15)
ax[0,1].set_xlabel("Cycles per year",
                size = 15, labelpad = 10)
ax[0,1].set_ylabel("Spectral Power",
                labelpad = 10, size = 15)
ax[0,1].tick_params(labelsize = 15)
# bottom left
ax[1,0].plot(f_mon, Mod_squared, 'k-')
ax[1,0].set_title("Periodogram in terms of month",
                pad = 10, size = 15)
ax[1,0].set_xlabel("Cycles per month",
                size = 15, labelpad = 10)
ax[1,0].set_ylabel("Spectral Power",
                labelpad = 10, size = 15)
ax[1,0].tick_params(labelsize = 15)
# bottom right
ax[1,1].plot(tau, Mod_squared, 'k-')
ax[1,1].set_title("Periodogram in terms of period",
                pad = 10, size = 15)
ax[1,1].set_xlabel("Period in Months (in log scale)",
                size = 15, labelpad = 10)
ax[1,1].set_ylabel("Spectral Power",
                labelpad = 10, size = 15)
ax[1,1].tick_params(labelsize = 15)
```

```
plt.savefig("fig0807.eps") # save figure
plt.show()
```

8.2.3 Energy Identity

DFT has the following identity:

$$|X|^2 = |\tilde{X}|^2, \tag{8.41}$$

i.e., the square of the Euclidean norm of a time series is equal to that of its DFT. This is Parseval's identity or Parseval's theorem, after French mathematician Marc-Antoine Parseval (1755–1836). You can easily verify the identify by the following computer code.

```
#R code to verify Parseval's identity
M = 8
X = rnorm(M) #Time series data
DFT_X = fft(X)/sqrt(M) #Compute DFT using FFT
t(X)%*%X - sum(Mod(DFT_X)^2)
#[1,] 2.220446e-15 #Approximately zero
```

```
#Python code to verify Parseval's identity
import numpy as np
from numpy.fft import fft, ifft
M = 8
X = np.random.normal(0, 1, M)#Time series data
DFT_X = fft(X)/np.sqrt(M) #Compute DFT using FFT
np.linalg.norm(X) - np.linalg.norm(DFT_X )
#4.440892098500626e-16
```

If X is a sample anomaly data vector of length M, then $|X|^2/M$ is the variance of X. Variance of data often measures the energy of a system corresponding to the data. Thus, Parseval's identity is also called energy identity in engineering, but it is more appropriate to identify it as variance identity in climate science, which is more interested in variance.

The identity shows that the same variance $|X|^2/M$ may be distributed in different frequencies in a variety of ways:

$$\frac{|X|^2}{M} = \frac{1}{M}\left(|\tilde{X}(1)|^2 + |\tilde{X}(2)|^2 + \cdots + |\tilde{X}(M)|^2\right). \tag{8.42}$$

The high peaks of the periodogram and Parseval's theorem imply that the system's variance contributions in a few frequencies or a range of frequencies are significantly more important than others. In the case of Tokyo temperature, the annual and semiannual cycles are significantly more important than other cycles. It can be computed that the sum of annual and semiannual variances is 93% of the total variance of all frequencies. The variations at higher frequencies may be due to random noise, or what is usually referred to as natural variability, which itself is of great interest. This example implies that the physical meaning of Parseval's theorem can help explain many signals in our lives, climate or others, such as voice tone, color, and seasonal variations of climate. For instance, the second harmonic in the Tokyo spectrum can be traced to the semiannual harmonic in the sunlight driver.

Given the same variance, Eq. (8.42) implies that one system may have more variance at a lower frequency $|\tilde{X}(1)|^2$, and another more variance at a higher frequency $|\tilde{X}(6)|^2$. Thus, if one increases the annual cycle by an amount of variance E and another increases the diurnal cycle by the same amount of variance E, they result in the same total variance increase. However, the two resulting systems have different changes, one with an enhanced annual cycle, and another with an enhanced daily cycle. Using hourly data, one may explore how the climate change is manifested in the climate variability at daily, annual, and other time scales. See the climate change application examples of Parseval's identity in Dillon et al. (2016).

The proof of Parseval's theorem is very simple. The iDFT formula (8.26) implies that

$$
\begin{aligned}
|X|^2 &= |U^H\tilde{X}|^2 \\
&= (U^H\tilde{X})^H(U^H\tilde{X}) \\
&= \tilde{X}^H(UU^H)\tilde{X} \\
&= \tilde{X}^H\tilde{X} \\
&= |\tilde{X}|^2.
\end{aligned} \tag{8.43}
$$

8.2.4 Periodogram of White Noise

The periodogram of white noise, denoted by $W_t \sim N(0,\sigma^2)$, $t = 1,2,\ldots,M$, is a constant. This statement can be easily derived following the definition of white noise and that of DFT (8.25):

$$
\begin{aligned}
\mathrm{E}[\tilde{X}\tilde{X}^H] &= \mathrm{E}\left[(U^HX)(U^HX)^H\right] \quad \text{(due to the DFT definition)} \\
&- U^H\mathrm{E}\left[XX^H\right]U \\
&= U^H\sigma^2IU \quad \text{(due to the definition of white noise)} \\
&= \sigma^2I, \tag{8.44}
\end{aligned}
$$

i.e.,

$$
\mathrm{E}[\tilde{X}(k)\tilde{X}(l)] = \sigma^2\delta_{kl}, \tag{8.45}
$$

where δ_{kl} is the Kronecker delta. This expression implies that the spectra of different frequencies (when $k \neq l$) of white noise are uncorrelated. The expected value of the periodogram (when $k = l$) is a constant σ^2, i.e.,

$$
\mathrm{E}\left[|\tilde{X}(k)|^2\right] = \sigma^2, \ k = 1,2,\ldots,M. \tag{8.46}
$$

Hence, the spectra of white noise are flat.

8.3 Fourier Transform in $(-\infty, \infty)$

The DFT defined on time t at time steps $1, 2, \ldots, M$ can be extended to a continuous function $x(t)$ with t in $(-\infty, \infty)$. This is the so-called Fourier transform defined as

$$\tilde{x} = \frac{1}{\sqrt{2\pi}} \int_{-\infty}^{\infty} x(t) e^{2\pi i f t} \, dt. \tag{8.47}$$

The symbol $\mathrm{FT}(x)$ is also used to denote $\tilde{x}$.

The inverse Fourier transform is

$$x(t) = \frac{1}{\sqrt{2\pi}} \int_{-\infty}^{\infty} \tilde{x}(t) e^{-2\pi i f t} \, df. \tag{8.48}$$

The energy identity of the FT in $(-\infty, \infty)$ is

$$\int_{-\infty}^{\infty} |x(t)|^2 \, dt = \int_{-\infty}^{\infty} |\tilde{x}(f)|^2 \, df. \tag{8.49}$$

The Fourier transform in $(-\infty, \infty)$ has an important property:

$$\mathrm{FT}\left(\frac{dx}{dt}\right) = -2\pi i f \, \mathrm{FT}(x), \tag{8.50}$$

which converts a derivative in the time domain into a complex function in the frequency domain. Thus, linear differential equations may be solved in an algebraic way. This method turns out to be very useful in electrical engineering and all the fields, such as climate science, that use linear differential equation models.

Example 8.1   Calculate the power spectral density (PSD) function of the following damped Brownian motion equation:

$$\frac{dx}{dt} + \frac{x}{\tau_0} = Z(t), \tag{8.51}$$

where τ_0 is the characteristic time scale, $Z(t) \sim N(0, \sigma^2)$ is white noise, t is time in $(-\infty, \infty)$, and x is the random variable in the damped Brownian motion. This models a continuous AR(1) process.

The Fourier transform of this differential equation yields

$$-2\pi i f \tilde{x} + \frac{\tilde{x}}{\tau_0} = \tilde{Z}. \tag{8.52}$$

This leads to

$$\tilde{x} = \frac{\tau_0 \tilde{Z}(f)}{1 - 2\pi i \tau_0 f}, \tag{8.53}$$

where

$$\tilde{Z} = \mathrm{FT}(Z). \tag{8.54}$$

The corresponding PSD for the continuous AR(1) process is

$$E\left(|\tilde{x}(f)|^2\right) = \frac{\tau_0^2 \sigma^2}{1 + 4\pi^2 \tau_0^2 f^2}. \tag{8.55}$$

8.4 Fourier Series for a Continuous Time Series on a Finite Time Interval $[-T/2, T/2]$

The SVD expression of a space-time data matrix has a counterpart for a piecewise continuous function $g(x,t)$:

$$g(x,t) = \sum_k g_k \psi_k(x) G_k(t), \tag{8.56}$$

where g_k corresponds to eigenvalue, $\psi_k(x)$ to spatial eigenvector, and $G_k(t)$ to the temporal eigenvector. A special case of this is that x has only one point, and the function $g(x,t)$ decays to a continuous time series $x(t)$, which can have the following expansion:

$$x(t) = \sum_{k=-\infty}^{\infty} \tilde{x}(k) \frac{1}{\sqrt{T}} e^{i2\pi kt/T}. \tag{8.57}$$

This infinite series is called the Fourier series over the time interval $[-T/2, T/2]$. The quantities $\tilde{x}(k)$ are called Fourier coefficients:

$$\tilde{x}(k) = \int_{-T/2}^{T/2} x(t) \frac{1}{\sqrt{T}} e^{i2\pi kt/T} \, \mathrm{d}t. \tag{8.58}$$

Although

$$G_k(t) = \frac{1}{\sqrt{T}} e^{i2\pi kt/T} \tag{8.59}$$

are orthonormal

$$\int_{-T/2}^{T/2} G_k(t) G_l^*(t) \mathrm{d}t = \delta_{kl}, \tag{8.60}$$

they are prescribed in advance, in contrast to SVD where the temporal eigenvectors $G_k(t)$ are determined by a space-time data matrix $g(x,t)$.

The infinite sequence

$$|\tilde{x}(k)|^2 \ (k = 0, \pm 1, \pm 2, \ldots)$$

is called the power spectrum of $x(t)$. This infinite sequence is in contrast to the finite periodogram

$$|\tilde{x}(k)|^2 \ (k = 0, 1, 2, \ldots, M-1)$$

of DFT for a finite discrete time series.

Similar to the energy identity of DFT for a finite discrete time series, the spectra of the continuous $x(t)$ also have an energy identity which is

$$\sum_{k=-\infty}^{\infty} |\tilde{x}(k)|^2 = \int_{-T/2}^{T/2} x^2(t)\, dt. \tag{8.61}$$

If $x(t)$ are anomalies of a climate variable, then the right-hand side of Eq. (8.61) is the variance of the sample variable, and hence the sum of the power spectra is equal to the variance of the anomalies.

Because $e^{i2\pi kt/T}$ are periodic functions in $t \in (-\infty, \infty)$, the Fourier expansion (8.57) is by default a periodic function over $(-\infty, \infty)$ with period equal to T. In practice, $x(t)$ may not be a continuous function. If there is a jump discontinuity, the series (8.57) converges to the midpoint of the jump, i.e., $(x(c_-) + x(c_+))/2$, where c is the discontinuity point and $x(c_-)$ is the x value on the left side of c, and $x(c_+)$ the x value on the right side of c. This statement is intuitively reasonable because sine and cosine are continuous functions, and it can be rigorously proved. The proof and the other detailed Fourier series theory are beyond the scope of this book and can be found from many advanced calculus textbooks, such as Khuri (2003).

Example 8.2   The following Fourier series

$$x(t) = \sum_{k=-\infty}^{\infty} \frac{8}{i\pi(2k-1)} e^{i\pi(2k-1)t}, \quad -1 < t < 1 \tag{8.62}$$

converges to a function as follows:

$$x(t) = \begin{cases} -4, & \text{if } -1 < t < 0 \\ 4, & \text{if } 0 < t < 1 \end{cases} \tag{8.63}$$

and $x(t+2) = x(t)$ when the time domain is extended to $t \in (-\infty, \infty)$. Thus, the period is $T = 2$.

Let

$$R_K(t) = \sum_{k=-K}^{K} \frac{8}{i\pi(2k-1)} e^{i\pi(2k-1)t}, \quad -1 < t < 1 \tag{8.64}$$

be a finite sum of the Fourier series (8.62). This is called the partial sum. Figure 8.8 shows the approximations to $x(t)$ by the real part of $R_K(t)$ for $K = 3, 10$, and 100. The figure shows that the approximations are excellent in the neighborhood of $t = -0.5$ and 0.5 where the function $x(t)$ is smooth and does not have much change, and the approximation is bad at the discontinuity point $t = 0$, and the end points, which are also discontinuous points when the function is periodically extended to infinity. This observation is generally true. The approximation is the best at the differentiable points (e.g., smooth points), the next best at the continuous but nondifferentiable points (e.g., sharp cusps), and the worst at discontinuous points. At the jump discontinuous points, the approximation oscillates very fast and has large errors. This is called the Gibbs phenomenon, named after American scientist Josiah

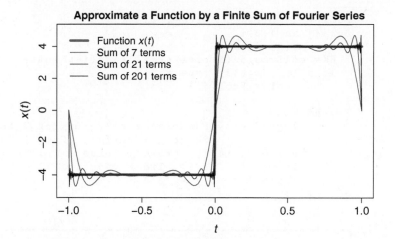

Figure 8.8   Approximation to a discontinuous function by a partial sum of Fourier series.

Willard Gibbs (1839–1903). Why the Gibbs phenomenon occurs is an important question in physics and is beyond the scope of this book.

Another question concerns the proper value K in order to have a good approximation. This depends on the smoothness of the function. If a function $x(t)$ is differentiable in the entire interval, then $|\tilde{x}(k)|$ is proportional to $1/k^3$ and hence does not need a large K to get a good approximation. We say that the partial sum $R_K(t)$ converges very fast. If $x(t)$ has some nondifferentiable points but is continuous everywhere in the interval, then $|\tilde{x}(k)|$ is proportional to $1/k^2$. Thus, the $R_K(t)$ convergence is slower. For a discontinuous function like (8.63), $|\tilde{x}(k)|$ is proportional to $1/k$, and the $R_K(t)$ convergence is very slow.

Figure 8.8 may be generated by the following computer code.

```
#R plot Fig. 8.8: Fourier series over [-1,1]
i  = complex(real = 0, imaginary = 1)
T = 2
t = seq(-T/2,T/2, len = 401)
#Define the original function x(t)
xt <- ( t >= -1 & t < 0) * (-4) +
  ( t <= 1 & t > 0) * (4)
#Plot the function x(t)
setEPS()
postscript("fig0808.eps", height=5, width=8)
par(mar = c(4, 4.5, 2, 0.5))
plot(t, xt, type = 'l',
     ylim = c(-5,5),
     xlab='t', ylab='x(t)',
main = 'Approximate a function by a finite sum of series',
     cex.lab = 1.5, cex.axis =1.4, cex.main =1.4,
     lwd = 5, col = 'red')

#Plot the partial sum of Fourier series from -K to K
J = c(3, 10, 100)
```

```
Fcol = c('brown', 'blue','black')
for (j in 1:3){
  k = -J[j]:J[j]
  RK= colSums(8/(i*pi*(2*k-1))*exp(i*pi*outer(2*k-1, t)))
  lines(t, Re(RK), type = 'l',
        col = Fcol[j])
}
legend(-1.05, 5.1,
       legend = c('Function␣x(t)', 'Sum␣of␣7␣terms',
                  'Sum␣of␣21␣terms', 'Sum␣of␣201␣terms'),
       col = c('red', 'brown', 'blue', 'black'),
       lty = c(1,1,1,1), lwd = c(5, 1,1,1),
       cex = 1.3, bty = 'n')
dev.off()
```

```
#Python plot Fig. 8.8: Partial sum of a Fourier series
i = complex(0,1)
T = 2
t = np.linspace(-T/2, T/2, 401)
# define the original function x(t)
# all (-) values = -4, all (+) values = 4
condlist = [t < 0, t > 0]
choicelist = [-4, 4]
xt = np.select(condlist, choicelist)
J = [3, 10, 100]
# plot x(t)
plt.plot(t,xt, color = 'red', linewidth = 6)
plt.title('Approximate␣x(t)␣by␣a␣finite␣sum␣of␣series',
          pad = 15)
plt.xlabel("t", labelpad = 15)
plt.ylabel("x(t)", labelpad = 10)
plt.xticks([-1.0,-0.5,0.0,0.5,1.0])
# plot the sums
Fcol = ('brown', 'blue', 'black')
for j in range(3):
    k = np.linspace(-J[j],J[j],2*J[j]+1)
    a = 8/(i*np.pi*(2*k-1))
    b = np.exp(i * np.pi * np.outer(2*k-1, t))
    RK = np.sum((a[:, np.newaxis] * b), axis=0)
    plt.plot(t, np.real(RK), color = Fcol[j], linewidth = 2)
    plt.legend(['Function␣x(t)','Sum␣of␣7␣terms',
     'Sum␣of␣21␣terms', 'Sum␣of␣201␣terms'], fontsize = 20)

plt.savefig("fig0808.eps") # save the figure
```

8.5 Chapter Summary

Spectral analysis is a kind of visualization of climate data or climate dynamics in an abstract way. From the point of view of waves, spectral analysis treats climate data from the perspective of light, sound, or music. You may say that the spectral analysis treats climate data as music notes and makes the climate data time series "sing" so that scientists can "hear" and detect climate signals (Lau et al. 1996). The essential components in the spectral analysis are period or frequency, amplitude, phase, and energy, as well as their variation in space and time. The spectral properties refer various kinds of relationships among these components.

This chapter has discussed the following materials:

(i) A mathematical expression of the simplest oscillation: A sine wave described in Eq. (8.1) may be regarded as the simplest periodic motion:

$$T(t) = A\sin(2\pi t/\tau + \phi). \tag{8.65}$$

It has amplitude A, period τ, and phase ϕ.

(ii) Discrete sine transform (DST): Treat the data X of a time series as a vector. Use sine function to make an orthogonal matrix Φ. Then, the matrix multiplication $S = \Phi^t X$ is the DST. We may regard S as music notes, and X as the motions of piano keys that play the music. Thus, producing DST S is like writing music notes, and producing the inverse iDST $X = \Phi S$ is like playing the corresponding music according to the music notes.

(iii) Discrete Fourier transform (DFT): DST can be extended to include both sine and cosine functions, or the complex function $e^{it} = \cos t + i\sin t$ where $i = \sqrt{-1}$. A unitary matrix U is made of the function e^{it} and is used to multiply the data vector X to make DFT.

(iv) Fast Fourier transform (FFT) and big data: Matrix multiplication is computationally expensive, requiring too much computer memory for big data. FFT is an algorithm to compute DFT and reduces the DFT computational complexity from $O(M^2)$ to $O(M\log(M))$, e.g., from $O(10^{12})$ to $O(6 \times 10^6)$, a difference 0.2 million times faster!

(v) A periodogram with the proper unit in its horizontal coordinate: For real-time series data, the time unit, in hours or days, must be properly considered. See Figure 8.7, its formulas, and its computer codes as examples for your own data. It is hard to remember the theory, but the code for this example can be helpful for you to generate one or two periodograms that can be used to explain climate cycles in terms of frequency or period.

(vi) Energy identity further shows the equivalence of a song and its corresponding music notes. Namely, it displays climate variations in time and its equivalent expression in terms of periodicity and amplitude.

(vii) DFT can be extended to continuous periodic functions in a finite interval $[-T/2, T/T]$ or a nonperiodic function in an infinite interval $(-\infty, \infty)$. These are called Fourier series or Fourier transform, correspondingly.

Computer codes and examples based on real climate data in this chapter may help you make spectral analysis on your own time series data. The analysis may make your data "sing"! Thus, spectral analysis allows you to explore your time domain climate data in the spectral space. In this way, you may better quantify many properties of climate variation, such as annual cycles, diurnal cycles, and the El Niño Southern Oscillation.

Note that we have only considered nonrandom functions on a discrete time domain with the exception of the white noise example, which used expectation values. In climate data analysis, we often estimate the spectrum for supposedly an infinitely long time series from the finite sample data. We play as if the finite time series were a realization from a population of infinitely many realizations. We need to bear in mind that the (sample) periodogram is computed from a finite data segment and is only an approximation to the spectrum from the ideal infinitely long stochastic time series. You can simulate this using short time series data of white noise and their sample periodogram. The ideal spectrum is flat, but the sample periodogram is not flat. Similar experiments can be done for AR(1) processes.

Modern statistical research over many decades has led us to numerous ways you can treat the data and obtain more accurate estimates of the true spectrum. Modern texts have covered these methods and there are many codes available from R or Python to assist. Our book does not exhaust all these methods.

References and Further Reading

[1] J. W. Cooley and J. W. Tukey, 1965: An algorithm for the machine calculation of complex Fourier series. *Mathematics of Computation*, 19, 297–301.

> This seminal work established the most commonly used FFT algorithm, known as the Cooley–Tukey algorithm. The main idea is to break a large DFT into many smaller DFTs that can be handled by a computer.

[2] M. E. Dillon, H. A. Woods, G. Wang et al., 2016: Life in the frequency domain: The biological impacts of changes in climate variability at multiple time scales. *Integrative and Comparative Biology*, 56, 14–30.

> This paper shows application examples of Parseval's identity in climate change studies, exploring climate changes at different frequencies using three 3-hourly data from GLDAS Version 2.0 data sets released by NASA GES DISC. R code is provided in the supplementary material of the paper.

[3] A. Khuri, 2003: *Advanced Calculus with Applications in Statistics*. 2nd ed., Wiley-Interscience.

> This book has a chapter devoted to the Fourier series and its statistics applications. It has rigorous proofs of convergence, continuity, and differentiability.

[4] K. M. Lau and H. Weng, 1996: Climate signal detection using wavelet transform: How to make a time series sing. *Bulletin of the American Meteorological Society*, 76, 2391–2402.

> This beautifully written paper has an attractive title showing the core spirit of spectral analysis: make time series data sing. In their spectral analysis, instead of sine and cosine functions, they used wavelet functions in their transformation matrices. Wavelets are an efficient way to treat the data when frequencies change with time.

[5] W. H. Press, H. William, S. A. Teukolsky et al., 2007: *Numerical Recipes: The Art of Scientific Computing*. 3rd ed., Cambridge University Press.

This is the most popular numerical analysis tool book beginning in the 1980s. It contains many kinds of numerical algorithms, including FFT, used in engineering and science.

[6] R. H. Shumway, D. S. Stoffer, 2016: *Time Series Analysis and Its Applications With R Examples*. 4th ed., Springer.

The R code and examples in chapter 4 of this book are helpful references.

[7] F. Zalkow and M. Müller, 2022: *Discrete Fourier Transform (DFT)*:

`www.audiolabs-erlangen.de/resources/MIR/FMP/C2/C2_DFT-FFT.html`

Access date: April 19, 2022.

This is the website of resources for the book by M. Müller entitled *Fundamentals of Music Processing Using Python and Jupyter Notebooks*, 2nd ed., Springer, 2021. The website is hosted at the International Audio Labs, Erlangen, Germany.

Exercises

8.1 Use a computer to plot six cosine functions, each of which has a different amplitude A, period τ, and phase ϕ. You can reference Figure 8.1 for this problem.

8.2 Plot the sum of the six cosine functions in the previous problem. You can reference Figure 8.2 for this problem.

8.3 (a) For $M = 10$ and on an $M \times M$ grid, make a pixel-color plot for the discrete sine transform (DST) matrix Φ defined by Eq. (8.4). You can reference Figure 8.4 for this problem.

(b) Make a similar plot for $M = 30$, and compare the two plots. *Hint: You may use the R command* `image(x, y, MatrixData)`, *or Python command* `ax.imshow(MatrixData)`.

8.4 This exercise is to test the idea of using DST to filter out noise. Because of the limitation of computer memory, the DST method can only be tested for a short sequence of data. You may choose $M = 51$ or smaller.

(a) Regard $y_s = 10\sin(t)$ as signal and $y_n \sim N(0, 3^2)$ as noise. Plot the signal y_s, noise y_n, and signal plus noise $y_s + y_n$ on the same figure for $0 \le t \le 10$. See Figure 8.9 as a reference.

(b) Make the DST of the data of signal plus noise $y_s + y_n$.

(c) Identify the DST component with the largest absolute value, and replace all the other components by zero. This is the filtering step, i.e., filtering out certain harmonics.

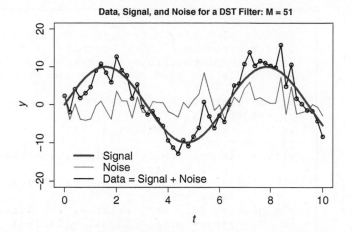

Figure 8.9 Data, signal, and noise for a discrete sine transform (DST) filter: $M = 51$ for 50 time steps in the time interval $[0, 10]$. The small black circles indicate data at discrete time t_m ($m = 1, 2, \ldots, M$) in the continuous time interval $0 \le t \le 10$.

(d) Make a reconstruction based on this modified DST vector with only one nonzero component. Plot this reconstruction and compare it with the signal in Step (b).

(e) Replace the DST component with the largest absolute value by zero, and make a reconstruction based on this modified DST vector. Plot this reconstruction and compare it with the noise in Step (b).

Figure 8.9 can be plotted by the following computer code.

```
#R plot Fig. 8.9: Data = Signal + Noise for Exercise 8.4
setwd('~/climstats')
setEPS() #Automatically saves the .eps file
postscript("fig0809.eps", height=5, width=7)
par(mar = c(4.5, 4, 2.5, 0.2))
seed(101)
M = 51
t = seq(0, 10, len = M)
ys = 10*sin(t)
yn = rnorm(M, 0, 3)
yd = ys + yn
plot(t, yd, type = 'o', lwd = 2,
     ylim = c(-20, 20),
     xlab = 't', ylab = 'y',
     main = 'Data, signal, and noise for a DST filter',
     cex.lab = 1.4, cex.axis = 1.4)
legend(0, -16, 'Data = Signal + Noise',
       lty = 1, bty = 'n', lwd = 2, cex = 1.4)
lines(t, ys, col = 'blue', lwd = 4)
legend(0, -10, 'Signal', cex = 1.4,
       lty = 1, bty = 'n', lwd = 4, col = 'blue')
lines(t, yn, col = 'brown')
legend(0, -13, 'Noise', cex = 1.4,
       lty = 1, bty = 'n', col = 'brown')
dev.off()
```

```
#Python plot Fig. 8.9: Data = Signal + Noise
from numpy.random import normal
fig, ax = plt.subplots(figsize = (12,7))
seed(102)# set seed to ensure the same simulation result
M = 51
t = np.linspace(0, 10, M)
ys = 10*np.sin(t)
yn = np.random.normal(0, 3, M)
yd = ys + yn
plt.plot(t, ys, color = 'blue', linewidth = 5,
         label = 'Signal')
plt.plot(t, yn, color = 'brown',linewidth = 1.5,
         label ='Noise')
plt.plot(t, yd, 'o-', color = 'k', linewidth = 3,
         label='Data = Signal + Noise')
plt.title('Data, signal, and noise for a DST filter')
plt.ylim(-20, 20)
plt.xlabel('t', fontsize = 20)
plt.ylabel('y', fontsize = 20)
plt.legend(loc='lower left', fontsize = 20)
plt.savefig("fig0809.eps") # save the figure
plt.show()
```

8.5 Use complex number operations to prove the following:

$$\left|1 - \lambda e^{-2\pi i f}\right|^2 = 1 - 2\lambda \cos(2\pi f) + \lambda^2. \tag{8.66}$$

8.6 For the following AR(1) process

$$X_t = \lambda X_{t-1} + W_t, \quad t = 0, \pm 1, \pm 2, \pm 3, \ldots, \tag{8.67}$$

where λ is a constant and $W_t \sim N(0, \sigma^2)$ are white noises with the same variance σ^2 for every time step t, show that the lag-m auto-covariances are

$$\gamma_m = \text{Cov}(X_t X_{t+m}) = \frac{\sigma^2 \lambda^m}{1 - \lambda^2}, \quad \text{when } |\lambda| < 1. \tag{8.68}$$

8.7 Prove the following equality:

$$\sigma^2 \sum_{m=-\infty}^{\infty} \frac{\lambda^m}{1 - \lambda^2} e^{-2\pi i m f} = \frac{\sigma^2}{\left|1 - \lambda e^{-2\pi i f}\right|^2} \quad \text{for } |\lambda| < 1 \tag{8.69}$$

This derivation is related to the power spectral density (PSD) for a discrete random process.

8.8 Show that

$$U^H U = I \tag{8.70}$$

is an identity matrix if U is defined by Eq. (8.20).

8.9 Plot the real part and the imaginary part of U defined by Eq. (8.20) on a grid similar to Figure 8.4 for $M = 30$. You can reference the computer code for Figure 8.4 for this problem.

8.10 Make a pixel-color plot for the real part and the imaginary part of U defined by Eq. (8.20) on a grid for $M = 30$. See the hint for Exercise 8.3 about the pixel-color plot.

8.11   For the data sequence

$$X = (\sin(0), \sin(1), \sin(2), \sin(3), \ldots, \sin(8), \sin(9), \sin(10)), \qquad (8.71)$$

use $\tilde{X} = U^H X$ to find the DFT of X, and use $U\tilde{X}$ to find the iDFT. Comment on the data X, as well as the results $\tilde{X}$ and iDFT $U\tilde{X}$.

8.12   Compute and plot the sample periodogram of the monthly mean surface air temperature Tavg of Chicago, USA, from January 1991 to December 2020. You can use the CSV data `ChicagoTavg.csv` from the book data file `data.zip` downloadable at the book website. Or you can download the data online from the Global Historical Climatology Network (GHCN) using the station ID: USC00111577. Please do the following.

(a)   Plot the Chicago data time series from January 1991 to December 2020, a figure similar to Figure 8.6.

(b)   Compute and plot the periodogram with the cycles per year as the horizontal axis (see Figure 8.7(b)).

(c)   Compute and plot the periodogram with the period in months as the horizontal axis (see Figure 8.7(d)).

(d)   Use text to make climate interpretations of these three figures in text (limited to 30–200 words).

8.13   Compute and plot the sample periodogram of the monthly mean surface air temperature Tavg of Berlin Dahlem, Germany, from January 1971 to December 2000. You can use the CSV data `BerlinGermanyTavg.csv` from the book data file `data.zip` downloadable at the book website. Or you can download the data online from the GHCN website using the station ID: GM000003319.

(a)   Plot the Berlin data time series from January 1971 to December 2000, a figure similar to Figure 8.6.

(b)   Compute and plot the periodogram with the cycles per year as the horizontal axis (see Figure 8.7(b)).

(c)   Compute and plot the periodogram with the period in months as the horizontal axis (see Figure 8.7(d)).

(d)   Use text to make climate interpretations of these three figures in text (limited to 30–200 words).

8.14   Do the same plotting and analysis for a location and a climate parameter (e.g., Tmax or precipitation) of your interest. Use a monthly climate data sequence from a data source of your choice, such as the data from a GHCN station or from a grid box of a Reanalysis climate model.

8.15   The monthly Central England (CE) surface air temperature from January 1659 is the longest thermometer-measuring record in the world. You can use the txt data `CentralEnglandT1659.txt` from the book data file `data.zip` downloadable at the book website. You can also download the data from the UK Met Office Hadley Center.

(a)   Plot the CE data time series from January 1659 to December 2021, a figure similar to Figure 8.6.

(b)   Compute and plot the periodogram with the cycles per year as the horizontal axis (see Figure 8.7(b)).

(c) Compute and plot the periodogram with the period in months as the horizontal axis (see Figure 8.7(d)).

(d) Use text to make climate interpretations of these three figures in text (limited to 30–200 words).

8.16 Verify Parseval's identity (8.41) for the monthly surface air temperature data X of the Central England from January 1659 to December 2021.

8.17 Use FFT to compute and then plot the sample periodogram of the daily mean surface air temperature Tmin of St. Paul, Minnesota, USA, from January 1, 1941 to December 31, 1949. You can use the txt data `StPaulStn.txt` from the book data file `data.zip` downloadable at the book website. Or you can download the data online from the Global Historical Climatology Network-Daily (GHCN-D) using the station ID: USW00014927.

(a) Plot the St. Paul data time series from January 1, 1941 to December 31, 1949, a figure similar to Figure 8.6.

(b) Plot the periodogram with the cycles per year as the horizontal axis (see Figure 8.7(b)).

(c) Plot the periodogram with the period in days as the horizontal axis (see Figure 8.7(d)).

(d) Use text to make climate interpretations of these three figures in text. Please comment on the seasonality of Tmin.

8.18 (a) Isolate the annual component from the FFT analysis of the St. Paul Tmin data from January 1, 1941 to December 31, 1949 in the previous problem. Here, the annual component refers the reconstructed FFT component as a function of time with a period approximately equal to 365 days. Plot the annual component as a function of time from January 1, 1941 to December 31, 1949.

(b) Compute and plot the monthly climatology of this dataset based on the 1941–1949 mean for each month in January, February, ..., December. You may compute the monthly data first, and then compute the ten-year mean. You can exclude February 29 from your computing. However, you can also write your code in such a way that the monthly climatology is computed directly from the daily data.

(c) Compare the figures from Steps (a) and (b) and comment on the annual cycles defined in different ways.

8.19 (a) Compute and plot the daily anomalies of the St. Paul Tmin from January 1, 1941 to December 31, 1949 based on the monthly climatology from the previous problem.

(b) Compute and plot the periodogram of the daily anomalies with the period as the horizontal axis (see Figure 8.7(d)). The period units are in days.

8.20 Evaluate the following integral:

$$\int_0^\infty \frac{\tau_0^2 \sigma^2}{1 + 4\pi^2 \tau_0^2 f^2} df, \tag{8.72}$$

where the integrand is the PSD function defined by (8.55). If the integral is regarded as the total power, explain how the total power depends on the time scale τ_0 and variance σ^2.

8.21 Plot the power spectral density (PSD) as a function of $f \in (-\infty, \infty)$ for three pairs of different values of τ_0 and σ^2:

$$p(f; \tau_0, \sigma^2) = \frac{\tau_0^2 \sigma^2}{1 + 4\pi^2 \tau_0^2 f^2}. \tag{8.73}$$

Plot all three curves on the same figure, and explain their differences.

8.22 For the following Fourier series

$$|\sin t| = \frac{2}{\pi} - \frac{4}{\pi} \sum_{k=1}^{\infty} \frac{1}{4k^2 - 1} \cos(2kt), \tag{8.74}$$

plot the function $f(t) = |\sin t|$ and the partial sums

$$R_K(t) = \frac{2}{\pi} - \frac{4}{\pi} \sum_{k=1}^{K} \frac{1}{4k^2 - 1} \cos(2kt) \tag{8.75}$$

for $K = 1, 2, 9$ over the domain $t \in [-2\pi, 2\pi]$ on the same figure. Comment on the accuracy of using the partial sum of a Fourier series to approximate a function.

8.23 (a) For the previous problem, plot $R_3(t)$. This is the result of removing the high-frequency components, i.e., filtering out all the frequencies $k \geq 4$. This procedure is a low-pass filter.

   (b) For the previous problem, plot $f(t) - R_3(t)$. This is the result of removing the low-frequency components, i.e., filtering out all the frequencies $k \leq 3$. This procedure is a high-pass filter.

8.24 Apply a low-pass filter and a high-pass filter to the monthly mean surface air temperature Tavg of Chicago, USA, from January 1971 to December 2020. You can use the CSV data `ChicagoTavg.csv` from the book data file `data.zip` downloadable at the book website. Or you can download the data online from the Global Historical Climatology Network (GHCN) using the station ID: USC00111577. Plot the data and the filtered results. You can choose your preferred cutoff frequency for your filters.

8.25 (a) Approximate the annual cycle of Chicago surface air temperature Tavg by its monthly climatology as the mean from 1971 to 2020. Plot the monthly climatology.

   (b) Approximate the annual cycle of Chicago surface air temperature Tavg by the harmonic component of period equal to 12 months. Plot this harmonic on the same figure as the monthly climatology in Step (a).

   (c) Compare the two curves and describe their differences in text (limited to 20–80 words).

8.26 A periodic function in $[0, \pi]$

$$f(t) = t(\pi - t) \text{ when } t \in [0, \pi] \tag{8.76}$$

is extended to $(-\infty, \infty)$ periodically with a period equal to π. This periodic function has the following Fourier series:

$$f(t) = \frac{8}{\pi} \sum_{k=1}^{\infty} \frac{\sin(2k-1)t}{(2k-1)^3}. \tag{8.77}$$

Plot the periodic function $f(t)$ and the partial sums

$$R_K(t) = \frac{8}{\pi} \sum_{k=1}^{K} \frac{\sin(2k-1)t}{(2k-1)^3} \tag{8.78}$$

for $K = 1, 2, 9$ over the domain $t \in [-2\pi, 2\pi]$ on the same figure.

8.27   (a)   A square-wave periodic signal with period equal to 2π is defined as follows:

$$f(t) = \begin{cases} 0 & \text{when} \quad t \in (-\pi, -\pi/4) \\ 1 & \text{when} \quad t \in [-\pi/4, \pi/4] \\ 0 & \text{when} \quad t \in (\pi/4, \pi]. \end{cases} \tag{8.79}$$

This periodic function is extended from $(-\pi, \pi]$ to $(-\infty, \infty)$ with a period equal to 2π. The Fourier series of this function is given as

$$f(t) = \frac{1}{4} + \frac{2}{\pi} \sum_{k=1}^{\infty} \frac{1}{k} \sin\left(\frac{k\pi}{4}\right) \cos(kt). \tag{8.80}$$

Plot the function $f(t)$ and the partial sums

$$R_K(t) = \frac{1}{4} + \frac{2}{\pi} \sum_{k=1}^{K} \frac{1}{k} \sin\left(\frac{k\pi}{4}\right) \cos(kt) \tag{8.81}$$

for $K = 3, 9, 201$ over the domain $t \in (-3\pi, 3\pi]$ on the same figure.

(b)   Compare the coefficients of the Fourier series and note the convergence rate of the series. A smoother function has a faster convergence $1/k^3$, and a discontinuous function has a very slow convergence rate $1/k$. Discuss your numerical results of convergence.

9   Introduction to Machine Learning

Machine learning (ML) is a branch of science that uses data and algorithms to mimic how human beings learn. The accuracy of ML results can be gradually improved based on new training data and algorithm updates. For example, a baby learns how to pick an orange from a fruit plate containing apples, bananas, and oranges. Another baby learns how to sort out different kinds of fruits from a basket into three categories without naming the fruits. Then, how does ML work? It is basically a decision process for clustering, classification, or prediction, based on the input data, decision criteria, and algorithms. It does not stop here, however. It further validates the decision results and quantifies errors. The errors and the updated data will help update the algorithms and improve the results.

Machine learning has recently become very popular in climate science due to the availability of powerful and convenient resources of computing. It has been used to predict weather and climate and to develop climate models. This chapter is a brief introduction to ML and provides basic ideas and examples. Our materials will help readers understand and improve the more complex ML algorithms used in climate science, so that they can go a step beyond only applying the ML software packages as a black box. We also provide R and Python codes for some basic ML algorithms, such as K-means for clustering, the support vector machine for the maximum separation of sets, random forest of decision trees for classification and regression, and neural network training and predictions.

Artificial intelligence (AI) allows computers to automatically learn from past data without human programming, which enables a machine to learn and to have intelligence. Machine learning is a subset of AI. Our chapter here focuses on ML, not general AI.

9.1 K-Means Clustering

A few toddlers at a daycare center may learn how to grab a few candies near them. It can be a point of fighting for a candy at a location that is not obviously closer to one than another. K-means method can help divide the candies among the toddlers in a fair way.

Based on the historical weather data over a country, can ML decide the climate regimes for the country? Can ML determine the ecoregions of a country? Can ML define the regimes of wild fire over a region? The K-means clustering method can be useful to answering these questions.

9.1.1 K-Means Setup and Trivial Examples

The aim of K-means clustering is to divide N points into K clusters so that the total within cluster sum of squares (tWCSS) is minimized. Here we use 2D data points to describe tWCSS and the K-means algorithm, although the K-means method can be formulated in higher dimensions. We regard the data $(\mathbf{x}_1, \mathbf{x}_2, \ldots, \mathbf{x}_N)$ as the 2D coordinates of N points. For example, we treat $\mathbf{x}_1 = (1.2, -7.3)$ as observational data. Assume that these N points can be divided into K clusters $(C_1, C_2, \ldots, C_K)$, where K is subjectively given by you, the user who wishes to divide the N points into K clusters by the K-means method. Let $\mathbf{x} \in C_i$, i.e., points within cluster C_i. The number of points in cluster C_i is unknown and is to be determined by the K-means algorithm. The total WCSS is defined by the following formula:

$$\text{tWCSS} = \sum_{i=1}^{K} \left(\sum_{\mathbf{x} \in C_i} |\mathbf{x} - \boldsymbol{\mu}_i|^2 \right), \qquad (9.1)$$

where

$$\boldsymbol{\mu}_i = \frac{1}{K_i} \sum_{\mathbf{x} \in C_i} \mathbf{x} \qquad (9.2)$$

is the mean of the points within the cluster C_i, as K_i is the number of points in C_i. Thus, we have K means, the name of the K-means method. The part in the parentheses in Eq. (9.1) is called the within cluster sum of squares (WCSS)

$$\text{WCSS}_i = \left(\sum_{\mathbf{x} \in C_i} |\mathbf{x} - \boldsymbol{\mu}_i|^2 \right). \qquad (9.3)$$

This is defined for each cluster, and tWCSS is the sum of these WCSS_i for all the K clusters.

Some publications use WCSS or Tot WCSS, instead of tWCSS, to express the right-hand side of Eq. (9.1). Be careful when reading literature concerning the WCSS definition. Some computer software for K-means may even use a different tWCSS definition.

Our K-means computing algorithm is to minimize the tWCSS by optimally organizing the N points into K clusters. This algorithm can assign each data point to a cluster. If we regard $\boldsymbol{\mu}_i$ as the centroid or center of cluster C_i, we may say that the points in cluster C_i have some kind of similarity, such as similar climate or similar ecological characteristics, based on certain criteria. In the following, we use two simple cases ($N = 2$ and $N = 3$) to illustrate the K-means method and its solutions.

(i) **Case $N = 2$:**

When $N = 2$, if we assume $K = 2$, then

$$\boldsymbol{\mu}_1 = \mathbf{x}_1, \quad \boldsymbol{\mu}_2 = \mathbf{x}_2 \qquad (9.4)$$

and

$$\text{tWCSS} = \sum_{i=1}^{2} \sum_{\mathbf{x} \in C_i} |\mathbf{x} - \boldsymbol{\mu}_i|^2 = 0. \qquad (9.5)$$

This is the global minimum for $K = 2$, and

$$\mathbf{x}_1 \in C_1, \ \ \mathbf{x}_2 \in C_2, \ \ \mu_1 = \mathbf{x}_1, \ \ \mu_2 = \mathbf{x}_2 \tag{9.6}$$

is the unique solution. Of course, this is a trivial solution and bears no meaning.

When $N = 2$, if we assume $K = 1$, then

$$\mu_1 = \frac{\mathbf{x}_1 + \mathbf{x}_2}{2} \tag{9.7}$$

and

$$\text{tWCSS} = \sum_{i=1}^{1} \sum_{\mathbf{x} \in C_i} |\mathbf{x} - \mu_i|^2 = \frac{1}{2} |\mathbf{x}_1 - \mathbf{x}_2|^2. \tag{9.8}$$

This is the unique minimum of tWCSS, and the solution is also trivial.

(ii) Case $N = 3$:

When $N = 3$, if we assume $K = 2$, then there are three different cases of clustering:

$$C_1 = \{\mathbf{x}_1, \mathbf{x}_2\}, \ C_1 = \{\mathbf{x}_1, \mathbf{x}_3\}, \ C_1 = \{\mathbf{x}_2, \mathbf{x}_3\}. \tag{9.9}$$

Here, the cases

$$C_1 = \{\mathbf{x}_1, \mathbf{x}_2\}, \ C_2 = \{\mathbf{x}_3\} \tag{9.10}$$

and

$$C_2 = \{\mathbf{x}_1, \mathbf{x}_2\}, \ C_1 = \{\mathbf{x}_3\} \tag{9.11}$$

are considered the same, since their difference is only a matter of which cluster is called C_1 or C_2.

For case $C_1 = \{\mathbf{x}_1, \mathbf{x}_2\}, C_2 = \{\mathbf{x}_3\}$, we have

$$\mu_1 = \frac{1}{2}(\mathbf{x}_1 + \mathbf{x}_2), \ \ \mu_2 = \mathbf{x}_3 \tag{9.12}$$

and

$$\text{tWCSS} = \frac{1}{2}|\mathbf{x}_1 - \mathbf{x}_2|^2. \tag{9.13}$$

The tWCSS for the other two cases of clustering ($C_1 = \{\mathbf{x}_1, \mathbf{x}_3\}$, $C_1 = \{\mathbf{x}_2, \mathbf{x}_3\}$) can be computed in a similar way. Then, the smallest tWCSS of the three determines the final solution, which is one of the three cases of clustering. A numerical example is given in the following.

Example 9.1 Find the two K-means clusters for the following three points:

$$P_1(1, 1), \ \ P_2(2, 1), \ \ P_3(3, 3.5). \tag{9.14}$$

This is a K-means problem with $N = 3$ and $K = 2$. The K-means clusters can be found by a computer command, e.g., kmeans(mydata, 2) in R. Figure 9.1 shows the two K-means clusters with their centers at

$$C_1(1.5, 1), \ \ C_2(3, 3.5) \tag{9.15}$$

and

$$\text{tWCSS} = 0.5. \tag{9.16}$$

Points P_1 and P_2, indicated by red dots, are assigned to cluster C_1, whose center is marked by a red cross. Point P_3, indicated by the sky blue dot, is assigned to cluster C_2, whose

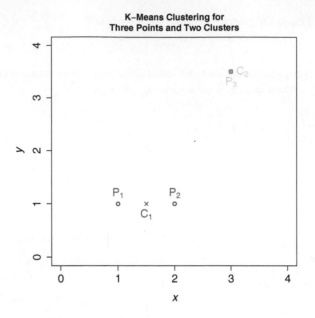

**K−Means Clustering for
Three Points and Two Clusters**

Figure 9.1   The K-means clustering results for three points ($N = 3$) and two clusters ($K = 2$).

center is marked with a blue cross and obviously overlaps with P_3. This result agrees with our intuition since P_1 and P_2 are together. The three points can have two other possibilities of combination: $C_1 = \{P_1, P_3\}$ and $C_1 = \{P_2, P_3\}$. The case $C_1 = \{P_1, P_3\}$ leads to tWCSS $=$ 5.125, and $C_1 = \{P_2, P_3\}$ to tWCSS $=$ 3.625. Both tWCSS are greater than 0.5. Thus, these two combinations should not be a solution of the K-means clustering. These numerical results may be produced by the following computer codes.

```
#R code: tWCSS calculation for N = 3 and K = 2
mydata <- matrix(c(1, 1, 2, 1, 3, 3.5),
                    nrow = N, byrow = TRUE)
x1 = mydata[1, ]
x2 = mydata[2, ]
x3 = mydata[3, ]

#Case C1 = (P1, P2)
c1 = (mydata[1, ] + mydata[2, ])/2
c2 = mydata[3, ]
tWCSS = norm(x1 - c1, type = '2')^2 +
  norm(x2 - c1, type = '2')^2 +
  norm(x3 - c2, type = '2')^2
tWCSS
#[1] 0.5

#Case C1 = (P1, P3)
c1 = (mydata[1, ] + mydata[3, ])/2
c2 = mydata[2, ]
norm(x1 - c1, type = '2')^2 +
  norm(x3 - c1, type = '2')^2 +
  norm(x2 - c2, type = '2')^2
#[1] 5.125
```

```
#Case C1 = (P2, P3)
c1 = (mydata[2, ] + mydata[3, ])/2
c2 = mydata[1, ]
norm(x2 - c1, type = '2')^2 +
  norm(x3 - c1, type = '2')^2 +
  norm(x1 - c2, type = '2')^2
#[1] 3.625

#The case C1 = (P1, P2) can be quickly found by
kmeans(mydata, 2)
#Clustering vector:
#[1] 1 1 2 #points P1, P2 in C1
```

```
#Python code: K-means computing for N = 3 and K = 2
N = 3
# create a 3 x 2 matrix of data
mydata = np.array([1,1,2,1,3,3.5]).reshape(3,2)

# first row of data
x1 = mydata[0,:]
# second row of data
x2 = mydata[1,:]
# third row of data
x3 = mydata[2,:]

# case C1 = (P1, P2)
c1 = (x1 + x2)/2
c2 = mydata[2,:]
tWCSS = np.linalg.norm(x1 - c1)**2 +\
np.linalg.norm(x2 - c1)**2 +\
        np.linalg.norm(x3 - c2)**2
print(tWCSS)
#0.5

# case C1 = (P1, P3)
c1 = (x1 + x3)/2
c2 = mydata[1,:]
print(np.linalg.norm(x1 - c1)**2 +\
np.linalg.norm(x3 - c1)**2 + \
        np.linalg.norm(x2 - c2)**2)

# case C1 = (P2, P3)
c1 = (x2 + x3)/2
c2 = mydata[0,:]
print(np.linalg.norm(x2 - c1)**2 +\
np.linalg.norm(x3 - c1)**2 + \
        np.linalg.norm(x1 - c2)**2)
#3,625

# case C1 = (P1, P2) can be quickly found by
kmeans = KMeans(n_clusters=2).fit(mydata)

print(kmeans.labels_)
#[0 0 1] #points P1, P2 in C1
```

```
#because Python index begins with 0 while R starts from 1

# number of data points
N = 3
# assume K clusters
K = 2
# create a 3 x 2 matrix of data
mydata =  np.array([1, 1, 2, 1, 3, 3.5]).reshape(3,2)

Kclusters = KMeans(n_clusters=K).fit(mydata)
# cluster means
print(Kclusters.cluster_centers_, "\n")
#[[1.5 1. ]
#[3.  3.5]]
# sum of squares by cluster
print(Kclusters.inertia_)
#0.5
```

Of course, this problem is trivial and can even be solved analytically by hand. The computer solution of the K-means clustering is necessary when more points are involved and when it is not obvious that a point belongs to a specific cluster.

Figure 9.1 may be generated by the following computer code.

```
#R plot Fig. 9.1: K-means for N = 3 and K = 2
setwd("~/climstats")
N = 3 #The number of data points
K = 2 #Assume K clusters
mydata = matrix(c(1, 1, 2, 1, 3, 3.5),
                nrow = N, byrow = TRUE)
Kclusters = kmeans(mydata, K)
Kclusters #gives the K-means results,
#e.g., cluster centers and WCSS
#Cluster means:
#[,1] [,2]
#1  1.5  1.0
#2  3.0  3.5
#Within cluster sum of squares by cluster:
#  [1] 0.5 0.0
Kclusters$centers
par(mar = c(4,4,2.5,0.5))
plot(mydata[,1], mydata[,2], lwd = 2,
    xlim =c(0, 4), ylim = c(0, 4),
    xlab = 'x', ylab = 'y', col = c(2, 2, 4),
    main = 'K-means clustering for
      three points and two clusters',
    cex.lab = 1.4, cex.axis = 1.4)
points(Kclusters$centers[,1], Kclusters$centers[,2],
      col = c(2, 4), pch = 4)
text(1.5, 0.8, bquote(C[1]), col = 'red', cex = 1.4)
text(3.2, 3.5, bquote(C[2]), col = 'skyblue', cex = 1.4)
text(1, 1.2, bquote(P[1]), cex = 1.4, col = 'red')
text(2, 1.2, bquote(P[2]), cex = 1.4, col = 'red')
text(3, 3.3, bquote(P[3]), cex = 1.4, col = 'skyblue')
```

```
#Python plot Fig. 9.1: SVM clustering for N=3 and K=2
#starting from setting up the plot
plt.figure(figsize = (8,8))
plt.ylim([0,4])
plt.yticks([0,1,2,3,4])
plt.xlim([0,4])
plt.xticks([0,1,2,3,4])
plt.title("K-means␣clustering␣for␣\n␣\
␣␣three␣points␣and␣two␣clusters", pad = 10)
plt.xlabel("x", labelpad = 10)
plt.ylabel("y", labelpad = 10)

# plot P1-P3
plt.scatter(mydata[:,0], mydata[:,1],
            color = ('r', 'r', 'dodgerblue'),
            facecolors='none', s = 80)
# plot C1 and C2
plt.scatter((Kclusters.cluster_centers_[0][0],
             Kclusters.cluster_centers_[1][0]),
            (Kclusters.cluster_centers_[0][1],
             Kclusters.cluster_centers_[1][1]),
            marker = 'X', color = ('dodgerblue', 'r'))

# add labels
plt.text(1.43, 0.8, "$C_1$", color = 'r', size = 14)
plt.text(3.1, 3.45, "$C_2$", color = 'dodgerblue', size =14)
plt.text(0.95,1.1, "$P_1$", color = 'r', size = 14)
plt.text(1.95,1.1, "$P_2$", color = 'r', size = 14)
plt.text(2.95,3.3, "$P_3$", color = 'dodgerblue', size = 14)

plt.show()
```

9.1.2 A K-Means Algorithm

From Subsection 9.1.1, we may have already developed a feeling that the K-means clustering method may roughly be divided into the following two steps:

- Assignment step: Assume a number K and randomly generate K points $\boldsymbol{\mu}_i^{(1)}$ ($i = 1, 2, \ldots, K$) as the initial K centroids for K clusters. Without replacement, for each point $\mathbf{x}$, compute the distance

$$d_i = \left| \mathbf{x} - \boldsymbol{\mu}_i^{(1)} \right| \quad i = 1, 2, \ldots, K. \tag{9.17}$$

Assign the point $\mathbf{x}$ to the cluster with the smallest distance. Then, assign the next point to a cluster until every point is assigned. Thus, the initial K clusters are formed: $C_i^{(1)}$ and each $C_i^{(1)}$ containing $K_i^{(1)}$ data points $(i = 1, 2, \ldots, K)$.

- Update step: Compute the centroids of the initial K clusters P:

$$\boldsymbol{\mu}_i^{(2)} = \frac{1}{K_i^{(1)}} \sum_{\mathbf{x} \in C_i^{(1)}} \mathbf{x} \tag{9.18}$$

and compute $tWCSS^{(2)}$

$$tWCSS^{(2)} = \sum_{i=1}^{K} \sum_{\mathbf{x} \in C_i^{(2)}} |\mathbf{x} - \boldsymbol{\mu}_i^{(2)}|^2. \tag{9.19}$$

These $\boldsymbol{\mu}_i^{(2)}$ $(i = 1, 2, \ldots, K)$ are regarded as the updated centroids. We then repeat the assignment step to assign each point to a cluster, again according to the rule of the smallest distance. Next, we compute $\boldsymbol{\mu}_i^{(3)}$ $(i = 1, 2, \ldots, K)$, and $tWCSS^{(3)}$.

The K-means algorithm continues these iterative steps of assignment and update until the centroids $\boldsymbol{\mu}_i^{(n)}$ $(i = 1, 2, \ldots, K)$ do not change from those in the previous assignments $\boldsymbol{\mu}_i^{(n-1)}$ $(i = 1, 2, \ldots, K)$. When the assignment centroids do not change, we say that the K-means iterative sequence has converged. It can be rigorously proven that this algorithm always converges.

R and Python have K-means functions. Many software packages are freely available for some modified K-means algorithms using different distance formulas. Climate scientists often just use these functions or packages without actually writing an original computer code for a specific K-means algorithm. For example, the R calculation in the previous subsection used the command kmeans(mydata, K) to calculate the cluster results for three points shown in Figure 9.1.

The K-means algorithm appears very simple, but it was invented only in the 1950s. Although the K-means iterative sequence here always converges, the final result may depend on the initial assignment of the centroid: $C_i^{(1)}$ $(i = 1, 2, \ldots, K)$, hence each run of the R K-means command kmeans(mydata, K) may not always lead to the same solution. Further, a K-means clustering result may not be a global minimum for tWCSS.

Thus, after we have obtained our K-means results, we should try to explain the results and see if they are reasonable. We may also run the code with different K values and check which value K is the most reasonable. The tWCSS score is a function of K. We may plot the function $tWCSS(K)$ and check the shape of the curve. Usually, $tWCSS(K)$ is a decreasing function, and further,

$$\Delta_K = tWCSS(K) - tWCSS(K+1) > 0, \quad K = 1, 2, \ldots \tag{9.20}$$

decreases fast for $K \leq K_e$, and then the decrease suddenly slows down or even increases at $K = K_e$. In this sense, the curve $tWCSS(K)$ looks like having an elbow at K_e. We would choose this K_e as the optimal K. See the elbow in the example of the next subsection.

Different applications may need different optimization criteria for the selection of the best number of clusters. When you work on your own data, try different criteria to choose the best number of clusters. For example, the Silhouette method uses the Silhouette score to measure the similarity level of a point with its own cluster compared to other clusters.

Again, a computer code for the K-means clustering may not always obtain the minimum tWCSS against all partitions. Sometimes, a computer yields a "local" minimum for $tWCSS(K)$.

In short, when using the K-means method to divide given points into clusters, you should run the K-means code multiple times and check the relevant solution results. Although most

K-means code yields a unique solution, multiple solutions sometimes appear, and then we need to examine which solution is the most reasonable for our purposes.

9.1.3 K-Means Clustering for the Daily Miami Weather Data

In this subsection, we present an example using the K-means method to cluster the observed daily weather data at the Miami International Airport in the United States.

9.1.3.1 K-Means Clustering for $N = 365, K = 2$

Figure 9.2(a) shows a scatter plot of the daily minimum surface air temperature Tmin [in °C] and the direction of the fastest 2-minute wind (WDF2) in a day [in degrees] of the Miami International Airport. The wind blowing from the north is zero degree, from the east is 90 degrees, and from the west is 270 degrees. The scatter plot seems to support two clusters or weather regimes: the lower regime showing the wind direction between 0 and 180 degrees, meaning the wind from the east, northeast, or southeast; and another with the wind direction between 180 degrees and 360 degrees, meaning that the wind came from the west, northwest, or southwest.

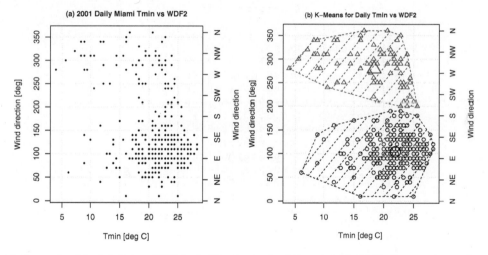

Figure 9.2   (a) Scatter plot of the daily Tmin vs WDF2 for the Miami International Airport in 2001. (b) The K-means clusters of the daily Tmin vs WDF2 data points when assuming $K = 2$.

We would like to use the K-means method to make the clustering automatically and justify our previous intuitive conclusion from the scatter diagram. This example, although simple, is closer to reality. You can apply the procedures described here to your own data for various clustering purposes, such as defining wildfire zones, ecoregions, climate regimes, and agricultural areas. You can find numerous application examples by an online search using keywords like "K-means clustering for ecoregions" or "K-means clustering for climate regimes."

The daily weather data for the Miami International Airport can be obtained from the Global Historical Climatology Network-Daily (GHCN-D), station code USW00012839. The daily data from 2001 to 2020 are included in the book's master data file `data.zip` from the book website `www.climatestatistics.org`. The file name of the Miami data is `MiamiIntlAirport2001_2020.csv`. You can access the updated data by an online search for "NOAA Climate Data Online USW00012839." Figure 9.2(b) shows the K-means clustering result for the daily Tmin vs WDF2 for the 2001 daily data at the Miami International Airport. The two clusters generated by the K-means method agree with our intuition by looking at the scatter diagram Figure 9.2(a). However, the K-means clustering result Figure 9.2(b) provides a clearer picture: a red cluster with each data point indicated by a small triangle, and a black cluster with each data point indicated by a small circle. The cluster boundaries are linked by dashed lines. The large vertical gap at Tmin around 5°C (in winter) between the two clusters indicates that the fastest 2-minute wind in a day has a WDF2 angle value around 50°, a northeast wind for the black cluster, and around 300°, a northwest wind for the red cluster.

Figure 9.2 may be generated by the following computer codes.

```
#R for Fig. 9.2: K-means clustering for 2001 daily weather
#data at Miami International Airport, Station ID USW00012839
setwd("~/climstats")
dat = read.csv("data/MiamiIntlAirport2001_2020.csv",
               header=TRUE)
dim(dat)
#[1] 7305    29
tmin = dat[,'TMIN']
wdf2 = dat[,'WDF2']
# plot the scatter diagram Tmin vs WDF2
setEPS() #Plot the data of 150 observations
postscript("fig0902a.eps", width=5, height=5)
par(mar=c(4.5, 4.5, 2, 4.5))
plot(tmin[2:366], wdf2[2:366],
     pch =16, cex = 0.5,
     xlab = 'Tmin [deg C]',
     ylab = 'Wind Direction [deg]', grid())
title('(a) 2001 Daily Miami Tmin vs WDF2',
     cex.main = 0.9, line = 1)
axis(4, at = c(0, 45, 90, 135, 180, 225, 270, 315, 360),
     lab = c('N', 'NE', 'E', 'SE', 'S', 'SW', 'W', 'NW', 'N'))
mtext('Wind Direction', side = 4, line =3)
dev.off()
#K-means clustering
K = 2 #assuming K = 2, i.e., 2 clusters
mydata = cbind(tmin[2:366], wdf2[2:366])
fit = kmeans(mydata, K) # K-means clustering
#Output the coordinates of the cluster centers
fit$centers
#1 18.38608 278.8608
#2 21.93357 103.9161
fit$tot.withinss # total WCSS
#[1] 457844.9 for # the value may vary for each run

#Visualize the clusters by kmeans.ani()
mycluster <- data.frame(mydata, fit$cluster)
```

```
names(mycluster)<-c('Tmin␣[deg␣C]',
                    'Wind␣Direction␣[deg]',
                    'Cluster')
library(animation)
par(mar = c(4.5, 4.5, 2, 4.5))
kmeans.ani(mycluster, centers = K, pch=1:K, col=1:K,
           hints = '')
title(main=
       "(b)␣K-means␣Clusters␣for␣Daily␣Tmin␣vs␣WDF2",
      cex.main = 0.8)
axis(4, at = c(0, 45, 90, 135, 180, 225, 270, 315, 360),
     lab = c('N', 'NE', 'E', 'SE', 'S', 'SW', 'W', 'NW', 'N'))
mtext('Wind␣Direction', side = 4, line =3)
```

```
#Python plot Fig. 9.2: K-means clustering for Tmin and WDF2
# K-means clustering for Fig. 9.2(b)
K = 2
# combine data into two column df
mydata = pd.concat([tmin[1:366],wdf2[1:366]], axis = 1)

# K-means clustering
fit = KMeans(n_clusters=K).fit(mydata)

# cluster center coordinates
print(fit.cluster_centers_, '\n')#enters of the two clusters
print(fit.inertia_) # total WCSS

#kmeans = KMeans(n_clusters=K, random_state=0)
mydata['cluster'] = \
kmeans.fit_predict(mydata[['TMIN', 'WDF2']])
colors = ['black', 'red']
mydata['color'] = \
mydata.cluster.map({0:colors[0], 1:colors[1]})

kmeans = KMeans(n_clusters=K, random_state=0)
mydata['cluster'] = \
kmeans.fit_predict(mydata[['TMIN', 'WDF2']])

colors = ['black', 'red']
mydata['color'] = \
mydata.cluster.map({0:colors[0], 1:colors[1]})

x1 = fit.cluster_centers_[0][0] #Cluster center
x2 = fit.cluster_centers_[1][0]
y1 = fit.cluster_centers_[0][1]
y2 = fit.cluster_centers_[1][1]

plt.title("(b)␣K-means␣for␣Tmin␣vs␣WDF2", pad = 10)
plt.xlabel(r"Tmin␣[$\degree$C]", labelpad = 10)
plt.ylabel(r"Wind␣Direction␣[$\degree$]", labelpad = 10)
plt.yticks([0,50,100,200,300])
# plot points
plt.scatter(mydata['TMIN'], mydata['WDF2'],
            s = 10, color = mydata['color']) #s is size
```

```
for i in mydata.cluster.unique():
    points = \
    mydata[mydata.cluster == i][['TMIN', 'WDF2']].values
    # get convex hull
    hull = ConvexHull(points)
    # get x and y coordinates
    # repeat last point to close the polygon
    x_hull = np.append(points[hull.vertices,0],
                       points[hull.vertices,0][0])
    y_hull = np.append(points[hull.vertices,1],
                       points[hull.vertices,1][0])
    # plot shape
    plt.fill(x_hull, y_hull, alpha=0.2, c=colors[i])
    # plot centers
plt.scatter([x1,x2],[y1,y2], color = ['red', 'black'],
            linewidth = 2, facecolors= 'none', s = 700)
plt.show()
```

9.1.3.2 Why $K = 2$ for the Miami Weather Data?

The K-means computer code for the Miami Tmin and WDF2 data shows two cluster centers at $C_1(18.4°C, 278.9°)$, and $C_2(21.9°C, 103.9°)$. The corresponding total WCSS is 457844.9. To justify our choice of $K = 2$, we compute tWCSS, the total WCSS, for different numbers of K, and then use the elbow rule to determine the best K value. Figure 9.3(a) shows the tWCSS scores for $K = 2, 3, \ldots, 8$. We can see that the elbow appears at $K = 2$, since total WCSS decrement suddenly slows down from $K = 2$. Thus, the elbow rule has justified our choice of $K = 2$.

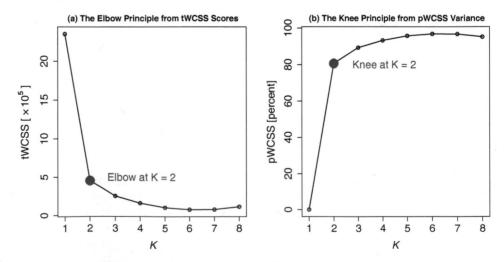

Figure 9.3   (a) tWCSS scores for different K. (b) pWCSS variances for different K.

Another method to choose K is the knee rule, which is determined by the variances explained by K clusters. When $K = 1$, tWCSS is the spatial variance of the entire data. When $K = N$, tWCSS $= 0$, where N is the total number of the data points. Usually, we have

$$\text{tWCSS}[1] \geq \text{tWCSS}[K] \geq \text{tWCSS}[N] \tag{9.21}$$

for any $1 \leq K \leq N$. Then,

$$\text{pWCSS}[K] = \frac{\text{tWCSS}[1] - \text{tWCSS}[K]}{\text{tWCSS}[1]} \times 100\% \tag{9.22}$$

is defined as the percentage of variance explained by the K clusters. Usually, pWCSS$[K]$ is an increasing function of K. When the increase rate suddenly slows down from K, a shape like a knee appears in the pWCSS$[K]$ curve. We then use this K as the optimal number of centers for our K-means clustering. This is the so-called the knee rule. Figure 9.3(b) shows that the increase of pWCSS$[K]$ dramatically slows down at $K = 2$. Thus, we choose $K = 2$, which provides another justification why $K = 2$.

Figure 9.3 may be generated by the following computer codes.

```
#R plot Fig. 9.3: tWCSS(K) and pWCSS(K)
twcss = c()
for(K in 1:8){
  mydata=cbind(tmax[2:366], wdf2[2:366])
  twcss[K] = kmeans(mydata, K)$tot.withinss
}
 twcss
par(mar = c(4.5, 6, 2, 0.5))
par(mfrow=c(1,2))
plot(twcss/100000, type = 'o', lwd = 2,
    xlab = 'K', ylab = bquote('tWCSS␣[␣x' ~  10^5 ~ ']'),
    main = '(a)␣The␣elbow␣principle␣from␣tWCSS␣scores',
    cex.lab = 1.5, cex.axis = 1.5)
points(2, twcss[2]/100000, pch =16, cex = 3, col = 'blue')
text(4, 5, 'Elbow␣at␣K␣=␣2', cex = 1.5, col = 'blue')
#compute percentage of variance explained
pWCSS = 100*(twcss[1]- twcss)/twcss[1]
plot(pWCSS, type = 'o', lwd = 2,
    xlab = 'K', ylab = 'pWCSS␣[percent]',
    main = '(b)␣The␣knee␣principle␣from␣pWCSS␣variance',
    cex.lab = 1.5, cex.axis = 1.5)
points(2, pWCSS[2], pch =16, cex = 3, col = 'blue')
text(4, 80, 'Knee␣at␣K␣=␣2', cex = 1.5, col = 'blue')
dev.off()
```

```
#Python plot Fig. 9.3: tWCSS(K) and pWCSS(K)
#Plot Fig. 9.3(a): Elbow principle
twcss = np.zeros(9)
for K in range(1,9):
    mydata = pd.concat([tmin[1:366],wdf2[1:366]], axis = 1)
    twcss[K] = KMeans(n_clusters=K).fit(mydata).inertia_
# remove K = 0 value from twcss
```

```
twcss = twcss[1:]
# plot elbow
plt.plot(np.linspace(1,8,8), twcss/100000, 'k', linewidth=2)
# plot points
plt.scatter([np.linspace(1,8,8)], [twcss/100000], color='k')

# plot elbow at K = 2
plt.scatter(2,twcss[1]/100000, color = 'blue', s = 500)
# add text
plt.text(2.2, 5.2, 'Elbow at K = 2', color='blue', size=20)
# add labels
plt.title("(a) The elbow principle from tWCSS scores",
          pad = 10)
plt.xlabel("K", labelpad = 10)
plt.ylabel(r"tWCSS [x $10^5$]", labelpad = 10)
plt.show()

#Plot Fig. 9.3(b): Knee principle
# compute percentage of variance explained
pWCSS = 100*(twcss[0]-twcss)/twcss[0]
# plot knee
plt.plot(np.linspace(1,8,8),pWCSS, 'k', linewidth = 2)
# add points
plt.scatter(np.linspace(1,8,8),pWCSS, color = 'k')

# plot knee at K = 2
plt.scatter(2,pWCSS[1], color = 'blue', s = 500)
# add text
plt.text(2.2, 76, 'Knee at K = 2', color = 'blue',
         size = 20)
# add labels
plt.title("(b) The knee princple from pWCSS variance",
          pad = 10)
plt.xlabel("K", labelpad = 10)
plt.ylabel("pWCSS [%]", labelpad = 10)
plt.show()
```

9.1.3.3 Steps of the K-Means Clustering Data Analysis

From the previous subsections, we may summarize that the K-means method for data analysis consists of the following steps:

(i) Data preparation: Write your data into an $N \times 2$ matrix, where N is the total number of data points, and 2 is for two variables, i.e., in a 2-dimensional space. The real value matrix must have no missing data. Although our previous examples are for 2-dimensional data, the K-means clustering can be used for higher dimensional data. On the Internet, you can easily find examples of the K-means for data of three or more variables.

(ii) Preliminary data analysis: You may wish to make some preliminary simple analyses before applying the K-means method. These analyses may include plotting a scatter diagram, plotting a histogram for each variable, and computing mean, standard deviation, and quantiles for each variable. If the data for the two variables have different units, you may convert the data into their standardized anomalies. The standardized anomaly computing procedure is called "scale" in the ML community. For example, the R command scale(x) yields the standardized anomalies of data sequence x, i.e, divided by the standard deviation after a subtraction of mean. Mathematically speaking, the scaling procedure is preferred when two variables have different units, because the $unit1^2 + unit2^2$ in tWCSS usually do not have a good interpretation. In contrast, standardized anomalies are dimensionless and have no units. Nondimensional data are often preferred when applying statistical methods. However, our goal of the K-means clustering is to find patterns without labeling the data, which is an unsupervised learning. The K-means results from scaled data and unscaled data may not be the same. As long as the K-means result is reasonable and interpretable, regardless of the scaled or unscaled data, we adopt the result. For example, in our Miami Tmin and WDF2 example in the previous subsection, we found the K-means clusters for both scaled and unscaled data, but we have chosen the result from the unscaled data from our view point of interpretation. Therefore, we keep in mind that the K-means is a result-oriented method.

(iii) Test K-means clustering: Apply a computer code of K-means to the prepared $N \times 2$ data matrix with a K value inspired from the scatter diagram or science background of the data. You may apply the code once to the scaled data and again to the unscaled data, and see if the results make sense. Run your code a few times and see if the results vary.

(iv) Determine the optimal K by the elbow rule: Compute the tWCSS scores for different K and find the elbow location of the $\text{tWCSS}(K)$ curve. Apply the K-means computer code for this optimal K value. Compare this K with the K in the previous step. Check the details of the clustering results, such as cluster assignment, tWCSS, and centers.

(v) Visualize the K-means clustering results: Use a computer code to plot the K clusters. Different visualization codes are available for your choice.

(vi) Interpret the final result: Identify the characteristics of the clusters and interpret them with text or more figures.

Applying the K-means computer code is very easy, but in the K-means data analysis, we should avoid the following common mistakes:

- Skipping the preliminary data analysis step: Some people may quickly apply the K-means computer code and treat the K-means results as unique, hence regard their plot of the clusters as the end of the analysis.
- Missing the interpretation: Interpretation step is needed, short or long. This is our purpose anyway.
- Forgetting to scale the data: We should always analyze the scaled data. At the same time, because of the interpretation requirement, we should also check the results from the unscaled data and see which result is more reasonable.
- Overlooking the justification of the optimal K: A complete K-means clustering should have this step, which helps us to explore all the possible clusters.

In addition to using the standard K-means computer code you can find from the Internet or that built in R or Python, you sometimes may wish to make special plots for your clustering results with some specific features, such as a single cluster on a figure with the marked data order, as shown in Figure 9.4. A convex hull is a simple method to plot a cluster. By definition, the convex hull of a set of points is the smallest convex polygon that encloses all of the points in the set. Below is the computer code to plot Figure 9.4 using the convex hull method.

Cluster 1 of Miami Tmin vs WDF2

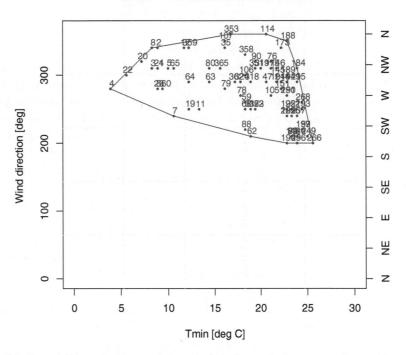

Figure 9.4 | Convex hull for Cluster 1. The number above each data point is the day number in a year, e.g., January 1 being 1, February 2 being 33, and December 31 being 365 in 2001. The day numbers help identify the seasonality of the points, for example the left-most tip of the cluster being Day 4, a specific winter day.

Figure 9.4 may be generated by the following computer codes.

```
#R plot Fig. 9.4: Convex hull for a cluster
setwd("~/climstats")
dat = read.csv("data/MiamiIntlAirport2001_2020.csv",
                header=TRUE)
dim(dat)
#[1] 7305   29
tmin = dat[,'TMIN']
wdf2 = dat[,'WDF2']
#K-means clustering
K = 2 #assuming K = 2, i.e., 2 clusters
mydata = cbind(tmin[2:366], wdf2[2:366]) #2001 data
clusterK = kmeans(mydata, K) # K-means clustering
mycluster <- data.frame(mydata, clusterK$cluster)
```

```
plotdat = cbind(1:N, mycluster)

par(mar = c(4.5, 6, 2, 0.5)) #set up the plot margin
i = 1 # plot Cluster 1
N = 365 #Number of data points
X = plotdat[which(mycluster[,3] == i), 1:3]
colnames(X)<-c('Day', 'Tmin␣[deg␣C]', 'WDF2␣[deg]')
plot(X[,2:3], pch = 16, cex = 0.5, col = i,
     xlim = c(0, 30), ylim = c(0, 365),
     main = 'Cluster␣1␣of␣Miami␣Tmin␣vs␣WDF2' )
grid(5, 5)
#chull() finds the boundary points of a convex hull
hpts = chull(X[, 2:3])
hpts1 = c(hpts, hpts[1]) #close the boundary
lines(X[hpts1, 2:3], col = i)
for(j in 1:length(X[,1])){
  text(X[j,2], X[j,3] + 8, paste("", X[j,1]),
       col = i, cex = 0.8)
} #Put the data order on the cluster
```

```
#Python plot Fig. 9.4: Convex hull for a cluster
K = 2
clusterK = KMeans(n_clusters=K).fit(mydata)
mydata['cluster'] = \
       kmeans.fit_predict(mydata[['TMIN', 'WDF2']])
mydata['day']= np.arange(1, 366)
subset = mydata[mydata['cluster']==0]
# set up plot
plt.title("Cluster␣1␣of␣Miami␣Tmin␣vs␣WDF2", pad = 10)
plt.xlabel(r"Tmin␣[$\degree$C]", labelpad = 10)
plt.ylabel(r"WDF2␣[$\degree$]", labelpad = 10)

plt.xlim([0,30])
plt.ylim([0,400])
plt.yticks([0,100,200,300])

# plot points
plt.scatter(subset["TMIN"],subset["WDF2"], color = 'k')
# add labels
for i in range(len(subset["TMIN"])):
    plt.text(subset["TMIN"].values[i],
             subset["WDF2"].values[i],
             subset["day"].values[i], size=12)
# add boundary
for i in subset.cluster.unique():
    points = \
      subset[subset.cluster == i][['TMIN', 'WDF2']].values
    # get convex hull
    hull = ConvexHull(points)
    # get x and y coordinates
    # repeat last point to close the polygon
    x_hull = np.append(points[hull.vertices,0],
                       points[hull.vertices,0][0])
    y_hull = np.append(points[hull.vertices,1],
```

```
                                    points[hull.vertices,1][0])
        # plots hape
        plt.fill(x_hull, y_hull, alpha=0.2, c=colors[i])
```

The K-means method cannot only be used for weather data clustering and climate classification, but also for many other purposes, such as weather forecasting, storm identification, and more. In many applications, K-means is part of a machine learning algorithm, and takes a modified version, such as the incremental K-means for weather forecasting and the weighted K-means for climate analysis.

9.2 Support Vector Machine

The K-means clustering can organize a suite of given data points into K clusters. The data points are not labeled. This is like organizing a basket of fruits into K piles according to the minimum tWCSS principle. After the clustering, we may name the clusters, say $1, 2, \ldots, K$, or apples, oranges, peaches, ..., grapes. Machine learning often uses the term "cluster label" in lieu of "cluster name." So, we say that we label a cluster when we name a cluster. The learning process for unlabeled data is called unsupervised learning. If a basket of fruits is sorted by a baby who cannot articulate the fruit names, he is performing an unsupervised learning.

In contrast, support vector machine (SVM) is in general a supervised learning, working on the labeled data, e.g., a basket of apples, oranges, peaches, and grapes. Based on the numerical data, SVM has two purposes. First, SVM determines the maximum "distances" between the sets of labeled data, say apples, oranges, peaches, and grapes. This is a training process and helps a computer learn the differences between different labeled sets, e.g., different fruits, different weather, or different climate regimes. Second, SVM makes a prediction, which, based on the trained model, predicts the labels (i.e., names) of the new and unlabeled data. Thus, SVM is a system to learn the maximum differences among the labeled clusters of data, and to predict the labels of the new unlabeled data. This process of training and prediction mimics a human's learning experience. For example, over the years, a child is trained by his parents to acquire knowledge of apples, oranges, and other fruits. At a certain point, the child is well trained and ready to independently tell (i.e., predict) whether a new fruit is an apple or an orange. During this learning process, the fruits are labeled (e.g., apples, oranges, etc.). His parents help input the data into his brain through a teaching and learning process. His brain was trained to maximize the difference between an apple and an orange. In the prediction stage, the child independently imports the data using his eyes, and makes the prediction whether a fruit is an apple or an orange. SVM mimics this process by maximizing the difference between two or more labeled categories or clusters in the training process, then, by using the trained model, SVM predicts the categories for the new data. The key is how to quantify the difference between any two categories.

In the following, we will illustrate these SVM learning ideas using three examples: (i) a trivial case of two data points, (ii) training and prediction for a system of three data points, and (iii) a more general case of many data points.

9.2.1 Maximize the Difference between Two Labeled Points

We formulate a two-point learning question as follows. We have an apple and an orange. We know all the data about the apple and orange. We teach a baby how to use the data to quantify the maximum difference between the apple and the orange. In the ML language, "apple" and "orange" are called labels. The data are called the training data, such as color and surface smoothness. The maximum difference will be used to predict whether a new fruit is an apple or an orange.

An abstract description of this may be put in the following way. Given the data for an object labeled A and also the data for another object labeled B, quantify the maximum difference between A and B. A simple example corresponding to this statement is that given two points $P_1(1,1)$ and $P_2(3,3)$, labeled $y_1 = -1$ and $y_2 = 1$, quantify the maximum difference between y_1 and y_2. For this simple problem, the obvious answer is that the two categories y_1 and y_2 are divided by the perpendicular bisector of the two points P_1 and P_2 as shown in Figure 9.5. The bisector is called the *separating hyperplane*. A hyperplane means a plane in a space of more than three dimensions. In a 2-dimensional space, it is a straight line. In a 3-dimensional space, it is just a regular flat plane. The real ML applications are often for higher dimensional data, e.g., a comprehensive description of an apple may need data of color, surface smoothness, stem size, skin thickness, and more, which form a p-dimensional vector.

The maximum margin of difference between the two categories is bounded by two lines parallel to the bisector, one line passing through P_1 and another P_2. One line is called the negative hyperplane, and the other the positive hyperplane. The data points on these hyperplanes are called *support vectors*. The support vector machine (SVM) has its name from these vectors.

The equation for the separating hyperplane (i.e., the purple line) is

$$x_1 + x_2 = 4, \tag{9.23}$$

that for the positive hyperplane (i.e., the blue dashed line) is

$$x_1 + x_2 = 6, \tag{9.24}$$

and that for the negative hyperplane (i.e., the red dashed line) is

$$x_1 + x_2 = 2. \tag{9.25}$$

The unit vector in the perpendicular bisector direction from P_1 to P_2 is

$$\mathbf{n} = \left(\frac{\sqrt{2}}{2}, \frac{\sqrt{2}}{2} \right). \tag{9.26}$$

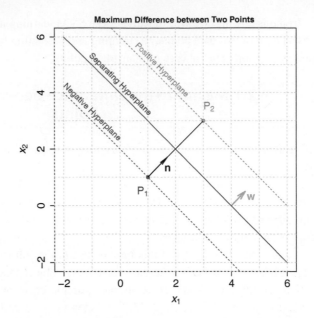

Figure 9.5 Quantify the maximum difference between two points.

This vector is a normal vector for the separating hyperplane. The equation of the separating hyperplane may be expressed in the normal vector form

$$\mathbf{n} \cdot \mathbf{x} = 2\sqrt{2}. \tag{9.27}$$

This can further be written in the following form

$$\mathbf{w} \cdot \mathbf{x} - b = 0, \tag{9.28}$$

where

$$\mathbf{w} = (0.5, 0.5) \tag{9.29}$$

and

$$b = 2. \tag{9.30}$$

An advantage of this form is that the equations for the positive and negative hyperplanes can be simply written in the form

$$\mathbf{w} \cdot \mathbf{x} - b = \pm 1, \tag{9.31}$$

and that for the separating hyperplane can be written as

$$\mathbf{w} \cdot \mathbf{x} - b = 0. \tag{9.32}$$

Here, $\mathbf{w}$ is called a normal vector and is related with the unit normal vector $\mathbf{n}$ as follows:

$$\mathbf{w} = -\frac{1}{\sqrt{2}} \mathbf{n}. \tag{9.33}$$

The distance between the positive and negative hyperplanes is

$$D_m = \frac{2}{|\mathbf{w}|} = 2\sqrt{2}. \tag{9.34}$$

This is the margin size. The maximum difference between P_1 and P_2 is quantified by the margin size D_m. The primary goal of SVM is to maximize D_m when training data are given.

The distance of the origin to the separating hyperplane is

$$D_o = -\frac{b}{|\mathbf{w}|} = 2\sqrt{2}. \tag{9.35}$$

These formulations of the hyperplanes and the quantities $\mathbf{w}, b, D_m$ can be used for the general SVM mathematical formulations, which are described in the next subsection.

Figure 9.5 may be generated by the following computer code.

```
#R plot Fig. 9.5: Maximum difference between two points
x = matrix(c(1, 1, 3, 3),
           ncol = 2, byrow = TRUE)#Two points
y= c(-1, 1) #Two labels -1 and 1
#Plot the figure and save it as a .eps file
setEPS()
postscript("fig0905.eps", height=7, width=7)
par(mar = c(4.5, 4.5, 2.0, 2.0))
plot(x, col = y + 3, pch = 19,
     xlim = c(-2, 6), ylim = c(-2, 6),
     xlab = bquote(x[1]), ylab = bquote(x[2]),
     cex.lab = 1.5, cex.axis = 1.5,
     main = "Maximum Difference between Two Points")
axis(2, at = (-2):6, tck = 1, lty = 2,
     col = "grey", labels = NA)
axis(1, at = (-2):6, tck = 1, lty = 2,
     col = "groy", labels = NA)
segments(1, 1, 3, 3)
arrows(1, 1, 1.71, 1.71, lwd = 2,
       angle = 9, length= 0.2)
text(1.71, 1.71-0.4, quote(bold('n')), cex = 1.5)
arrows(4, 0, 4 + 0.5, 0 + 0.5, lwd = 3,
       angle = 15, length= 0.2, col = 'green' )
text(4.7, 0.5 -0.2, quote(bold('w')),
     cex = 1.5, col = 'green')
x1 = seq(-2, 6, len = 31)
x20 = 4 - x1
lines(x1, x20, lwd = 1.5, col = 'purple')
x2m = 2 - x1
lines(x1, x2m, lty = 2, col = 2)
x2p = 6 - x1
lines(x1, x2p, lty = 2, col = 4)
text(1-0.2,1-0.5, bquote(P[1]), cex = 1.5, col = 2)
text(3+0.2,3+0.5, bquote(P[2]), cex = 1.5, col = 4)
text(1-0.2,1-0.5, bquote(P[1]), cex = 1.5, col = 2)
text(0,4.3, 'Separating Hyperplane',
     srt = -45, cex = 1.2, col = 'purple')
text(1.5, 4.8, 'Positive Hyperplane',
     srt = -45, cex = 1.2, col = 4)
text(-1, 3.3, 'Negative Hyperplane',
     srt = -45, cex = 1.2, col = 2)
dev.off()
```

```
#Python plot Fig. 9.5: Maximum difference between two points
# define the two points
x = np.array([1,1,3,3]).reshape(2,2)
# define the two labels
y = (-1,1)
# set up plot
plt.figure(figsize = (10,10))
plt.title("Maximum␣Difference␣between␣Two␣Points", pad = 10)
plt.xlabel("$x_1$", labelpad = 10)
plt.ylabel("$x_2$", labelpad = 10)
plt.xlim(-2.2,6.2)
plt.xticks([-2,-1,0,1,2,3,4,5,6],[-2,'',0,'',2,'',4,'',6])
plt.ylim(-2.2,6.2)
plt.yticks([-2,-1,0,1,2,3,4,5,6],[-2,'',0,'',2,'',4,'',6])
plt.grid()
# plot points
plt.scatter(x[0],x[0], color = 'r')
plt.scatter(x[1],x[1], color = 'dodgerblue')
# connect points
plt.plot([1,3],[1,3], 'k')
# add dashed lines
plt.plot([0,6],[6,0], color ='dodgerblue', linestyle ='--')
plt.plot([-2,6],[6,-2], color = 'purple', linestyle = '--')
plt.plot([-2,4],[4,-2], 'r', linestyle = '--')
# plot arrows
plt.arrow(1,1,0.5,0.5, head_width = 0.12, color='k')
plt.arrow(4,0,0.3,0.3, head_width = 0.12, color='limegreen')
# add text
plt.text(0.5,4.3, "Positive␣Hyperplane", rotation = -45,
         size = 14, color = 'dodgerblue')
plt.text(3.2,3.2, "$P_2$", size = 15, color = 'dodgerblue')
plt.text(-1.2,3.8, "Separating␣Hyperplane", rotation = -45,
         size = 14, color = 'purple')
plt.text(4.4,0.1, "w", size = 15, color = 'limegreen')
plt.text(1.5,1.2, "n", size = 15)
plt.text(-2,2.7, "Negative␣Hyperplane", rotation = -45,
         size = 14, color = 'r')
plt.text(0.7, 0.6, "$P_1$", size = 15, color = 'r')
plt.show()
```

9.2.2 SVM for a System of Three Points Labeled in Two Categories

Let us progress from a two-point system to a three-point system and build an SVM using a computer code. Figure 9.6 shows the three points labeled in two categories. The two blue points are $P_1(1,1), P_2(2,1)$ which are in the first category corresponding to the categorical value $y = 1$. The red point $P_3(3,3.5)$ is in the second category corresponding to $y = -1$. Training these data is to maximize the margin size D_m, which results in the following SVM parameters.

$$w = (-0.28 - 0.69), \quad b = 2.24, \quad D_m = 2.69. \tag{9.36}$$

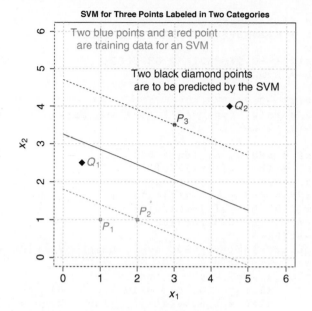

Figure 9.6 SVM training from three data points, and the prediction by the trained SVM.

The support vectors are $P_2(2,1)$ and $P_3(3,3.5)$.

We acquire two new data points $Q_1(0.5,2.5)$ and $Q_2(4.5,4.0)$. We wish to use the trained SVM to predict whether $Q_1(0.5,2.5)$ corresponds to 1 (i.e., the first category) or -1 (i.e., the second category). The SVM prediction results are shown in Figure 9.6: $Q_1(0.5,2.5)$ belongs to the first category and $Q_2(4.5,4.0)$ to the second category. Geometrically, this prediction result in Figure 9.6 is obvious and has nothing surprising. The significance of this example is its systematic use of the SVM method and its computer algorithm. Although this system is simple and has only three points, solving this problem by hand is very difficult. The following computer code solves this problem and generates Figure 9.6.

```
#R plot Fig. 9.6: SVM for three points
#training data x
x = matrix(c(1, 1, 2, 1, 3, 3.5),
           ncol = 2, byrow = TRUE)
y = c(1, 1, -1) #two categories 1 and -1
plot(x, col = y + 3, pch = 19,
     xlim = c(-2, 8), ylim = c(-2, 8))
library(e1071)
dat = data.frame(x, y = as.factor(y))
svm3P = svm(y ~ ., data = dat,
            kernel = "linear", cost = 10,
            scale = FALSE,
            type = 'C-classification')
svm3P #This is the trained SVM
xnew = matrix(c(0.5, 2.5, 4.5, 4),
           ncol = 2, byrow = TRUE) #New data
predict(svm3P, xnew) #prediction using the trained SVM
# 1  2
# 1 -1
```

```
# Find hyperplane, normal vector, and SV (wx + b = 0)
w = t(svm3P$coefs) %*% svm3P$SV
w
#[1,] -0.2758621 -0.6896552
b = svm3P$rho
b
#[1] -2.241379
2/norm(w, type ='2') #maximum margin of separation
#[1] 2.692582

x1 = seq(0, 5, len = 31)
x2 = (b - w[1]*x1)/w[2]
x2p = (1 + b - w[1]*x1)/w[2]
x2m = (-1 + b - w[1]*x1)/w[2]
x20 = (b - w[1]*x1)/w[2]
#plot the SVM results
setEPS()
postscript("fig0906.eps", height=7, width=7)
par(mar = c(4.5, 4.5, 2.0, 2.0))
plot(x, col = y + 3, pch = 19,
     xlim = c(0, 6), ylim = c(0, 6),
     xlab = bquote(x[1]), ylab = bquote(x[2]),
     cex.lab = 1.5, cex.axis = 1.5,
     main = 'SVM for three points labeled in two categories')
axis(2, at = (-2):8, tck = 1, lty = 2,
     col = "grey", labels = NA)
axis(1, at = (-2):8, tck = 1, lty = 2,
     col = "grey", labels = NA)
lines(x1, x2p, lty = 2, col = 4)
lines(x1, x2m, lty = 2, col = 2)
lines(x1, x20, lwd = 1.5, col = 'purple')
xnew = matrix(c(0.5, 2.5, 4.5, 4),
              ncol = 2, byrow = TRUE)
points(xnew, pch = 18, cex = 2)
for(i in 1:2){
  text(xnew[i,1] + 0.5, xnew[i,2] , paste('Q',i),
       cex = 1.5, col = 6-2*i)
}
text(2.2,5.8, "Two blue points and a red point
are training data for an SVM ",
     cex = 1.5, col = 4)
text(3.5,4.7, "Two black diamond points
          are to be predicted by the SVM",
     cex = 1.5)
dev.off()
```

```
#Python plot Fig. 9.6: SVM for three points
# define x for ease of plotting
x = np.array([1,1,2,1,3,3.5])
# redefine x for svm
X = [[1, 1],[2,1],[3,3.5]]
y = [1, 1, -1]
# define SVM
```

```
svm3P = svm.SVC(kernel='linear')
# train SVM
svm3P.fit(X, y)
# predict new data using the trained SVM
print('The prediction is: ',
      svm3P.predict([[0.5, 2.5],[4.5,4]]), '\n')
# find hyperplane, normal vector, and SV (wx + b = 0)
w = svm3P.coef_[0]
print('w is: ',w, '\n')
b = -svm3P.intercept_ #intercept
print('b is: ', b, '\n')
# find the maximum margin of separation
print(2/np.linalg.norm(w))
x1 = np.linspace(0,5,5)
x2 = (b - w[0]*x1)/w[1]
x2p = (1 + b - w[0]*x1)/w[1]
x2m = (-1 + b - w[0]*x1)/w[1]
x20 = (b - w[0]*x1)/w[1]
# set up plot
plt.figure(figsize = (10,10))
plt.title("SVM for three points labeled in two categories",
          pad = 10)
plt.xlabel("$x_1$", labelpad = 10)
plt.ylabel("$x_2$", labelpad = 10)
plt.xlim(-0.2,5.2)
plt.ylim(-0.2,6.2)
plt.grid()
# plot original points
plt.scatter([x[::2]],[x[1::2]],
            color=['dodgerblue','dodgerblue','red'])
# add lines
plt.plot(x1,x2p, color = 'dodgerblue', linestyle = '--')
plt.plot(x1, x2m, color = 'red', linestyle = '--')
plt.plot(x1, x20, color = 'purple')
# plot diamonds
plt.scatter([0.5,4.5], [2.5,4], color = 'k', marker = 'D')
# add text
plt.text(0.65,2.45,"$Q_1$", size = 16, color = 'dodgerblue')
plt.text(4.65,3.95, "$Q_2$", size = 16, color = 'r')
plt.text(0.3,5.4, "Two blue points and a red point \n \
are training data for an SVM",
         size = 16, color = 'dodgerblue')
plt.text(1.7,4.5, "Two black diamond points \n \
          are to be predicted by the SVM",
         size = 16, color = 'k')
plt.show()
```

9.2.3 SVM Mathematical Formulation for a System of Many Points in Two Categories

The previous quantification of maximum separation between two points or three points can be generalized to the separation of many points of two categories labeled $(1, -1)$, or $(1, 2)$, or (A, B), or another way. What is a systematic formulation of mathematics to make the best separation, by maximizing the margin? How can we use a computer code to implement the separation? This subsection attempts to answer these questions.

We have n training data points:

$$(\mathbf{x}_1, \mathbf{y}_1), (\mathbf{x}_2, \mathbf{y}_2), \ldots, (\mathbf{x}_n, \mathbf{y}_n), \tag{9.37}$$

where $\mathbf{x}_i$ is a p-dimensional vector (i.e., using p parameters to describe every data point in a category), and $\mathbf{y}_i$ is equal to 1 or -1, a category indicator, $i = 1, 2, \ldots, n$.

The separating hyperplane has its linear equation as follows:

$$\mathbf{w} \cdot \mathbf{x} - b = 0. \tag{9.38}$$

Our SVM algorithm is to find the p-dimensional normal vector $\mathbf{w}$ and a real-valued scaler b so that the distance

$$D_m = \frac{2}{|\mathbf{w}|} \tag{9.39}$$

between the positive and negative hyperplanes

$$\mathbf{w} \cdot \mathbf{x} - b = \pm 1 \tag{9.40}$$

is maximized.

The maximization is for the domain in the p-dimensional space below the negative hyperplane and above the positive hyperplane, i.e.,

$$\mathbf{w} \cdot \mathbf{x}_i - b \geq 1 \text{ when } y_i = 1 \tag{9.41}$$

and

$$\mathbf{w} \cdot \mathbf{x}_i - b \leq 1 \text{ when } y_i = -1. \tag{9.42}$$

This definition of the maximization implies that no training data points are located between the positive and negative hyperplanes. Thus, the solution of the maximization of $D_m = 2/|\mathbf{w}|$ must occur at the boundary of the domain, i.e., the positive and negative hyperplanes. The points on these hyperplanes are called the support vectors (SV). R or Python SVM code can find these SVs.

Figure 9.7 shows an example of an SVM training with 20 training data points, and an SVM prediction for 3 data points. The data are given below.

```
    X1  X2 y
1    1 6.0 1
2    2 8.0 1
3    3 7.5 1
4    1 8.0 1
5    4 9.0 1
6    5 9.0 1
7    3 7.0 1
8    5 9.0 1
9    1 5.0 1
```

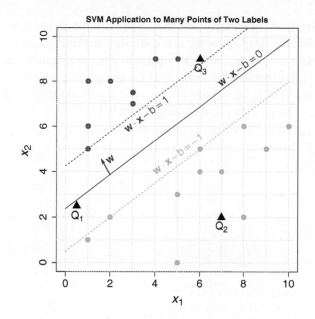

Figure 9.7   SVM training from 20 data points, and the prediction of 3 new data points by the trained SVM.

```
10   5 3.0 2
11   6 4.0 2
12   7 4.0 2
13   8 6.0 2
14   9 5.0 2
15  10 6.0 2
16   5 0.0 2
17   6 5.0 2
18   8 2.0 2
19   2 2.0 2
20   1 1.0 2
```

The new data to be predicted are

```
[1,]   0.5   2.5
[2,]   7.0   2.0
[3,]   6.0   9.0
```

The following computer code can do the SVM calculation and plot Figure 9.7. We acquire three new data points $Q_1(0.5, 2.5)$, $Q_2(7.5, 2.0)$, and $Q_3(6.0, 9.0)$. We wish to use the trained SVM to find out which point belongs to which category. The SVM prediction result is shown in Figure 9.7: $Q_1(0.5, 2.5)$ and $Q_2(7.5, 2.0)$ belong to Category 1, and $Q_3(6.0, 9.0)$ to Category 2. Again, this prediction result is obvious to us when the new data points are plotted, our eyes can see the points, and our brains can make the decisions based on our trained intelligence. The SVM applications in the real world often involve data of

higher dimensions, in which case the visual decision is impossible, and requires us to predict the category for the new data without plotting. The prediction is based on the training
data, new data, and SVM algorithm without visualization. For example, the R prediction
of this problem can be done by the command predict(svmP, xnew) where xnew is the
new data in a 3×2 matrix, and svmP is the trained SVM. This command is included in the
computer code for generating Figure 9.7.

```
#R plot Fig. 9.7: SVM for many points
#Training data x and y
x = matrix(c(1, 6, 2, 8, 3, 7.5, 1, 8, 4, 9, 5, 9,
             3, 7, 5, 9, 1, 5,
             5, 3, 6, 4, 7, 4,   8, 6, 9, 5, 10, 6,
             5, 0, 6, 5, 8, 2, 2, 2, 1, 1),
           ncol = 2, byrow = TRUE)
y= c(1, 1, 1, 1, 1, 1,
     1, 1, 1,
     2, 2, 2, 2, 2, 2 ,
     2, 2, 2, 2, 2)

library(e1071)
dat = data.frame(x, y = as.factor(y))
svmP = svm(y ~ ., data = dat,
           kernel = "linear", cost = 10,
           scale = FALSE,
           type = 'C-classification')
svmP
#Number of Support Vectors:  3
svmP$SV #SVs are #x[9,], x[17,], x[19,]
#9   1   5
#17  6   5
#19  2   2

# Find SVM parameters: w, b, SV (wx+c=0)
w = t(svmP$coefs) %*% svmP$SV
# In essence this finds the hyperplane that separates our points
w
#[1,] -0.39996 0.53328
b = svmP$rho
b
#[1] 1.266573
2/norm(w, type ='2')
#[1] 3.0003 is the maximum margin
x1 = seq(0, 10, len = 31)
x2 = (b - w[1]*x1)/w[2]
x2p = (1 + b - w[1]*x1)/w[2]
x2m = (-1 + b - w[1]*x1)/w[2]
x20 = (b - w[1]*x1)/w[2]

#plot the svm results
setEPS()
postscript("fig0907.eps", height=7, width=7)
par(mar = c(4.5, 4.5, 2.0, 2.0))
plot(x, col = y + 9, pch = 19, cex =1.5,
     xlim = c(0, 10), ylim = c(0, 10),
     xlab = bquote(x[1]), ylab = bquote(x[2]),
     cex.lab = 1.5, cex.axis = 1.5,
```

```
         main = 'SVM␣application␣to␣many␣points␣of␣two␣labels')
axis(2, at = 0:10, tck = 1, lty = 3,
     col = "grey", labels = NA)
axis(1, at = 0:10, tck = 1, lty = 3,
     col = "grey", labels = NA)
lines(x1, x2p, lty = 2, col = 10)
lines(x1, x2m, lty = 2, col = 11)
lines(x1, x20, lwd = 1.5, col = 'purple')
thetasvm = atan(-w[1]/w[2])*180/pi
thetasvm
#[1] 36.8699 #36.9 deg angle of the hyperplane
#linear equations for the hyperplanes
delx = 1.4
dely = delx * (-w[1]/w[2])
text(5 + 2*delx, 6.5 + 2*dely, bquote(bold(w%.%x) - b == 0),
     srt = thetasvm, cex = 1.5, col = 'purple')
text(5 - delx, 7.6 - dely, bquote(bold(w%.%x) - b == 1),
     srt = thetasvm, cex = 1.5, col = 10)
text(5, 4.8, bquote(bold(w%.%x) - b == -1),
     srt = thetasvm, cex = 1.5, col = 11)
#normal direction of the hyperplanes
arrows(2, 3.86, 2 + w[1], 4 + w[2], lwd = 2,
       angle = 15, length= 0.1, col = 'blue' )
text(2 + w[1] + 0.4, 4 + w[2], bquote(bold(w)),
     srt = thetasvm, cex = 1.5, col = 'blue')

#new data points to be predicted
xnew = matrix(c(0.5, 2.5, 7, 2, 6, 9),
              ncol = 2, byrow = TRUE)
points(xnew, pch = 17, cex = 2)
predict(svmP, xnew) #Prediction
#1 2 3
#2 2 1
for(i in 1:3){
  text(xnew[i,1], xnew[i,2] - 0.4 ,
       paste('Q',i), cex = 1.5)
}
dev.off()
```

```
#Python for Fig. 9.7: SVM for a system of many points
# define x for plotting
x = np.array([1,6,2,8,3,7.5,1,8,
              4,9,5,9,3,7,5,9,1,5,
              5,3,6,4,7,4,8,6,9,5,
              10,6,5,0,6,5,8,2,2,2,1,1])
# define y
y = [1,1,1,1,1,1,1,1,1,2,2,2,2,2,2,2,2,2,2,2]
# redefine x for SVM
X = [[1,6],[2,8],[3,7.5],[1,8],
            [4,9],[5,9],[3,7],[5,9],[1,5],
            [5,3],[6,4],[7,4],[8,6],[9,5],
            [10,6],[5,0],[6,5],[8,2],[2,2],[1,1]]
```

```
svmP = svm.SVC(kernel='linear')
# train SVM
svmP.fit(X, y)

# predict new data using trained SVM
print('The support vectors are: ',
      svmP.support_vectors_, '\n')

# find hyperplane, normal vector, and SV (wx + b = 0)
w = -svmP.coef_[0]
print('w is: ',w, '\n')

# rho is the negative intercept in R
# intercept is the positive intercept in Python
b = svmP.intercept_
print('b is: ',b, '\n')

# find the maximum margin of separation
print(2/np.linalg.norm(w))
x1 = np.linspace(0,10,31)
x2 = (b - w[0]*x1)/w[1]
x2p = (1 + b - w[0]*x1)/w[1]
x2m = (-1 + b - w[0]*x1)/w[1]
x20 = (b - w[0]*x1)/w[1]

thetasvm = math.atan(-w[0]/w[1])*180/np.pi
# degree angle of the hyperplane
print(thetasvm)

delx = 1.4
dely = delx * (-w[0]/w[1])

# new predicted data points
svmP.predict([[0.5,2.5],[7,2],[6,9]])

#Python plot Fig. 9.7: SVM for a system of many points
#The plotting part of  Fig. 9.7
plt.figure(figsize = (10,10))#set up plot
plt.xlim(-0.2,10.2)
plt.ylim(-0.2,10.2)
plt.title("SVM application to many points of two labels",
          pad = 10)
plt.xlabel("$x_1$", labelpad = 10)
plt.ylabel("$x_2$", labelpad = 10)

# plot points
color = ['r','r','r','r','r','r','r','r','r','forestgreen',
         'forestgreen','forestgreen','forestgreen',
   'forestgreen','forestgreen','forestgreen','forestgreen',
         'forestgreen','forestgreen','forestgreen',]
plt.scatter(x[::2],x[1::2], color = color)
# plot newly predicted points
plt.scatter([0.5,7,6],[2.5,2,9], marker = "^",
            color = 'k', s = 90)
```

```
# plot lines
plt.plot(x1,x2p, color = 'red', linestyle = '--')
plt.plot(x1,x2m, color = 'forestgreen', linestyle = '--')
plt.plot(x1,x20,color = 'purple')

# add text
plt.text(5+2*delx,6.5+2*dely,'$w_\cdot_x_-_b_=_0$',
         color = 'purple', rotation = thetasvm, size = 15)
plt.text(5-delx, 7.6-dely, '$w_\cdot_x_-_b_=_1$',
         color = 'red', rotation = thetasvm, size = 15)
plt.text(5, 4.8,'$w_\cdot_x_-_b_=_-1$',
    color = 'forestgreen', rotation = thetasvm, size = 15)
plt.text(1.8,4.3,'w', color = 'blue', size = 15)
plt.text(0.3,2.1,'$Q_1$', color = 'k', size = 15)
plt.text(6.8,1.6,'$Q_2$', color = 'k', size = 15)
plt.text(5.8,8.6,'$Q_3$', color = 'k', size = 15)

# add normal direction of the hyperplanes
plt.arrow(2,3.86,w[0],w[1],color = 'blue', head_width = 0.1)
plt.show()
```

SVM maximizes $2/|\mathbf{w}|$, i.e., minimizes $|\mathbf{w}|$. The computer code shows that $\mathbf{w} = (-0.39996, 0.53328)$. This optimization is reached at the three SVs $P_9(1,5)$, $P_{17}(6,5)$, and $P_{19}(2,2)$. We can verify that the end points of these three SVs are on the two hyperplanes, by plugging these coordinates into the following two linear equations:

$$\mathbf{w} \cdot \mathbf{x} - b = \pm 1. \tag{9.43}$$

Reversely, these equations have three parameters w_1, w_2, b, which can be determined by three support vectors (SVs).

The linear SVM separation can have nonlinear extensions, such as circular or spherical hypersurfaces. Nonlinear SVM is apparently needed when the data cannot be easily separated by linear hyperplanes. Both R and Python codes have function parameters for the nonlinear SVM. Another SVM extension is from two classes to many classes. Multi-class and nonlinear SVM are beyond the scope of this book. The interested readers are referred to the designated machine learning books, such as those listed at the section of References and Further Reading at the end of this chapter.

9.3 Random Forest Method for Classification and Regression

Random forest (RF) is a popular machine learning algorithm and belongs to the class of supervised learning. It uses many decision trees trained from the given data, called the training data. The training data usually include categorical data, such as weather types (e.g., rainy, cloudy, and sunny) as labels, and numeric data, such as historic instrument observations (e.g., atmospheric pressure, humidity, wind speed, wind direction, and air temperature). These data form many logical decision trees that decide under what sets of conditions the weather is considered rainy, cloudy, or sunny. Many decision trees form a forest. Multiple decision nodes and branches for each tree are determined and trained by using a set of random sampling procedures. The random sampling and the decision trees lead to the name random forest. The training step results in an RF model that is a set of decision trees. In the prediction step, you submit your new data to the trained RF model whose each decision tree makes a vote for a category based on your new data. If the category "tomorrow is rainy" receives the most votes, then you have predicted a rainy day tomorrow. In this sense, RF is an ensemble learning method. Because of the nature of ensemble predictions, an RF algorithm should ensure the votes among the trained trees have as little correlation as possible.

This is a layman's description of RF for classification, when the RF objective is for categorical data, or called factor in data types. RF can also be used to fill in the missing numeric real values. The missing values, or missing data in climate science, are often denoted by NA, NaN, or -99999. RF makes a prediction for the missing data. The prediction is an ensemble mean from the trained RF trees. This is an RF regression, which produces numerical results.

9.3.1 RF Flower Classification for a Benchmark Iris Dataset

We use the popular iris flower dataset (Fisher 1936) frequently used for RF teaching as an example to explain RF computing procedures and to interpret RF results. Sir Ronald A. Fisher (1890–1962) was a British statistician and biologist. The Fisher dataset contains three iris species: *setosa*, *versicolor*, *virginica*. The data are the length and width of sepal and petal of a flower. The first two rows of the total 150 rows of the dataset are as follows.

```
#     Sepal.Length Sepal.Width Petal.Length Petal.Width Species
#1         5.1         3.5         1.4          0.2     setosa
#2         4.9         3.0         1.4          0.2     setosa
```

Figure 9.8 shows the entire dataset by connected dots. Figure 9.8 may be generated by the following computer code.

```
#R plot Fig. 9.8: R.A. Fisher data of three iris species
setwd('~/climstats')
data(iris) #read the data already embedded in R
dim(iris)
#[1] 150    5
iris[1:2,] # Check the first two rows of the data
#     Sepal.Length Sepal.Width Petal.Length Petal.Width Species
#1         5.1         3.5         1.4          0.2     setosa
#2         4.9         3.0         1.4          0.2     setosa

str(iris) # Check the structure of the data
#'data.frame':  150 obs. of  5 variables:
#$ Sepal.Length: num  5.1 4.9 4.7 4.6 5 5.4 4.6 5 4.4 4.9 ...
#$ Species      : Factor w/ 3 levels "setosa","versicolor",...
```

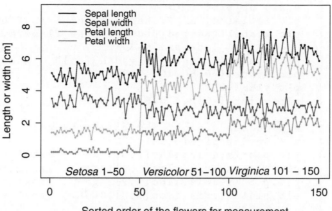

Figure 9.8 The R. A. Fisher iris dataset of sepal length, sepal width, petal length, and petal width for flower species (Fisher 1936).

```
setEPS() #Plot the data of 150 observations
postscript("fig0908.eps", width=7, height=5)
par(mar= c(4.5, 4.5, 2.5, 0.2))
plot(iris[,1], type = 'o', pch = 16, cex = 0.5, ylim = c(-1, 9),
     xlab = 'Sorted order of the flowers for measurement',
     ylab = 'Length or width [cm]',
     main = 'R. A. Fisher data of iris flowers', col = 1,
     cex.lab = 1.3, cex.axis = 1.3)
lines(iris[,2], type = 'o', pch = 16, cex = 0.5, col=2)
lines(iris[,3], type = 'o', pch = 16, cex = 0.5, col=3)
lines(iris[,4], type = 'o', pch = 16, cex = 0.5, col=4)
legend(0, 9.5, legend = c('Sepal length', 'Sepal width',
                          'Petal length', 'Petal width')),
       col = 1:4, lty = 1, lwd = 2, bty = 'n',
       y.intersp = 0.8, cex = 1.2)
text(25, -1, 'Setosa 1-50', cex = 1.3)
text(75, -1, 'Versicolor 51-100', cex = 1.3)
text(125, -1, 'Virginica 101 - 150', cex = 1.3)
dev.off()
```

```
#Python plot Fig. 9.8: R.A. Fisher data of iris species
iris = datasets.load_iris()
# notice "target" is equivalent to "species"
iris = pd.DataFrame(data=np.c_[iris['data'],iris['target']],
            columns= iris['feature_names'] + ['target'])

# check the structure of the data
print(iris.info(), '\n')

# check the first two rows of the data
iris.iloc[0:2,]
```

```
# set up plot
plt.title("R.␣A.␣Fisher␣data␣of␣iris␣flowers", pad = 10)
plt.xlabel("Sorted␣order␣of␣the␣flowers␣for␣measurement",
           labelpad = 10)
plt.ylabel("Length␣or␣width␣[cm]", labelpad = 10)
plt.xticks([0,50,100,150])
plt.ylim(-2,9)
plt.yticks([0,2,4,6,8])

# plot the data of 150 observations
x = np.linspace(0,149,150)
plt.plot(x,iris.iloc[:,0],'ko-', label = 'Sepal␣length')
plt.plot(x,iris.iloc[:,1],'ro-', label = 'Sepal␣width')
plt.plot(x,iris.iloc[:,2],'go-', label = 'Petal␣length')
plt.plot(x,iris.iloc[:,3],'bo-', label = 'Petal␣width')

# add legend
plt.legend()

# add text
plt.text(13,-1, "Setosa␣1-50", size = 15)
plt.text(57,-1, "Versicolor␣51-100", size = 15)
plt.text(107,-1, "Virginica␣101-150", size = 15)

plt.show()
```

The first 50 are *setosa*, which are the smallest flower, and the last 50 for *virginica* that have the largest petals. Given the data of sepal and petal sizes, one may correctly detect their corresponding flower species. However, not all the flowers have the same size. This makes the species detection more difficult. The RF algorithm can make the detection with a small error. We make an RF experiment with this dataset. We randomly select 120 rows of data as the training data to obtained a train RF model with 500 decision trees, use the remaining 30 rows as the new data for our detection. These decision trees vote on each of the 30 rows of the new data. The most species votes a row receives from the trained RF trees is the RF detected result. By default, the decision trees in an RF algorithm are built randomly so the RF result from each run based on the same data may be slightly different. The following computer code shows the RF run and its result of this experiment for Fisher's iris data. The code also generates Figure 9.9.

```
#R code: RF prediction using the Fisher iris data
#install.packages("randomForest")
library(randomForest)
#set.seed(8)  # run this line to get the same result
#randomly select 120 observations as training data
train_id = sort(sample(1:150, 120, replace = FALSE))
train_data = iris[train_id, ]
dim(train_data)
#[1] 120   5
#use the remaining 30 as the new data for prediction
new_data = iris[-train_id, ]
dim(new_data)
#[1] 30   5
```

```
#train the RF model
classifyRF = randomForest(x = train_data[, 1:4],
                y = train_data[, 5], ntree = 800)
classifyRF #output RF training result
#Type of random forest: classification
#Number of trees: 800
#No. of variables tried at each split: 2

#OOB estimate of  error rate: 4.17%
#Confusion matrix:
#              setosa versicolor virginica class.error
#setosa           41         0         0  0.00000000
#versicolor        0        34         2  0.05555556
#virginica         0         3        40  0.06976744

plot(classifyRF) #plot the errors vs RF trees Fig. 9.9a

classifyRF$importance #classifyRF$ has many outputs
#             MeanDecreaseGini
#Sepal.Length      8.313131
#Sepal.Width       1.507188
#Petal.Length     31.075960
#Petal.Width      38.169763
varImpPlot(classifyRF) #plot the importance result Fig. 9.9b

#RF prediction for the new data based on the trained trees
predict(classifyRF, newdata = new_data[,1:4])
# 2         4         10        11        12        14
#setosa    setosa    setosa    setosa    setosa    setosa

#It got two wrong: 120 versicolor and 135 versicolor

#Another version of the randomForest() command
anotherRF = randomForest(Species ~ .,
                    data = train_data, ntree = 500)
```

```
#Python code: RF prediction with the Fisher iris data
from sklearn.ensemble import RandomForestClassifier
from sklearn.metrics import confusion_matrix
from sklearn.metrics import accuracy_score
from sklearn import metrics
# randomly select 120 observations as training data
train_id = random.sample(range(0, 150), 120)
train_data = iris.iloc[train_id,:]
print(train_data.shape)# dimensions of training data
# use the remaining 30 as the new data for prediction
new_data = iris.drop(train_id)
print(new_data.shape)# dimensions of new data
# train the RF model
classifyRF = RandomForestClassifier(n_estimators=800,
                              oob_score=True)
training = classifyRF.fit(train_data.iloc[:,:4],
          train_data.iloc[:,4])
```

```
predictions = classifyRF.predict(new_data.iloc[:,:4])
print(predictions)
#[0. 0. 0. 0. 0. 0. 0. 0. 0. 1. 1. 1. 2.
#1. 2. 1. 1. 2. 1. 1. 2. 2. 2. 2.
# 2. 2. 2. 2. 2. 2.]
#0 = setosa, 1 = versicolor, 2 = virginica
confusion_M = confusion_matrix(new_data.iloc[:,4],
                                 predictions)
print(confusion_M)
#[[ 9  0  0] #all correct predictions for setosa
# [ 0  8  3]  #3 wrong predictions for versicolor
# [ 0  0 10]] #all correct predictions for virginica
accuracy_score(new_data.iloc[:,4],
                                 predictions)
#0.9

#Python lot Fig. 9.9(a): RF mean accuracy
n_estimators = 100
classifyRF = RandomForestClassifier(n_estimators,
                                    oob_score=True)
RFscore = []
for i in range(1, n_estimators + 1):
    classifyRF.set_params(n_estimators=i)
    classifyRF.fit(train_data.iloc[:,:4],
                train_data.iloc[:,4])
    RFscore.append(classifyRF.score(train_data.iloc[:,:4],
                train_data.iloc[:,4]))
plt.plot(RFscore)
plt.title("Random forest score of the mean accuracy")
plt.xlabel("Number of estimators")
plt.ylabel("OOB score")
plt.show()
```

```
#Python plot Fig. 9.9(b): RF importance plot
# Python shows the importance results differently from R
plt.plot(np.linspace(1,4,4),
        classifyRF.feature_importances_,
        'ko', markersize = 15)
plt.title("Importance plot of RF model", pad = 10)
plt.ylabel("Mean decrease in impurity", labelpad = 12)
plt.xticks([1,2,3,4], ['Sepal.Length','Sepal.Width',
    'Petal.Length', 'Petal.Width'], rotation = 90)
plt.yticks(classifyRF.feature_importances_,
        ['0.01','0.11','0.39','0.48'])
plt.grid()
plt.show()
```

The following explains the RF output data and figures.

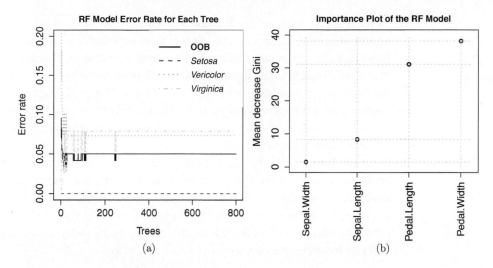

Figure 9.9   (a) RF errors vs trees. (b) The RF's importance plot.

- Confusion matrix: This matrix displays correct and wrong classifications based on the training data. The columns are truth and rows are RF classifications. The diagonals are corrected classifications, and off diagonals are the wrong ones. The 41 *setosas* are all correctly classified. Among the 37 *versicolors*, 34 are correctly classified, but 3 are wrongly classified as *virginica*. Among the 37 *versicolors*, 34 are correctly classified, but 3 are wrongly classified as *virginica*.

- Classification error: The last column in the confusion matrix is the classification error which is equal to the sum of the off-diagonal elements in that row divided by the sum of the entire row, i.e., the ratio of the number of wrong classifications to the total number of classifications. The errors show how good the trained RF model is.

- OOB estimate of error rate: OOB stands for out-of-bag, a term for a statistical sampling process that puts a part of the data for training the RF model and the remainder for validation. This remainder is referred to as the OOB set of data. The OOB error rate is defined as the ratio of the number of wrong decisions based on the majority vote to the size of OOB set.

- Number of variables tried at each split: This is the number of variables randomly sampled for growing trees, and is denoted by `mtry`. The Fisher iris data have four variables ($p = 4$). It is recommended that $\texttt{mtry} = \sqrt{p}$ for RF classification, and $= p/3$ for RF regression.

- Number of trees: We want to build enough trees for RF that the RF errors shown in Figure 9.9(a) stabilized. Too few trees may yield results of large differences at different runs.

- RF errors vs trees: The colored lines show errors of different types of validation procedures. As RF users, we pay attention to whether these errors are stabilized. The detailed error definitions are beyond the scope of this chapter.

- Mean decrease Gini scores and the importance plot (Fig. 9.9(b)): A higher mean decrease Gini (MDG) score corresponds to more importance of the variable in the model. In our example, the petal width is the most important variable. This agrees with our intuition from Figure 9.8. The figure shows that the petal width has clear distinctions among the three species and little variance within species. The petal length also has clear distinctions, but has larger variances, and is the second most important variable. The sepal width has almost no distinctions among the species and has the smallest MDG score, 1.5. It is the least important variable.
- RF prediction: Finally, RF predictions were made for the 30 rows of new data. The prediction results for the first six rows of the new data are shown here. These are the final results. Among the 30 rows, RF correctly predicted 28: 120 and 135 should be *virginica*, but RF predicted *versicolor* for both. In weather forecasting, these can be the public weather outlook of the 9th day from today (e.g., sunny, rainy, or cloudy) for 30 different locations in a country.

9.3.2 RF Regression for the Daily Ozone Data of New York City

Similar to Fisher's iris data, the daily air quality data, the file name `airquality`, of New York City (NYC) from May 1, 1973 to September 30, 1973 (153 days) is another benchmark data for the RF algorithms. The dataset contains the following weather parameters: ozone concentration in the ground-level air, cumulative solar radiation, average wind speed, and daily maximum air temperature. When the ozone concentration is higher than 86 parts per billion (ppb), the air is considered unhealthy. The ozone concentration is related to solar radiation, wind, and temperature. The following list provides more information on this `airquality` dataset:

- The ozone data in ppb observed between 1300 and 1500 hours at Roosevelt Island of New York City. Of the 153 days, 37 had no data and are denoted by NA. The data range is 1–168 ppb. The maximum ozone level of this dataset is 168 ppb, occurred on August 26, 1973 when the cumulative solar radiation was high at 238 Lang, the average wind speed low at 3.4 mph, and Tmax moderate at 81°F.
- Cumulative solar radiation from 0800 to 1200 hours with units in Langley (Lang or Ly) (1 Langley = 41, 868 Watt·sec/m^2, or 1 Watt/m^2 = 0.085985 Langley/hour) in the lightwave length range 4,000–7,700 Angstroms at Central Park in New York City. The data range is 7–334 Lang.
- Average wind speed in miles per hour (mph) at 0700 and 1000 hours at LaGuardia Airport, less than 10 km from Central Park. The data range is 1.7–20.7 mph.
- Maximum daily temperature Tmax in °F at La Guardia Airport. The data range is 56–97 °F. The daily Tmax data can also be downloaded from the NOAA NCEI website by an online search using the words `NOAA Climate Data Online LaGuardia Airport`.

While RF analysis for this `airquality` dataset can be done for many purposes, our example is to fill the 37 missing ozone data with RF regression results. RFs use the four aforementioned parameters and their data to build decision trees. Figure 9.10(a) shows the original 116 observed data points in black dots and lines, and the RF regression estimates

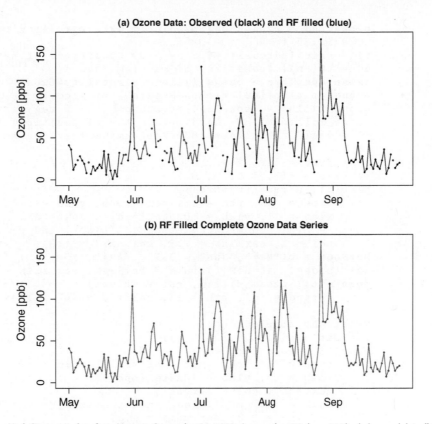

Figure 9.10   (a) New York City ozone data from May 1 to September 30, 1973. Among the 153 days, 116 had observed data (black dots and lines) and 37 had missing data. The missing data are filled by RF regression results (blue dots and lines). (b) The complete ozone data time series as a continuous curve with dots when the missing data are replaced by the RF regression.

for the 37 missing data in the blue dots and lines. A single blue dot means only an isolated day of missing data. A blue line indicates missing data in successive days. For better visualization, Figure 9.10(b) shows the complete NYC daily ozone time series that join the observed data with the result data from an RF regression.

Figure 9.10 may be generated by the following computer code.

```
#R plot Fig. 9.10: RF regression for ozone data
library(randomForest)
airquality[1:2,] #use R's RF benchmark data "airquality"
#   Ozone Solar.R Wind Temp Month Day
#1    41     190  7.4   67     5   1
#2    36     118  8.0   72     5   2
dim(airquality)
#[1] 153    6
ozoneRFreg = randomForest(Ozone ~ ., data = airquality,
            mtry = 2, ntree = 500, importance = TRUE,
            na.action = na.roughfix)
#na.roughfix allows NA to be replaced by medians
#to begin with when training the RF trees
```

```r
ozonePred = ozoneRFreg$predicted #RF regression result
t0 = 1:153
n1 = which(airquality$Ozone > 0) #positions of data
n0 = t0[-n1] #positions of missing data
ozone_complete = ozone_filled= airquality$Ozone
ozone_complete[n0] = ozonePred[n0] #filled by RF
ozone_filled = ozonePred #contains the RF reg result
ozone_filled[n1] <- NA #replace the n1 positions by NA
t1 = seq(5, 10, len = 153) #determine the time May - Sept

par(mfrow = c(2, 1))
par(mar = c(3, 4.5, 2, 0.1))
plot(t1, airquality$Ozone,
     type = 'o', pch = 16, cex = 0.5, ylim = c(0, 170),
     xlab = '', ylab = 'Ozone [ppb]', xaxt="n",
 main = '(a) Ozone data: Observed (black) and RF filled (blue)',
     col = 1, cex.lab = 1.3, cex.axis = 1.3)
MaySept = c("May","Jun", "Jul", "Aug", "Sep")
axis(side=1, at=5:9, labels = MaySept, cex.axis = 1.3)
points(t1, ozone_filled, col = 'blue',
       type = 'o', pch = 16, cex = 0.5)#RF filled data

#Plot the complete data
par(mar = c(3, 4.5, 2, 0.1))
plot(t1, ozone_complete,
     type = 'o', pch = 16, cex = 0.5, ylim = c(0, 170),
     xlab = '', ylab = 'Ozone [ppb]',
     xaxt="n", col = 'brown',
     main = '(b) RF filled complete ozone data series',
     cex.lab = 1.3, cex.axis = 1.3)
MaySept = c("May","Jun", "Jul", "Aug", "Sep")
axis(side=1, at=5:9, labels = MaySept, cex.axis = 1.3)
```

```python
#Python plot Fig. 9.10: RF regression for ozone data
from sklearn.ensemble import RandomForestRegressor
# Read in the airquality data and create a copy
airquality = pd.read_csv("data/airquality.csv",
                         index_col = 'Unnamed: 0')
airquality_copy = airquality.copy()
print(airquality.head())#print the first 5 rows of data
print(np.shape(airquality))#data matrix dim (153, 6)
# Create list of integers from 1 to 153
t0 = list(np.linspace(1,153,153, dtype=int))
# positions of recorded data
n1 = np.where(airquality["Ozone"] > 0)
n1 = n1[0].tolist()
n1 = [x+1 for x in n1]
# positions of unknown 'NaN' data
n0=[]
for x in t0:
    if x not in n1:
        n0.append(x)
# Replace NA with medians in order to train the RF trees
airquality['Solar.R'].fillna(
    value=airquality['Solar.R'].median(), inplace=True)
```

```python
airquality['Wind'].fillna(
    value=airquality['Wind'].median(), inplace=True)
airquality['Temp'].fillna(
    value=airquality['Temp'].median(), inplace=True)
airquality['Day'].fillna(
    value=airquality['Day'].median(), inplace=True)
airquality['Month'].fillna(
    value=airquality['Month'].median(), inplace=True)
airquality['Ozone'].fillna(
    value=airquality['Ozone'].median(), inplace=True)
# Create or features X and our target y
X = airquality[['Solar.R', 'Wind', 'Temp', 'Month', 'Day']]
y = airquality[['Ozone']]
# split our data into training and test sets
X_train = X.loc[n1,:]
X_test = X.loc[n0,:]
y_train = y.loc[n1,:]
y_test = y.loc[n0,:]
#create the RF Regressor model with 500 esitmators
ozoneRFreg = RandomForestRegressor(n_estimators=500,
                    oob_score=True, max_features = 2)
# fit model to our training set
ozoneRFreg.fit(X_train, y_train.values.ravel())
#create our prediction
ozonePrediction = ozoneRFreg.predict(X_test)
#create data frame with our results
result = X_test
result['Ozone'] = y_test
result['Prediction'] = ozonePrediction.tolist()
# create data frame with our predictions
ozone_filled = pd.DataFrame(None, index = np.arange(153),
                    columns = ["Prediction"] )
ozone_filled.loc[n0, 'Prediction'] = result["Prediction"]
ozone_filled.index += 1

#Python plot Fig. 9.10(a)
t1 = np.linspace(5,10,153)
mfrow = [2,1]
mar = [3, 4.5, 2, 0.1]
# original known data
ar = pd.read_csv("data/airquality.csv",
                    index_col = 'Unnamed:_0')
plt.plot(t1, ar["Ozone"],
        'ko-', markersize = 3)
plt.ylim(0, 170)
#unknown Predicted data
plt.plot(t1, ozone_filled['Prediction'],
        'bo-', markersize = 3)
plt.ylabel("Ozone_[ppb]", labelpad = 12)
MaySept = ["May","Jun", "Jul", "Aug", "Sep"]
plt.xticks([5,6,7,8,9], MaySept)
plt.title("(a)_Observed_(black)_and_RF_filled_(blue)")
plt.show()
```

```
#Python plot Fig. 9.10(b)
# combine unknown and known data into one dataframe
result_copy = result.copy()
result_copy = result_copy.rename(
    columns = {'Ozone':'y_tested', 'Prediction':'Ozone'})
ozone_complete = airquality_copy[
    "Ozone"].fillna(result_copy["Ozone"])
#plot the combined data
t1 = np.linspace(5,10,153)
mfrow = [2,1]
mar = [3, 4.5, 2, 0.1]
plt.plot(t1, ozone_complete, 'ro-', markersize = 3)
plt.ylim(0, 170)
plt.ylabel("Ozone [ppb]", labelpad = 12)
MaySept = ["May","Jun", "Jul", "Aug", "Sep"]
plt.xticks([5,6,7,8,9], MaySept)
plt.title("(b) RF-filled complete ozone data series")
plt.show()
```

9.3.3 What Does a Decision Tree Look Like?

We have learned that the RF prediction is the result of majority votes from the trained decision trees in a random forest. What, then, does a decision tree look like? Figure 9.11 shows a decision tree in the RF computing process for the training data of iris flowers. The gray box on top shows the percentage of each species in the 120 rows of training data. This particular set of training data was randomly selected from the entire R. A. Fisher dataset which has 150 rows. Of the 120 rows of training data, 41 rows are *setosa* (34%), 42 rows *versicolor* (35%), and 37 rows (31%) *virginica*. The first branch of the decision tree grows from a condition

$$\text{Petal Length} < 2.5 \text{ [cm].} \tag{9.44}$$

If "yes," this iris is *setosa*. All 41 *setosa* flowers (34% of the entire training data) have been detected. This decision is clearly supported by the real petal length data (the green line) shown in Figure 9.8, which shows that only *setosa* has petal length less than 2.5.

If "no," then two possibilities exist. This allows us to grow a new branch of the tree. The condition for this branch is

$$\text{Petal Width} < 1.8 \text{ [cm].} \tag{9.45}$$

This condition determines whether a flower is *versicolor* or *virginica*. After the isolation of 41 *setosa* flowers, the remaining 79 flowers are 42 *versicolor* (53%) and 37 *virginica* (47%). The "yes" result of condition (9.45) leads to *versicolor*. This result is 93%

A Decision Tree for the RF Training Data

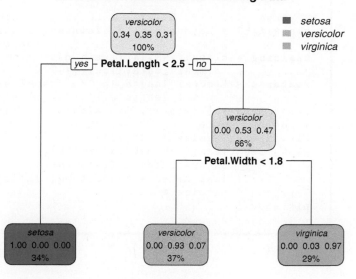

Figure 9.11   A trained tree in the random forest for the iris data of R. A. Fisher (1936).

correct and 7% incorrect. The "no" result implies *virginica*. This conclusion is 97% correct, and 3% incorrect. These conclusions are supported by the real petal width data (the blue line) shown in Figure 9.8. The petal width of *versicolor* iris flowers is in general less than 1.8 [cm]. However, the data have some fluctuations in both the *versicolor* and *virginica* sections in Figure 9.8. These fluctuations lead to the errors of decision.

The entire RF process for our iris species example had grown 800 such decision trees using the 120 rows of training data. Different trees use different branch conditions. RF algorithms grow these trees following various kinds of optimization principles, which involve some tedious mathematics not covered here.

Figure 9.11 may be generated by the following computer code.

```
#R plot of Fig. 9.11: A tree in a random forest
#install.packages('rpart')
library(rpart)
#install.packages('rpart.plot')
library(rpart.plot)

setwd('~/climstats')
setEPS() #Plot the data of 150 observations
postscript("fig0911.eps", width=6, height=6)
par(mar = c(0, 2, 1, 1))
iris_tree = rpart( Species ~. , data = train_data)
rpart.plot(iris_tree,
   main = 'A decision tree for the RF training data')
dev.off()
```

```
#Python plot Fig. 9.11: A tree in a random forest
classifyRF = RandomForestClassifier(n_estimators=1,
                               oob_score=True)
training = classifyRF.fit(train_data.iloc[:,:4],
            train_data.iloc[:,4])
features = ['sepal length (cm)', 'sepal width (cm)',
            'petal length(cm)', 'petal width (cm)']
class_name = ['setosa', 'versicolor', 'virginica']
# set up figure
plt.figure(figsize = (15,15))
plot_tree(classifyRF.estimators_[0],
        feature_names = features,
        class_names = class_name, filled = True,
        fontsize = 12)
plt.show()
```

9.4 Neural Network and Deep Learning

A neural network (NN) consists of a series of data fitting according to the training data and an application of the fitted NN model for the test data. The NN outputs categorical predictions, such as a sunny or rainy day, or numerical predictions as a regression. NN is also known as an artificial neural network (ANN) or simulated neural network (SNN). It is a popular machine learning method, and is a fundamental building block of the deep learning algorithms that often involve the data fitting of multiple layers, referred to as hidden layers.

Why is it called a neural network? How does a data fitting process have anything to do with "neural" and/or "network"? Artificial neurons were first proposed by Warren Sturgis McCulloch (1898–1969), an American neurophysiologist, and Walter Harry Pitts (1923–1969), an American logician, in their 1943 paper entitled "A logical calculus of ideas immanent in nervous activity." This mathematical paper used the terms "neuron," "action," "logic expression," and "net," which are among the keywords in modern NN writings. The ten theorems of McCulloch and Pitts' paper formulate a suite of logic expressions based on data. The word "neuron" was more a graphic indication for actions and logic expressions than a biological reference. A layman's understanding of NN machine learning is often incorrectly articulated in terms of biological neurons and a biological neural network. Therefore, ANN may be a more appropriate term for NN to avoid confusion.

This book attempts to briefly introduce NN so that you can use our R or Python code for your data and objectives, interpret your NN computing results, and understand the basic principles of mathematical formulations of an NN algorithm. However, we do not attempt to derive mathematical details of the NN theory. This section has two subsections: a simple NN example of a decision system, and an example of NN prediction using the benchmark data of Fisher's iris flowers.

9.4.1 An NN Model for an Automized Decision System

9.4.1.1 An Overall Idea of a Neural Network Decision System for Recruitment

A simple example is as follows. A senior human resource manager of an IT company recruited three new employees from among six candidates. Hiring decisions were based on the technical knowledge scores (TKS) and communication skills scores (CSS) given in Table 9.1: TKS and CSS were the interview results from company's working groups. A junior human resource manager wishes to use an NN model to help make recruitment decisions consistent with those made by the senior manager.

Table 9.1	TKS and CSS data and recruitment decisions				
TKS	20	10	30	20	80	30
CSS	90	20	40	50	50	80
Decision	Hire	Reject	Reject	Reject	Hire	Hire

The junior manager has received the following scores for three new job applicants A, B, and C, whose TKS and CSS scores are as follows: TKS (30, 51, 72), and CSS (85, 51, 30), respectively. The senior manager's ruling seems to suggest that a candidate is recruited when either TKS score or CSS score is high. Thus, this junior manager has an easy decision for Candidate A whose CSS 85 is high, so to be recruited. However, the decision for Candidate B is difficult as neither TKS 51 nor CSS 51 are high, but both are higher than those of the rejected candidates, and further the sum of TKS and CSS is 102, lower than the minimum of the totals of the recruited, 110. Should the junior manager hire Candidate B? The decision for Candidate C is also difficult. Candidate C seems weaker than the recruited case TKS 30 and CSS 80, but not too much weaker. Should the junior manager hire Candidate C? The following NN computer code may learn from the data of the senior manager and suggest an NN decision that may serve as a useful reference for the junior manager.

9.4.1.2 An NN Code, Results, and Their Interpretation

```
#R code for NN recruitment decision and Fig. 9.12
#Ref: https://www.datacamp.com/tutorial/neural-network-models-r
TKS = c(20,10,30,20,80,30)
CSS = c(90,20,40,50,50,80)
Recruited = c(1,0,0,0,1,1)
#make a data frame for the NN function neuralnet
df = data.frame(TKS, CSS, Recruited)

require(neuralnet) # load 'neuralnet' library
# fit neural network
set.seed(123)
nn = neuralnet(Recruited ~ TKS + CSS, data = df,
```

```
                 hidden = 5,  act.fct = "logistic",
                 linear.output = FALSE)
plot(nn) #Plot Fig 9.12: A neural network

TKS=c(30,51,72) #new data for decision
CSS=c(85,51,30) #new data for decision
test=data.frame(TKS,CSS)
Predict=neuralnet::compute(nn,test)
Predict$net.result #the result is probability
#[1,]  0.99014936
#[2,]  0.58160633
#[3,]  0.01309036

# Converting probabilities into decisions
ifelse(Predict$net.result > 0.5, 1, 0) #threshold = 0.5
#[1,]     1
#[2,]     1
#[3,]     0

#print bias and weights
nn$weights[[1]][[1]]
#print the last bias and weights before decision
nn$weights[[1]][[2]]
#print the random start bias and weights
nn$startweights
#print error and other technical indices of the nn run
nn$result.matrix #results data
```

```
   #Python code for the NN recruitment decision and Fig. 9.12
   #imports to build neural network model and visualize
   #You need to install the following Python environments:
   #tensorflow and keras-models
   #!pip3 install ann_visualizer
   #!pip3 install keras-models
   #from a terminal: pip install tensorflow
   from ann_visualizer.visualize import ann_viz
   #from graphviz import Source
   #from tensorflow.keras import layers
   from keras.models import Sequential
   from keras.layers import Dense

   TKS = [20,10,30,20,80,30]
   CSS = [90,20,40,50,50,80]
   Recruited = [1,0,0,0,1,1]
   #combine multiple columns into a single set
   df = pd.DataFrame({'TKS': TKS,'CSS': CSS,
        'Recruited': Recruited})
   df.head()
   X = df[["TKS", "CSS"]]
   y = df[["Recruited"]]
   #random.seed(123).
   #The model is random due to too little training data

   nn = Sequential()
```

```
nn.add(Dense(5, input_dim = 2, activation = "sigmoid"))
nn.add(Dense(1,activation = "sigmoid"))
#nn.add(Dense(2,activation = "sigmoid"))
nn.compile(optimizer = "adam", loss = "BinaryCrossentropy",
          metrics = "BinaryAccuracy")
#fit neural network to data
nn.fit(X,y)
#visualize the neural network
ann_viz(nn, title = "Recruitment Decision")
#output the nn model weights and biases
print(nn.get_weights())

TKS = [30,51,72] #new data for decision
CSS = [85,51,30] #new data for decision
test = pd.DataFrame({'TKS': TKS,'CSS': CSS})
prediction = nn.predict(test)
print(prediction)
#[[0.70047873]
# [0.66671777]
# [0.38699177]]
#Converting probabilities into decisions
for i in range(3):
  if prediction[i] >= 0.5:
    print("Hire")
  else:
    print("Reject")
#Hire
#Hire
#Reject
```

Figure 9.12 shows a simple neural network plotted by the previous computer code. The black numbers are called weights w_{ij} that are multiplied by the input data x_i for neuron j. The blue numbers are called biases b_j, associated with neurons j, indicated by the circles pointed to by a blue arrow. The last circle on the right is the result, or called the output layer. The first two circles on the left indicate input data, and form the input layer. The weights and biases are mathematically aggregated in the following way:

$$z_j = \sum_{i=1}^{n} w_{ij}x_i + b_j, \qquad (9.46)$$

where z_j are for fitting an activation function at neuron j when all the training data are used. In the computer code, the activation function is logistic. A logistic function is defined as

$$g(z) = \frac{1}{1 + \exp(-k(z - z_0))}, \qquad (9.47)$$

where k is called the logistic growth rate, and z_0 is the midpoint. This function can also be written as

$$g(z) = \frac{1}{2} + \frac{1}{2}\tanh\left(-\frac{k(z - z_0)}{2}\right), \qquad (9.48)$$

A Neural Network with One Hidden Layer

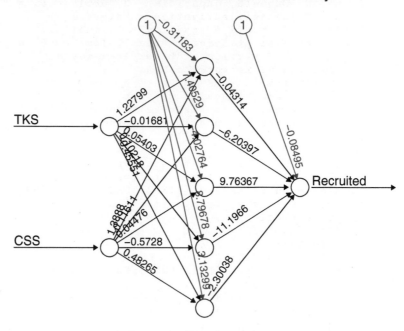

Error: 0.000309   Steps: 135

Figure 9.12   A simple neural network of five neurons in a hidden layer for an NN hiring decision system.

where the tanh function is defined as

$$\tanh(t) = \frac{\exp(t) - \exp(-t)}{\exp(t) + \exp(-t)}. \tag{9.49}$$

A curve for a logistic function is shown in Figure 9.13. The curve has properties $f(-\infty) = 0$ and $f(\infty) = 1$. This makes the logistic activation function useful for categorical assignment: False for 0 and True for 1.

Figure 9.13 can be plotted by the following computer code.

```
#R plot Fig. 9.13: Curve of a logistic function
y = seq(-2, 4, len = 101)
k = 3.2
y0 = 1
setEPS() #Automatically saves the .eps file
postscript("fig0913.eps", height=5, width=7)
par(mar = c(4.2, 4.2, 2, 0.5))
plot(y, 1/(1 + exp(-k*(y - y0))),
     type = 'l', col = 'blue', lwd =2,
     xlab = 'y', ylab = 'f(y)',
     main = 'Logistic function for k = 3.2, y0 = 1.0',
     cex.lab = 1.3, cex.axis = 1.3)
dev.off()
```

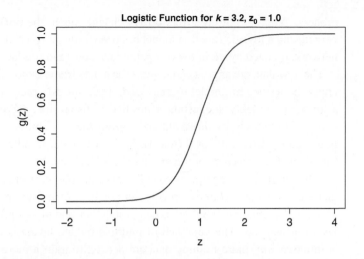

Figure 9.13 Logistic function with growth rate $k = 3.2$ and midpoint $y_0 = 1.0$.

```
#Python plot Fig. 9.13: Curve of a logistic function
# define variables
y = np.linspace(-2,4,101)
k = 3.2
y0 = 1

# plot figure
plt.plot(y, 1/(1+ np.exp(-k*(y-y0))), 'b', linewidth = 2.5)
# add labels
plt.title('Logistic␣function␣for␣k␣=␣3.2,␣z0␣=␣1.0',
          pad = 10)
plt.xlabel("z", labelpad = 10)
plt.ylabel("g(z)", labelpad = 10)
plt.show()
```

When the growth rate $k = 1$ and midpoint $y_0 = 0$, the logistic function is often referred to as a sigmoid function. Some Python or R code uses sigmoid to represent the activation function.

The idea of data-based decision processes is that you integrate all the data, develop a model through data fitting with a specified optimization, and apply the model to new data to generate predictions. The neural network integrates data by assigning each datum a weight, as in Eq. (9.46), and assigning each weighted sum a bias. The weighted data z_j and the decision data, 0 or 1, from the training dataset, form a series of data pairs (z, d). These data pairs are used for the logistic data fitting many times to train a neural network model. Each time, the NN algorithm will try to optimize its data fitting, such as minimizing mean square errors (MSE) of the fitting. This optimization process, called back-propagation in an NN algorithm, updates the weights and biases. This makes NN learning an optimization process. When the fitting errors do not change much, the optimization process has converged, and a logistic function model has been trained. You can now plot the trained neural network, denoted by nn in the computer code here, and produce Figure 9.12.

The new data are aggregated together as a data frame denoted by test in our code. You apply the trained nn model to test and obtain the NN prediction result as probabilities, a higher probability suggesting a positive decision, i.e., recruited. You may use 0.5 as your hiring probability threshold and convert the probability result into a decision result, indicated by 1 (recruit) or 0 (reject). In this example, Candidates A and B are hired and Candidate C is rejected, as suggested by this NN model.

When running the computer code for this example of hiring data, you may have noticed that each run has a different prediction result. This makes the prediction unreliable. To improve the situation, we need more training data and more optimizations by using several layers of neurons. The approach of multiple hidden layers is a type of deep learning. It is intuitive why more training data are needed to train a more reliable model, since few data cannot cover all possibilities. In the next subsection, we will present an example of NN deep learning using more than six groups of training data. Specifically, we will use 75 groups of the Fisher iris flower data.

9.4.1.3 Dissect NN Results and Errors

Logistic regression fits weighted data to a logistic function. Logistic regression can be a standalone statistics course or a chapter in a statistics book. Interested readers are referred to those books and courses. In addition, there are other types of activation functions, such as linear, rectified linear, leaky rectified linear, exponential linear, and SoftPlus. You can find these functions from the specialized ML books or online.

Usually a software package of neural network can output the weights and bias, such as using the following R command:

```
nn$weights[[1]][[1]]
```

in the R NN package neuralnet. R can also output the initial assignment of weights and bias:

```
nn$weights[[1]][[2]]
```

The initial weights and biases are randomly assigned, which makes the NN result different for different runs when having insufficient amounts of data, or an insufficient number of hidden layers.

An NN computer package can also output many other modeling results, such as an R command:

```
nn$result.matrix
```

These output data can be used to analyze the quality of your trained model, so, it is unfair to say that NN is a blackbox. Yet, analyzing a trained model for a complex NN is very difficult and requires some nontrivial mathematical preparation.

Many NN users may not have the mathematical ability to analyze the trained model based on mathematical theories. Instead, they apply the model to the new data and see if the NN prediction makes sense based on their experience or common sense. If not, train the model again with different parameters, or even feed more training data.

9.4.2 An NN Prediction of Iris Species

We wish to use the Fisher iris data to show an example of NN with a sufficiently large size of training data and a highly reliable result. The Fisher iris data is a 150×5 matrix with three iris species: 50 *setosa*, 50 *virginica*, and 50 *versicolor*, as described in the random forest section, Section 9.3 (Fisher 1936). We wish to use a percentage of the data as the training data and the rest as the test data. The following computer code is for 50% of the data to be the training data, i.e., $p = 0.5$ in the code. We choose 10 neurons and 10 hidden layers. The NN prediction result shows that NN correctly predicted all the 27 *setosa* irises, got 19 correct among the 20 *versivolor* irises but got one *versicolor* incorrectly as *virginica*, and got 26 correct among the 28 *virginica* irises but got 2 incorrectly as *versicolor*. Since the 75 groups of training data are plenty, compared to the 6 groups of training data for the hiring decision, the trained NN model for the iris flowers does not vary much for different runs. The prediction results are quite reliable. Using more neurons and hidden layers may also help achieve reliable results. Although the results from different runs of the NN code do not change much, small differences still exist. For example, you may get a result of predicting 27 *virginica*.

When you change $p = 0.8$, you will have 120 rows of training data and 30 rows of test data, as we used in Section 9.3 for the random forest example. This increase of training data size helps further stabilize prediction results, i.e., little difference between the results of the different runs. You can make some runs of the code on your own computer, and see how many incorrect predictions are there. Our example runs show only one or two incorrect predictions among the 30 rows of test data. For example, in one run, the 9 *setosa* and 12 *versicolor* irises were correctly predicted, and the 9 *virginica* irises had 8 correct predictions and 1 wrong. The level of the NN prediction accuracy seems similar to that of the random forest prediction for this particular dataset.

```
#R NN code for the Fisher iris flower data
#Ref: https://rpubs.com/vitorhs/iris
data(iris) #150-by-5 iris data
#attach True or False columns to iris data
iris$setosa = iris$Species == "setosa"
iris$virginica = iris$Species == "virginica"
iris$versicolor = iris$Species == "versicolor"
p = 0.5 # assign 50% of data for training
train.idx = sample(x = nrow(iris), size = p*nrow(iris))
train = iris[train.idx,] #determine the training data
test = iris[-train.idx,] #determine the test data
dim(train) #check the train data dimension
#[1] 75  8

#training a neural network
library(neuralnet)
```

```
#use the length, width, True and False data for training
iris.nn = neuralnet(setosa + versicolor + virginica ~
                        Sepal.Length + Sepal.Width +
                        Petal.Length + Petal.Width,
                    data = train, hidden=c(10, 10),
                    rep = 5, err.fct = "ce",
                    linear.output = F, lifesign = "minimal",
                    stepmax = 1000000, threshold = 0.001)

plot(iris.nn, rep="best") #plot the neural network

#Prediction for the rest data
prediction = neuralnet::compute(iris.nn, test[,1:4])
#prediction$net.result is 75-by-3 matrix
prediction$net.result[1:2,]
#  [,1]          [,2]             [,3]
#2     1 3.531638e-09 6.745692e-137
#4     1 3.527872e-09 6.756521e-137

#find the column that contains the max of a row
pred.idx <- apply(prediction$net.result, 1, which.max)
pred.idx
#2  4  6  8  9 10
#1  1  1  1  1  1

#Assign 1 for setosa, 2 for versicolor, 3 for virginica
predicted <- c('setosa', 'versicolor', 'virginica')[pred.idx]
predicted[1:6] #The prediction result
#[1] "setosa" "setosa" "setosa" "setosa" "setosa" "setosa"

#Create confusion matrix: table(prediction,observation)
table(predicted, test$Species)
#predicted       setosa versicolor virginica
#setosa            27          0          0
#versicolor         0         19          2
#virginica          0          1         26
```

```
    #Python NN code for the iris flower prediction
    from ann_visualizer.visualize import ann_viz
    from graphviz import Source
    from keras.models import Sequential
    from keras.layers import Dense
    from sklearn.model_selection import train_test_split

    iris = pd.read_csv("data/iris.csv") #150-by-5 iris data
    #attach True or False columns to iris data
    iris['setosa'] = iris['class'].map(lambda x:
                            x == "Iris-setosa")
    iris['versicolor'] = iris['class'].map(lambda x:
                            x == "Iris-versicolor")
    iris['virginica'] = iris['class'].map(lambda x:
                            x == "Iris-virginica")
```

```
X = iris.iloc[:,0:4]#legnth and width data
y = iris.iloc[:,5:8]#species data
y["setosa"] = y["setosa"].astype(int)
y["versicolor"] = y["versicolor"].astype(int)
y["virginica"] = y["virginica"].astype(int)

#Randomly generate training and test datasets
X_train, X_test, y_train, y_test = train_test_split(X, y,
                                        test_size=0.5)
nn = Sequential() #define neural network nn
nn.add(Dense(10, activation='relu', input_dim = 4))
nn.add(Dense(10, activation='relu', input_dim = 10))
nn.add(Dense(3, activation='sigmoid'))
nn.compile(optimizer = "sgd", loss = "binary_crossentropy",
            metrics = ["accuracy"])
nn.fit(X_train, y_train, epochs = 5)#train nn
prediction = nn.predict(X_test) #use nn to predict
#output prediction result
p = prediction
for i in range(0, p.shape[0]):
    p[i] = np.where(p[i] == np.max(p[i]), True, False)
np.hstack((p, y_test))
#array([[1., 0., 0., 1., 0., 0.],
#        [1., 0., 0., 1., 0., 0.],
#        [1., 0., 0., 1., 0., 0.],
#        [1., 0., 0., 1., 0., 0.],
#        [0., 0., 1., 0., 0., 1.],
#        [0., 0., 1., 0., 0., 1.],
#        [1., 0., 0., 1., 0., 0.],
#        [0., 0., 1., 0., 0., 1.],
#        [1., 0., 0., 1., 0., 0.],
#        [0., 0., 1., 0., 1., 0.],
#The left 3 columns are the predictions results, and
#the right 3 columns are the validation data.
#The first 9 predictions are correct and the 10th is wrong
```

9.5 Chapter Summary

This chapter has described four methods of machine learning: K-means, support vector machine (SVM), random forest (RF), and neural network (NN). In the description, we have used the following datasets: the daily weather data at the Miami International Airport, the daily air quality data of New York City, and R. A. Fisher's iris flower data of species. We have provided both R and Python codes for these algorithms and their application examples. The following provides a brief summary of our descriptions about the four methods.

(i) K-means clustering: Simply speaking, this clustering method can fairly divide N identical candies on a table for K kids according to the candy locations. It is normally used as an unsupervised learning method. The K-means algorithm minimizes the total within cluster sum of squares (tWCSS) through iterations. As an example, we applied the K-means clustering method to the data of the daily minimum surface air temperature Tmin [in °C] and the direction of the fastest 2-minute wind in a day WDF2 [in degrees] of the Miami International Airport. Our tests suggested that the data have two clusters, one corresponding to the prevailing wind from the east, and another from the west.

(ii) Support vector machine for the maximum separation of different sets: The SVM algorithm is built on the principle of the maximum differences among the labeled groups of data. The maximum separation of different groups is measured by the distance between positive and negative hyperplanes. SVM can also predict the labels of the new unlabeled data. SVM is usually used as a tool of supervised learning.

(iii) Random forest of decision trees: An RF model is a set of decision trees which form a "forest of decisions." It is usually used as a tool of supervised learning. It has two steps: training and prediction. RF can be used for both classification and regression. A benchmark RF example is used in our RF classification description: the separation and prediction of the R. A. Fisher iris species dataset. We have also included an example on RF regression using the daily air quality data of New York City.

(iv) Neural network of multiple data fitting: An NN model is a series of data fitting for a given activation function. NN can also make classifications and numerical predictions as a regression. The logistic function is a popular activation function, a smooth transition between category 0 to category 1. Because randomness is involved in the NN algorithm, different runs of R or Python code may yield different results. We have included an example of using the NN model to help make a hiring decision using the existing data. To compare with the RF model, we have also applied the NN model to the R. A. Fisher flower dataset of iris species.

References and Further Reading

[1] B. Boehmke and B. Greenwell, 2019: *Hands-on Machine Learning with R*. Chapman and Hall.

> This machine learning book has many R code samples and examples. The book website is `https://bradleyboehmke.github.io/HOML/`.

[2] J. M. Chambers, W. S. Cleveland, B. Kleiner, and P. A. Tukey, 1983: *Graphical Methods for Data Analysis*. Wasworth International Group.

> This book contains the daily ozone and weather data from May 1, 1973 to September 30, 1973 of New York City. This dataset serves as a benchmark dataset in random forest algorithms. John M. Chambers is a Canadian statistician, who developed the S programming language, and is a core member of the R programming language project.

[3] R. A. Fisher, 1936: The use of multiple measurements in taxonomic problems. *Annals of Eugenics*, 7(2), 179–188.

> This paper contains the iris data frequently used in the teaching of machine learning. The current name of the journal *Annals of Eugenics* is *Annals of Human Genetics*. The author, Ronald Aylmer Fisher (1890–1962), was a British mathematician, statistician, biologist, and geneticist. Modern description of this dataset and its applications in statistics and machine learning can be found from numerous websites, such as
>
> `www.angela1c.com/projects/iris_project/the-iris-dataset/`

[4] A. Géron, 2017: *Hands-on Machine Learning with Scikit-Learn and TensorFlow: Concepts: Tools, and Techniques to Build Intelligent Systems*. O'Reilly.

> This machine learning book has many Python code examples.

[5] W. McCulloch and W. Pitts, 1943: A logical calculus of ideas immanent in nervous activity. *Bulletin of Mathematical Biophysics* 5, 115–133.

> This seminal work is considered the first paper to have created modern NN theory. The paper is highly mathematical, contains ten theorems, and includes a figure of neurons.

[6] T. Hastie, R. Tibshirani, and J. H. Friedman, 2017: *The Elements of Statistical Learning: Data mining, Inference, and Prediction.* 2nd ed., Springer.

> This is a very famous machine learning book and includes in-depth descriptions of the commonly used ML algorithms.

Exercises

9.1 Use the K-means method to conduct a cluster analysis for the following five points in the 2D space: $P_1(1,1), P_2(2,2), P_3(2,3), P_4(3,4), P_5(4,4)$. Assume $K = 2$. Plot a figure similar to Figure 9.1. What is the final tWCSS equal to?

9.2 Given the following three points: $P_1(1,1), P_2(2,2), P_3(2,3)$, and given $K = 2$, conduct a K-means cluster analysis following the method presented in Sub-section 9.1.1. Plot your K-means clustering result.

9.3 Use the K-means method to conduct a cluster analysis for the daily Tmin and WDF2 data of the Miami International Airport in 2015, following the method presented in Sub-section 9.1.3. Here, Tmin is the daily minimum temperature, and WDF2 denotes the direction [in degrees] of the fastest 2-minute wind in a day. You can obtain the data from the NOAA Climate Data Online website, or from the file

`MiamiIntlAirport2001_2020.csv`

in the `data.zip` file downloadable from the book website

`www.climatestatistics.org`

9.4 Use the K-means method to conduct a cluster analysis for the daily Tmax and WDF2 data of the Miami International Airport in 2015.

9.5 Use the K-means method to conduct a cluster analysis for the daily Tmin and Tmax data of the Miami International Airport in 2015.

9.6 Use the K-means method to conduct a cluster analysis for the daily Tmin and PRCP data of the Miami International Airport in 2015. Here, PRCP denotes the daily total precipitation.

9.7 Identify two climate parameters of your interest, find the corresponding data online, and conduct a K-means cluster analysis similar to the previous problem for your data.

9.8 Following the method presented in Subsection 9.2.1, conduct an SVM analysis for the following five points in a 2D space: $P_1(1,1), P_2(2,2), P_3(2,3), P_4(3,4), P_5(4,4)$. The first three points are labeled 1 and the last two are labeled 2. What are w, b, and D_m? What points are the support vectors?

9.9 Two new points are introduced to the previous problem: $Q_1(1.5, 1)$ and $Q_2(3, 3)$. Use the SVM trained in the previous problem to find out which point belongs to which category.

9.10 From the Internet, download the historical daily data of minimum temperature (Tmin) and average wind speed (AWND) for a month and a location of your interest. Label your data as 1 if the daily precipitation is greater than 0.5 millimeter, and 0 otherwise. Conduct an SVM analysis for your labeled data.

9.11 SVM forecast: From the Internet, download the historical daily data of minimum temperature and sea level pressure for a month and a location of your interest. Label your data as 1 if the total precipitation of the next day is greater than 0.5 millimeter, and 0 otherwise. Conduct an SVM analysis for your labeled data. Given the daily data of minimum temperature and sea level pressure of a day in the same month but a different year, use your trained SVM model to forecast whether the next day is rainy or not, i.e., to determine whether the next day is 1 or 0.

9.12 In order to improve the accuracy of your prediction, can you use the data of multiple years to train your SVM model? Perform numerical experiments and show your results.

9.13 The first 50 in the 150 rows of the R. A. Fisher iris data are for *setosa*, rows 51–100 are for *versocolor*, and 101–150 are for *virginica*. Use the data rows 1–40, 51–90, and 101–140 to train an RF model, following the method in Subsection 9.3.1. Use the RF model to predict the species of the remaining data of lengths and widths of petal and sepal. You can download the R. A. Fisher dataset `iris.csv` from the book website or use the data already built in R or Python software packages.

9.14 For the 150 rows of the R. A. Fisher iris data, use only 20% of the data from each species to train your RF model. Select another 10% of the iris data of lengths and widths as the new data for prediction. Then use the RF model to predict the species of the new data. Discuss the errors of your RF model and your prediction.

9.15 Plot a decision tree like Figure 9.11 for the previous exercise problem.

9.16 For the same R. A. Fisher dataset, design your own RF training and prediction. Discuss the RF prediction accuracy for this problem.

9.17 RF forecast: From the Internet, download the historical daily data of minimum temperature and sea level pressure for a month and a location of your interest. Label your data as 1 if the total precipitation of the next day is greater than 0.5 millimeter, and 0 otherwise. Conduct an RF analysis for your labeled data. Given the daily data of minimum temperature and sea level pressure of a day in the same month but a different year, use your trained RF model to forecast whether the next day was rainy or not. Is your forecast accurate? How can you improve your forecast?

9.18 Find a climate data time series with missing data and fill in the missing data using the RF regression method described in Subsection 9.3.2.

9.19 The first 50 in the 150 rows of the R. A. Fisher iris data are for *setosa*, rows 51–100 are for *versocolor*, and rows 101–150 are for *virginica*. Use the data rows 1–40, 51–90, and 101–140 to train an NN model, following the method in Subsection 9.4.2. Use the NN model to predict the species of the remaining data of lengths and widths of petal and sepal.

9.20 For the 150 rows R. A. Fisher iris data, use only 20% of the data from each species to train your NN model. Select another 10% of the iris data of lengths and widthes as the new data for prediction. Then use the NN model to predict the species of the new data. Discuss the errors of your NN model and your prediction.

9.21 NN forecast: From the Internet, download the historical daily data of minimum temperature and sea level pressure for a month and a location of your interest. Label your data as 1 if the total precipitation of the next day is greater than 0.5 millimeter, and 0 otherwise. Conduct an NN analysis for your labeled data. Given the daily data of minimum temperature and sea level pressure of a day in the same month but a different year, use your trained NN model to forecast whether the next day is rainy or not. Is your forecast accurate? Compare your NN forecast with your RF and SVM forecasts.

9.22 Design a machine learning project for yourself or others. What are your training data? What are your test data? What is your training model? What is your training model error? How would you assess your prediction error?

Index

labeled data, 346
LaGuardia Airport, 366
Langley, 366
lapse rate, 121
least square, 135
Leptokurtic, 9
likelihood function, 135
linear
   combination, 291
   equations, 173
   regression, 10, 122
   transformation, 174
   trend line, 15
linear algebra, 173
linearly dependent, 175
lognormal distribution, 64
Ly, 366

machine learning (ML), 329
Madison, Wisconsin, 65
Mann–Kendall test (MK), 116, 147, 148
marginal distributions, 53
mathematical expectation, 52
matrix, 165
   diagonal, 168
   identity, 168
   inversion, 172
   orthogonal, 172
   principal component, 209
   rank, 177
   transpose, 168
Mauna Loa Observatory, 244
McCulloch, Warren Sturgis, 372
mean, 7
Miami International Airport, 337, 338
MK-test, 147, 148
mode mixing, 224
modulus, 294
moving average (MA), 6, 264
multiple linear regression, 149

$N(\mu, \sigma^2)$, 59
NASA, 19, 120
NCAR, 307
NCEI, 2, 366
NCEP, 307
NCEP/NCAR Reanalysis, 308
negative hyperplane, 348
netCDF, 16, 19
Network Common Data Form, 16
neural network, 372
New York City (NYC), 366
NN, 372
NOAA, 2
NOAA Climate at a Glance, 27
NOAAGlobalTemp, 2, 62, 157
   annual time series, 31

data, 82, 97
   gridded, 16
nonlinear fitting, 152
normal distribution, 47, 59
North's rule of thumb, 223
Nullschool, 27

Omaha, Nebraska, 104, 284
orthogonal functions, 290
   matrix, 172
   polynomial, 159–160
orthonormal, 170
   condition, 213
   property, 172
Ozone concentration, 366

p-value, 87, 90, 91, 103, 107, 108, 145, 148
PACF, 278
Panoply, 19
Paraview, 25
Parseval's theorem, 312
Parseval, Marc-Antoine, 312
PC, 193, 208
PDF, 42, 52
period, 288, 292
periodic function, 290, 316
periodogram, 292, 306
phase, 288
Pitts, Walter Harry, 372
platykurtic, 9
Plotly, 20
PMF, 43
Poisson distribution, 56, 58
polynomial fitting, 157
   third order, 158
population in statistics, 73
population mean, 74
power of statistical inference, 91
power spectra, 292, 306
   spectral density, 334
principal component, 208
probability distribution, 47
probability function, 42
PSD, 314, 327

Q-Q plot, 12, 145
quantiles, 7, 8

R-squared, 133
random field, 215
random forest, 359
random process, 262
   continuous, 262
   discrete, 262
random residuals, 249
random variable, 35
random walk, 257, 263